VASCULAR BIOCHEMISTRY

Vascular Biochemistry

Edited by

PETER ZAHRADKA

St. Boniface General Hospital
Research Center
351 Tache Avenue
R2H 2A6, Winnipeg, Manitoba
Canada

JEFFREY WIGLE

Institute of Cardiovascular Sciences
St. Boniface General Hospital
351 Tache Avenue
R2H 2A6, Winnipeg, Manitoba
Canada

GRANT N. PIERCE

Division of Stroke & Vascular Disease
St. Boniface General Hospital
Research Center
Faculty of Medicine, University of Manitoba
R2H 2A6, Winnipeg, Manitoba
Canada

Reprinted from *Molecular and Cellular Biochemistry*, Volume 246 (2003)

Springer Science+Business Media, LLC

Library of Congress Cataloging-in-Publication Data

A C.I.P. Catalogue record for this book is available from the Library of Congress

ISBN 978-1-4020-7398-4 ISBN 978-1-4615-0298-2 (eBook)
DOI 10.1007/978-1-4615-0298-2

Printed on acid-free paper

Molecular and Cellular Biochemistry:

An International Journal for Chemical Biology in Health and Disease

CONTENTS VOLUME 246, Nos. 1 & 2, April 2003

VASCULAR BIOCHEMISTRY

Drs. Peter Zahradka, Jeffrey Wigle and Grant N. Pierce

Molecular and Cellular Biochemistry **246:** 1, 2003.

Preface

Our vasculature is more than a simple conduit for carrying blood, it is a complex and dynamic organ system that affects every system in our bodies. The absolute requirement for a functioning circulatory system also means that its health impacts directly on our well being. It is for these reasons that there is so much interest directed at understanding how the vasculature functions in both healthy and disease states.

The articles in this volume examine all aspects of vascular biochemistry. Not only do they touch on the various tissue components present in the vascular wall, they also examine the hemodynamic and metabolic activities associated with its function. Why is this so important? It is well established that cardiovascular disease is the number one killer in the world. In the United States, 710,000 individuals died from heart disease in 2000, while across the world the death rate from heart disease is 17 million (1/3 of global deaths). Indeed, 12% of Americans have been diagnosed with heart disease, and this number increases to 35% for individuals over the age of 65. More significantly, coronary artery disease accounts for half of all cardiovascular conditions, and is the primary cause of heart attack and stroke. In addition, approximately 20% of individuals are hypertensive. These statistics clearly indicate the impact of vascular health on the general population. As well, there is a marked increase in the incidence of hypertension and coronary artery disease for diabetics. In fact, there is considerable support for the view that the cardiovascular consequences of diabetes occur as a result of vascular hyperreactivity. Since the incidence of diabetes is expected to increase from 4% to 5.4% of the total population between now and 2025, understanding the factors that influence vascular health will be necessary to develop the clinical tools needed to prevent a comparable increase in cardiovascular deaths.

The papers included in this volume represent research that is directed towards understanding vascular function. As well, many articles have examined the vasculature under conditions which mimic the various disease states described above. Atherosclerosis is considered in relation to lipid/cholesterol metabolism, infection, racial profiles and oxidation state. At the cellular level, the contribution of smooth muscle cell proliferation, gene expression and the activity of enzymes catalyzing cholesterol synthesis to atherosclerosis are considered. Similarly, hypertension is discussed in terms of vasoreactivity and endocrine responses. The significance of diabetes and autoimmune responses were also studied by contributors to this volume. As such, this volume presents a broad range of topics associated with vascular function.

The 27 manuscripts contained in this volume were contributed by scientists who had attended the XVII World Congress of the International Society for Heart Research held in Winnipeg, Canada in July, 2001. By participating in a Congress with almost 2000 attendees, the authors were able to provide new information for researchers that could have considerable impact on the future health of many individuals. The breadth of the work is also notable, since each of the pathological conditions associated with vascular disease typically has multiple causes. We therefore believe that the following articles will provide new insights into the mechanisms that control vascular function, as well as therapies designed to treat vascular disease.

Peter Zahradka, Jeffrey Wigle and Grant N. Pierce
St. Boniface General Hospital Research Centre
Winnipeg, Manitoba
Canada

Molecular and Cellular Biochemistry **246:** 3–11, 2003.

Osteoglycin expression and localization in rabbit tissues and atherosclerotic plaques

Borja Fernández,[1] Andreas Kampmann,[2] Frederic Pipp,[1] René Zimmermann[2] and Wolfgang Schaper[1]

[1]*Max-Planck-Institute, Department of Experimental Cardiology, Bad Nauheim;* [2]*Kerckhoff-Clinic, Vascular Genomics, Bad Nauheim, Germany*

Abstract

The localization of osteoglycin (OG), one of the corneal keratan sulfate proteoglycans, was studied in different normal rabbit tissues, as well as in atherosclerotic lesions, by means of *in situ* hybridization and immunohistochemistry. OG was associated with the vasculature of all the organs analyzed. Normal aortas showed abundance of the protein in the adventitia and focally in the media. Peripheral vessels showed OG localized only in the adventitia. OG mRNA was restricted to vascular smooth muscle cells, pericytes, and fibroblasts in aorta and skeletal muscle. In striated muscle, OG was abundant and distributed in foci around muscles and vessels, whereas in visceral muscle, the protein was homogeneously distributed throughout the extracellular matrix. In all the other organs studied, OG was only associated with the vasculature, with the exception of the lung and liver. In these two organs, the protein accumulated also around cartilage, alveoli, and hepatic duct. In atherosclerotic lesions, OG mRNA was down-regulated in the media and up-regulated in the activated endothelium and thick neo-intima, whereas the protein accumulated in the front edge of migrating smooth muscle cells. We conclude that OG is a basic component of the vascular extracellular matrix. OG also plays a role in atherosclerosis, and might be useful for therapeutic interventions. In addition, the possible involvement of OG in maintaining physical properties of tissues is discussed. (Mol Cell Biochem **246**: 3–11, 2003)

Key words: osteoglycin, keratan sulfate proteoglycans, mRNA and protein localization, rabbit tissues, vasculature

Introduction

Osteoglycin (OG), also called mimecan, is a keratan sulfate proteoglycan (KSPG) first isolated from bovine bone [1]. Later, it was found to be identical to an unidentified KSPG highly abundant in the cornea [2]. The OG core protein belongs to the small leucine-rich repeat protein (SLRP) family together with 20 other evolutionary conserved small interstitial proteoglycans [3]. SLRPs share three conserved domains [4]: (1) An amino-terminal domain containing negatively charged glycosaminoglycan chains, probably involved in binding to cell surface and extracellular matrix proteins. (2) A cysteine-free central domain with 8–10 tandem repeats of leucine-rich regions responsible for binding to type I collagen. (3) A carboxyl end domain that contains two cysteine residues with unknown function. The biological roles of SLRPs are most probably related to protein–protein interaction [4]. They are considered key regulators of collagen fibrillogenesis in skin and cornea [5]. In addition, members of the subfamily of SLRPs that includes OG have been shown to bind growth factors, especially members of the TGFβ superfamily, via the core protein [6].

KSPGs were initially characterized in three different tissues: cornea, cartilage and brain [7]. However, proteoglycans with similar antigenic and physical properties were also extracted from many different tissues, but with a limited distribution of keratan sulfate epitopes [8]. KSPGs are 10–20% as abundant in connective tissues as in the cornea [8], where they are responsible for corneal transparency by regulating collagen fibril structure [5, 9, 10]. Lumican, one of the cor-

Address for offprints: B. Fernández, Max-Planck-Institute, Department of Experimental Cardiology, Benekstrasse 2, D-61231 Bad Nauheim, Germany (E-mail: borjaf@kerckhoff.mpg.de)

neal KSPGs plays a similar regulatory role for fibrillogenesis in the skin [11]. No experimental evidence exists about the functions of KSPGs in tissues other than cornea and skin.

The distribution of OG, like other KSPGs, is widespread in many body tissues. It is highly abundant in cornea, sclera, and cultured keratinocytes, but it has also been detected at variable amounts in different organs [2, 12, 13]. However, the localization of OG in these tissues is mostly unknown. Only two studies have addressed OG protein or mRNA localization in tissues: OG protein has been detected in rat chondrocytes, osteocytes, and osteoblasts [14], whereas OG mRNA has been shown in normal and diseased arteries [15].

In the present study we have explored the localization of OG mRNA and protein in different organs, with special attention on vascular tissues. In addition, we have extended the previous observations of OG expression in rabbit atherosclerotic lesions. Our study is the first to describe the localization of OG in different organs, which may contribute to understanding the physiological functions of OG and KSPGs in health and disease.

Materials and methods

All animal studies were approved by the Bioethical Committee of the District of Darmstadt, Germany. Animals were handled in accordance with the American Physiological Society guidelines for animal welfare and the Guide for care and use of laboratory animals published by the US National Institutes of Health.

Normal tissues: aorta, skeletal muscle, heart, uterus, intestine, lung, liver, kidney, pancreas, spleen and brain were obtained from New Zealand White rabbits, and atherosclerotic lesions from Watanabe rabbits. All the animals were killed by an anesthetic overdose. For Northern and Western blot analyses, the tissues were rapidly collected, frozen in liquid Nitrogen, and stored at –80°C until use. For *in situ* hybridization, immunohistochemistry, and immunofluorescence the tissues were embedded in Tissue-teck mounting media (Sakura Finitek), and cryopreserved in liquid Nitrogen-cooled methyl butane (Merk).

Cloning of OG

OG was identified as clone AK47b2 in a Differential-Display-Reverse-Polymerase-Chain-Reaction (DDRT-PCR) screen for genes, which are differentially expressed during growth of arteries in rabbits (A. Kampmann *et al.*, paper in preparation). This clone showed homology to human OG (89%, Genbank Acc. No. NM014057). The full length cDNA (AK74b2 full length; submitted to Genbank) was amplified using the SMART 5′ RACE Kit (Clontech) and used as a probe for all further studies.

Northern blot

Total RNA was isolated from rabbit aorta and skeletal muscle according to the method of Chomczynski and Sacchi [16]. Northern blot hybridization was performed according to standard procedures [17] using the full length OG cDNA probe.

Western blot

Protein isolation and Western blot analysis were performed according to standard procedures [16]. In short, tissue was homogenized in extraction buffer (0.1 M Tris-HCl pH 8.0; 0.01 M EDTA; 0.04 M DTT; 10% (w/v) SDS) using an ultrasonic device (Sonoplus GM 70, Bandelin) and quantified using the Bio-rad assay. Equal amounts of proteins were then separated on precast 4–12% Tris-Gylcine gels (NOVEX). A polyclonal rat anti-OG antibody raised against a synthetic peptide from the N-terminus of the deduced rabbit OG amino acid sequence was used. Immunodetection was performed with an enzyme-linked chemiluminescence detection system (ECL-System; Amersham Pharmacia Biotech) and a peroxidase-conjucated goat anti-rat IgG secondary antibody (Santa Cruz).

In situ *hybridization*

Non-radioactive *in situ* hybridization was performed with digoxigenin-labeled mRNA probes, generated by the *in vitro* transcription method from the clone AK74b2 (full length). The RNA was labeled using the 'DIG RNA Labeling Mix' (Roche Diagnostic) as instructed by the supplier. The hybridization and detection protocols followed the 'Nonradioactive *in situ* hybridization application manual' (Roche Diagnostic). Briefly, cryosections were fixed in 4% paraformaldehyde, and washed first with phosphate buffer saline (PBS) and then with 2 × SSC (300 mM NaCl; 30 mM sodium citrate; pH 7.4). Prehybridization consisted of incubation with 4 × SSC + 50% formamide at 37°C for 20 min. Hybridization was performed overnight at 45°C with antisense or sense probes. After several 2× and 1 × SSC washing steps, sections were incubated with RNase A at 37°C for 30 min. Stringent washes consisted of: 2 × SSC + 50% formamide at 50°C; 1 × SSC at 37°C; 0.5 × SSC at room temperature, 10 min each. mRNA probes were detected with anti-digoxigenin antibodies conjugated with alkaline phosphatase as instructed.

Immunohistochemistry

Immunoperoxidase was performed in 5 or 10 μm cryosections. The sections were fixed with cooled acetone, air-dried and washed with PBS. After quenching the endogenous peroxidase activity with 3% hydrogen peroxide, the sections were washed in PBS, immersed in blocking solution (0.1% bovine serum albumin; 0.4% glycine in PBS) and incubated overnight with the rat polyclonal antibody against OG. After several PBS washing steps, the sections were incubated with a peroxidase-conjugated anti-rat IgG (Santa Cruz Biotechnology). Peroxidase was detected by incubation with 3,3′ diaminobenzidine (Sigma). The omission of the first antibody served as a negative control. In some sections, the nuclei were counterstained with hematoxilin. Photomicrographs were obtained with a Leica DMLD microscope.

In some cases, *in situ* hybridization with OG mRNA probes, or immunoperoxidase with OG antibodies were combined with immunofluorescence with vascular smooth muscle α-actin antibodies (FITC-conjugated anti-α-actin, *Sigma*) in consecutive cryosections. After the blocking step, the sections were incubated overnight with anti-α-actin antibodies and washed several times with PBS. Nuclei were counterstained with DAPI (Sigma).

Results

Northern blot analysis of tissue from rabbit aorta and skeletal muscle using clone AK74b2 (full length) revealed a single band of approximately 3.2 kb (Fig. 1A), corresponding well to the published mRNA size for rat OG [15]. Western blot results showed a major band at approximately 34 kDa in most organ tissues (Fig. 1B). This value corresponds to the predicted molecular weight of rat OG [15] and also to the predicted molecular weight of rabbit OG/AK74b2 (full length) (data not shown). Cardiac ventricles, uterus, and lung showed the strongest signals. Skeletal muscles, kidney, and brain also showed specific signals. In the liver and spleen, the signal was almost undetectable. The existence of multiple faint bands at higher molecular weight recognized by the antibody points to possible glycosylated forms of the protein.

In situ hybridization of normal rabbit aortas with antisense mRNA probes coding for OG revealed positive signals in the nuclei of smooth muscle cells of the arterial media and in adventitial fibroblasts (Figs 2A and 2B). The signal was homogeneously distributed in the media and was usually more abundant in the adventitia, whereas the endothelial cells were negative. Immunoperoxidase using anti-OG antibodies revealed focal deposits of the protein in the media and adventitia (Figs 2C and 2D). The localization of OG was extracellular, accumulating in the medial lamellae. The staining in adventitia was significantly more robust than in the media, showing a ring of immunoreactive material in the outer layer of the vessel. The endothelial layer and the subendothelial space were always devoid of protein.

In the quadriceps muscle, OG mRNA was located in the smooth muscle cells of the arteries, similar to the localization observed in the aorta (Fig. 3A). In addition, strong staining was found in tissue fibroblasts (Figs 3A and 3C), and in pericytes surrounding capillaries and small arterioles (Fig. 3B). Endothelial cells of capillaries, arteries and veins, as well as skeletal myocytes were negative. Immunoperoxidase staining of similar sections of the quadriceps muscle revealed a widespread distribution of OG protein in the extracellular matrix surrounding each muscle fiber (Fig. 4A). The signal decreased in intensity from the perimysium to the endomysium, with focal and patchy deposits of immunoreactive signal around individual myocytes (Fig. 4A). Strong OG staining was

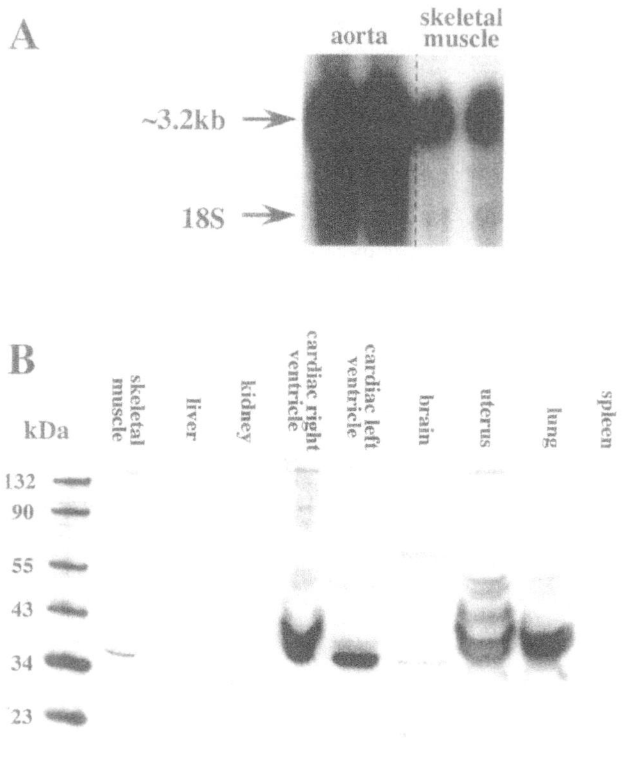

Fig. 1. (A) Northern blot analysis of OG in aorta and skeletal muscle. Hybridization to a single band of ≈ 3.2 kb is found. For reference, the size of the 18S rRNA is indicated. (B) Western blot showing the amount of OG protein in different rabbit organs. The polyclonal anti-OG antibodies recognized a major band of ≈ 34 kDa. The multiple faint bands recognized by the antibody point to possible glycosylated forms of the protein. Note that the secondary antibody used detects an epitope from the molecular weigh standard.

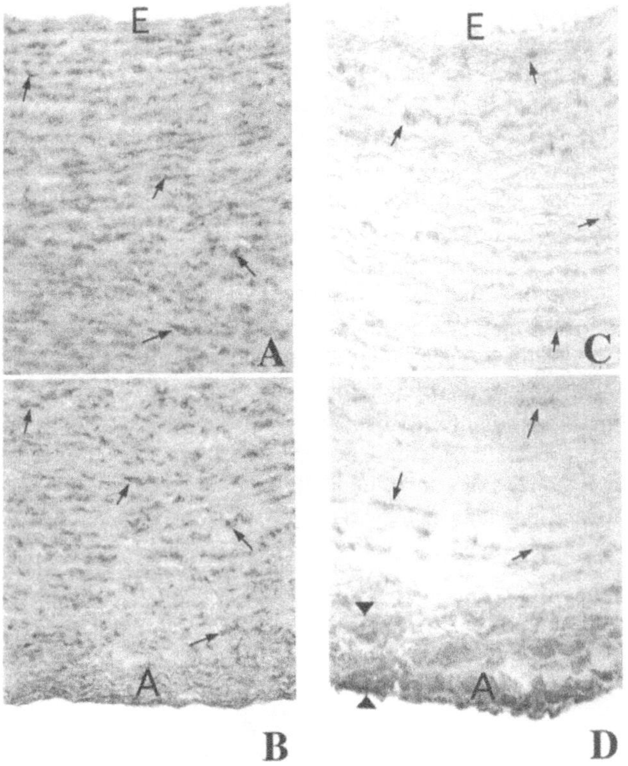

Fig. 2. In situ hybridization (A, B) and immunohistochemistry (C, D) of a normal rabbit aorta. A and C show the luminal part of the artery. B and D show the abluminal part. Arrows point to signals for OG mRNA (A, B), and OG protein (C, D) in the arterial media. The adventitia is indicated between arrowheads in D. A – adventitia; E – endothelium. (A–D: ×150).

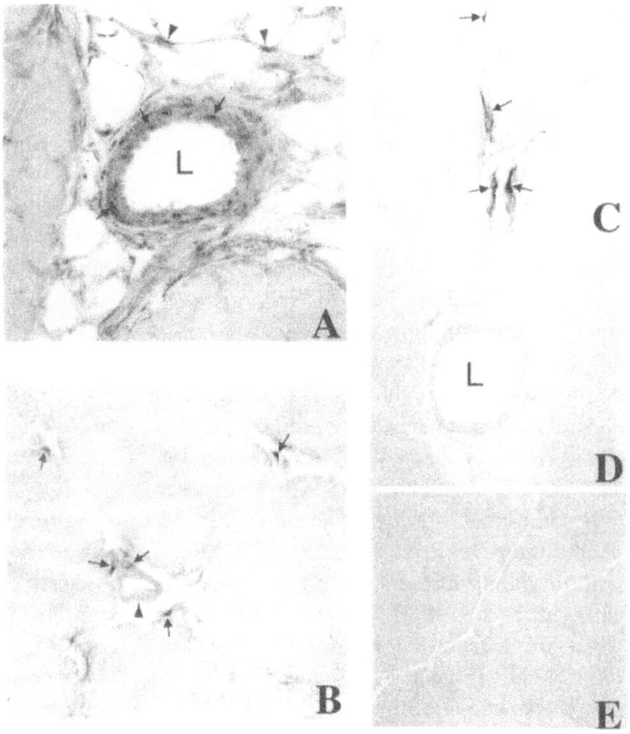

Fig. 3. In situ hybridization with OG mRNA antisense (A–C) and sense (D, E) probes of the quadriceps muscle. (A) Arrows indicate positive nuclei of arterial smooth muscle cells. Perivascular fibroblasts are also positive (arrowheads). (B) Positive pericytes (arrows) surround several capillaries and a small arteriole (arrowhead). Note that arteriolar endothelium, capillaries, and myocytes are negative. (C) Positive interstitial fibroblasts (arrows) are distributed among negative myocytes. (D, E) Incubation of the sections with a sense mRNA probe for OG resulted in no signal. L – lumen. (A: ×150; B: ×400; C: ×300; D: ×75; E: ×100).

observed around vessels. The protein accumulated in the adventitia of arteries and veins, but was always absent from the media and the endothelium (Fig. 4B).

In the heart, the pattern of OG immunostaining was found to be similar to that of skeletal muscles (Figs 4C and 4D). The protein was located in the extracellular matrix surrounding cardiomyocytes. The strongest staining was found in the subendocardium and subepicardium (Fig. 4C), as well as in the adventitia of the coronary vessels (Fig. 4D). The semilunar valves were also found to be immunoreactive (not shown).

In the uterus, the localization of OG was similar to that in the intestine (Fig. 5). However, it was found to be different compared to heart and skeletal muscles (compare Figs 4A and 4C with Figs 5A and 5C). Instead of patchy foci of immunoreactive material, a light brown staining was homogeneously distributed throughout the muscular, epithelial and vascular areas of both organs (Fig. 5). The protein intermingled with smooth muscle cell layers (Figs 5A–5C), and surrounded vessels of every size in the vascular areas (Figs 5B and 5D). OG also occupied the extracellular space around the intestinal and endometrial epithelia (Figs 5A–5C). In these two organs, no anatomical or histological structure was found to be more reactive than others to OG antibodies.

In the lung, the extracellular space beneath the bronchial epithelium was the richest in terms of OG immunoreactive content (Fig. 6A). The extracellular matrix surrounding the bronchial cartilage was also strongly immunoreactive. Dispersed foci of the protein were detected around the alveoli. The pulmonary vasculature showed a pattern of staining similar to that of the heart and skeletal muscle. However, the main pulmonary arteries and veins showed marked differences. In the pulmonary artery (Fig. 6B), OG concentrated in the adventitia, and was absent from the typically thick media of the vessel. In addition, a rim of the protein was found along the subendothelial surface. In the pulmonary vein, (Fig. 6C) focal deposits of the protein were present in the medial lamellae, like in the aorta (Figs 2C and 2D) and vena cava (not shown). In the subendothelial space, OG formed a thick layer, not found in any other vessel studied.

In the liver, OG was associated with two histological structures: the hepatic duct, and the hepatic vasculature (Fig. 6D). In the hepatic duct, OG strongly concentrated around the epithelium. In the hepatic vasculature, the distribution of OG was similar to other organs. The portal vein, like other big veins showed focal depositions of OG not only in the adventitia, but in the media as well. (Fig. 6E).

In the kidney (Fig. 6F), brain, pancreas and spleen (not shown), the distribution of OG was exclusively restricted to the adventitia of the vessels. In these three organs, the parenchyma was completely devoid of any immunoreactive signal, and the distribution of OG around the vasculature was similar to that in other organs like the heart and skeletal muscle (Fig. 6F).

Given the prevalence of OG distribution in the vascular system, we studied OG mRNA and protein in the aortas of Watanabe rabbits with different degrees of atherosclerotic plaque development. Early stage is characterized by the formation of a thin neo-intima layer beneath the endothelium (Figs 7A and 7B). At this stage, the neo-intima consists of loosely arranged mesenchymal-like cells with no immunoreactivity to smooth muscle α-actin. While neo-intimal cells did not show any expression of OG mRNA, an ectopic expression was consistently detected in the endothelial cells covering the neo-intimal progression (Fig. 7A).

More advanced lesions showed a thick neo-intima layer composed of α-actin negative cells (Figs 7C–7E). OG mRNA expression disappeared from the endothelium, whereas the inner portion of the neo-intima (closer to the endothelium) acquired a strong staining (Fig. 7C). In addition, the inner part of the media that was composed of normally arranged α-actin positive smooth muscle cells without signs of elastica disruption (Fig. 7D), showed a consistent downregulation of OG mRNA (Fig. 7C). Immunohistochemical staining of OG in adjacent sections revealed accumulation of the protein in the intima layer, just beneath the location of OG expressing cells (Figs 7C and 7E).

End-stage lesions were characterized by the presence of a plaque composed of α-actin negative fibroblast-like cells (Figs 7F and 7G). The fibroblast-like cells were devoid of OG expression, whereas a thin rim of α-actin-negative cells surrounding the plaque showed clear positive signals (Fig. 7F).

Discussion

We used *in situ* hybridization and immunohistochemistry to examine for the first time the localization of OG mRNA and protein in rabbit organs. Our results confirm previous studies [2, 12, 13] demonstrating abundance of the protein in different tissues, and show a heterogeneous distribution of OG in different organs.

Despite differences in distribution among organs, the presence of OG in the vasculature is a constant histological feature in all the organs studied. *In situ* hybridization revealed that vascular smooth muscle cells, pericytes, and adventitial fibroblasts are the sources of OG in vessels, whereas endothelial cells do normally not express OG mRNA. Immunohistochemistry showed accumulation of the secreted protein in the adventitia of all arteries and veins. In peripheral vessels, OG is restricted to the adventitia, whereas in big arteries and veins, foci of OG were distributed also throughout the media. The endothelium and the subendothelial space are free of protein. Although the mRNA and protein localization of OG in the vasculature showed a consistent and constant pattern, the pulmonary vessels were found special in terms of its amount and distribution. The pulmonary arteries were the only normal big arteries with OG in the intima layer but not in the media, and the pulmonary veins also showed OG in the thick intima. Our immunohistochemical staining revealed accumulations of OG protein around big and medium size vessels. Yet, microvessels (arterioles, venules and capillaries) were usually not stained. This observation can be interpreted (1) as a restriction of the histological distribution of the protein or (2) as a limitation of the method due to low concentrations of the antigen. The later is more likely, because *in situ* hybridization revealed expression of OG mRNA by pericytes surrounding arterioles and capillaries. In any case, we can conclude that the proteoglycan OG is a basic component of the vascular extracellular matrix in all organs of the body.

There is only one study that showed OG mRNA expression by smooth muscle cells in the normal media of neonatal and adult aortas [15]. Other members of the LRSPs family have also been detected in vascular extracellular matrices: byglican is associated with intramural blood vessels of the uterus [4], and Decorin accumulates in the adventitia of mammary gland vessels [4]. In addition, Thieszen and Rosenquist [18] found that decorin interacts with collagen type III in the chick embryonic cardiac outflow tract. Our results, together with all these data strengthen the hypothesis that OG together with other SLRP family members are important in developing and maintaining the mature vascular extracellular matrix.

OG mRNA expression has been shown to be regulated in diseased arteries [15]. Therefore, we examined the localization of OG mRNA and protein in the aortas of Watanabe rabbits, a well-known model for atherosclerosis. In the early stages of intima formation, a consistent ectopic expression of OG was found in endothelial cells. During neo-intima progression, OG mRNA was up-regulated in the inner neo-intima and down-regulated in the inner media, whereas the

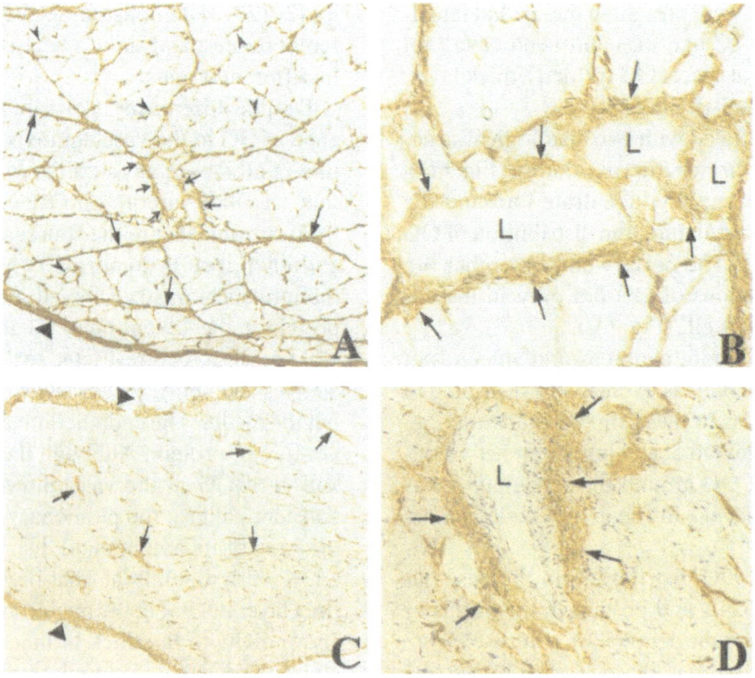

Fig. 4. OG immunohistochemical staining of rabbit quadriceps muscle (A, B) and heart (C, D). (A) A thick layer of immunoreactive OG surrounds the quadriceps muscle in the perimysium (big arrowhead). A smaller but continuous layer surrounds myofibers in the epimysium (big arrows). Focal deposits can be observed around individual myocytes in the endomysium (small arrowheads). Small arrows point to the adventitia of arteries. (B) Higher magnification of the arteries shown in A. OG accumulates in the adventitia (arrows). (C) In a papillary muscle of the heart, a thick layer of OG accumulates in the epicardium (arrowheads). Thick and focal deposits of OG are visible in the myocardial interstitium (arrows). (D) A coronary artery is surrounded by abundant immunodetectable OG. L – lumen. (A: ×90; B: ×300; C: ×50; D: ×200)

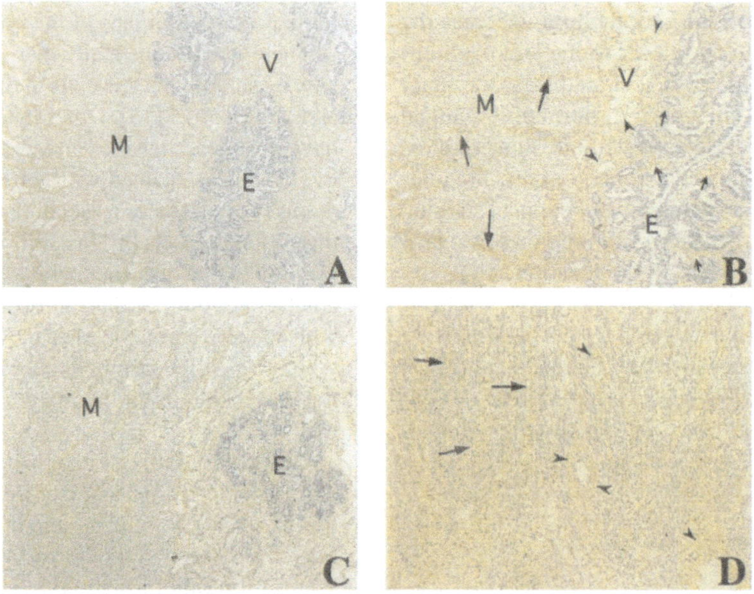

Fig. 5. OG immunohistochemical staining of rabbit intestine (A, B) and uterus (C, D). A light and homogeneous staining can be observed in the muscular (M), epithelial (E), and vascular (V) areas of both, intestine and uterus. Big arrows point to OG around smooth muscle cells, small arrows around epithelial cells, and arrowheads around vessels. (A, C: ×50; B, D: ×100).

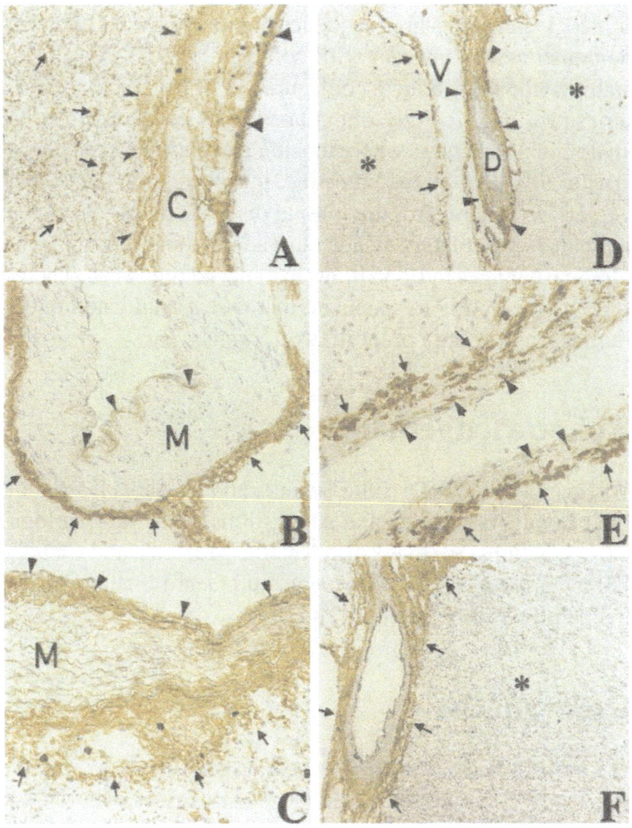

Fig. 6. OG immunohistochemical staining of rabbit lung (A–C), liver (D, E), and kidney (F). (A) OG forms a thick layer beneath the bronchial epithelium (big arrowheads). Small arrowheads point to OG surrounding cartilaginous tissue. Focal deposits of the protein associate with the lung alveoli (arrows). (B) A pulmonary artery shows strong OG immunorreactivity in the adventitia (arrows). OG forms a thin but detectable rim beneath the arterial endothelium (arrowheads). (C) In a pulmonary vein, OG accumulates in the adventitia (arrows), media (M), and intima (arrowheads). (D) The liver parenchyma is devoid of OG staining (asterisks). Arrowheads and arrows point to OG deposits around the hepatic duct and portal vein respectively. (E) In the portal vein, OG is present in the adventitia (arrows) and media (arrowheads). F: the parenchyma of the kidney is free of OG (asterisk), whereas the adventitia of an artery shows a strong staining (arrows). C – cartilage; D – hepatic duct; M – media; V – portal vein. (A, D, F: ×50; B, C, E:

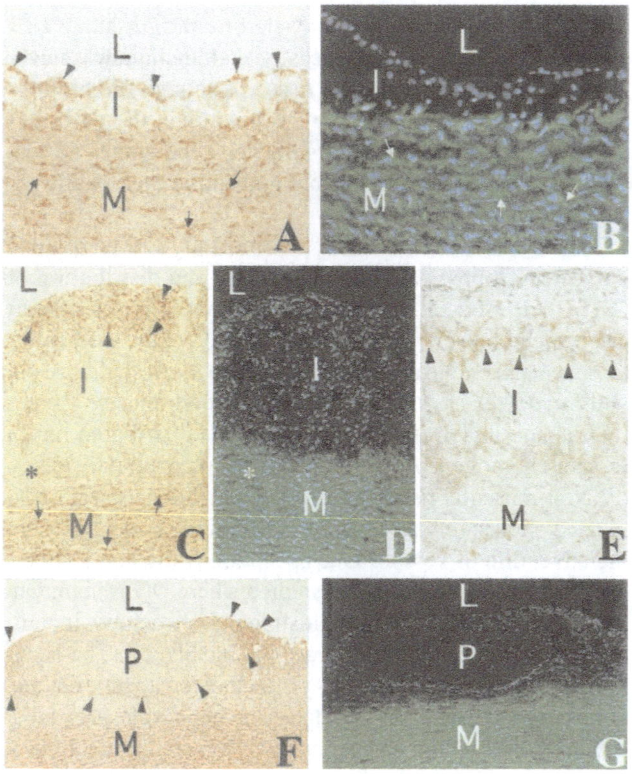

Fig. 7. In situ hybridization for OG (A, C, F), immunohistochemistry for OG (E), and immunofluorescence for vascular smooth muscle a-actin (B, D, G) of Watanabe atherosclerotic lesions. (A, B) Consecutive sections of an early lesion. Alpha-actin-positive smooth muscle cells of the media (arrows in B) show a normal OG mRNA expression pattern (arrows in A; compare with Fig. 1A). OG mRNA is also clearly detectable in the endothelium (arrowheads), but not in the neo-intima (I). (C–E): Consecutive sections of an advanced lesion. A thick neo-intima with α-actin-negative cells (I in D) shows OG mRNA expression only in the cells close to the lumen (arrowheads in C). The α-actin-positive smooth muscle cells at the inner part of the media (asterisk in D) show no OG mRNA signal (asterisk in C). OG protein accumulates in the neo-intima, just beneath the OG expressing cells (arrowheads in E). (F, G) Consecutive sections of an end stage lesion. The cells of the fibrous plaque are devoid of α-actin protein and OG mRNA (P in G and F respectively). The cells surrounding the plaque show some OG expression (arrowheads in F). I – neo-intima; L – lumen; M – media; P – plaque. (A, B: ×150; C–E: ×75; F, G: ×50).

protein accumulated in the intima. End stage atherosclerotic lesions showed no OG expression in the fibrous plaque, but an up-regulation in the cells surrounding the plaque was observed. Our results mainly support those from Shanahan and collaborators [15] obtained in injured carotid and atherosclerotic coronary arteries. The localization of OG mRNA in the neo-intima of advanced lesions is of special interest. Both, the study of Shanahan and ours revealed expression of OG in neo-intimal cells close to the lumen, whereas medial smooth muscle cells close to the neo-intima showed a clear down-regulation. In addition, our immunohistochemical study indicates that the secreted protein accumulates beneath the OG expressing cells, forming a lamina of immunoreactive material (Fig. 7E). These results suggest that the protein forms a border for the smooth muscle cells (not expressing OG) migrating from the media (Figs 7C–7E). Whether OG is implicated in the migration of medial smooth muscle cells during the progression of vascular disease remains to be confirmed. Another relevant finding of our study is the consistent expression of OG mRNA by activated endothelial cells as an early event during atherosclerotic progression. This expression pattern is ectopic because endothelium was never found positive in any normal tissue. Although the biological significance of this finding is unknown, the early expression in a relatively accessible cell type makes OG an attractive candidate for possible preventive therapeutic interventions.

The organs analyzed in the present study can be divided into three groups, based on the amount and distribution of OG in their extracellular matrices. (1) Striated muscles (heart and skeletal muscle): abundant foci of OG in the extracellular space, not exclusively associated with the vasculature but with myocytes as well. (2) Visceral smooth muscle tissues (uterus and intestine): abundance of OG distributed not in foci, but homogeneously around all the extracellular spaces. (3) Non-muscular tissues (kidney, brain, and pancreas): OG located exclusively around the vasculature, associated with the adventitia of vessels. One exception is the lung. This is the only non-muscular organ studied where OG is abundant and associated with different histological structures including the epithelium, the cartilage, and the alveoli. The accumulation of OG in the portal duct is another remarkable feature. Our study confirms previous data showing a widespread distribution of SLRPs in many organs. However, the histological localization of only three members of the protein family is currently known in a limited number of organs. Biglycan protein is abundant in the epidermis and follicular epithelium of the mammary glands [4]. In the uterus, biglycan surrounds the glandular epithelium and intramural blood vessels, similar to the distribution of OG showed in the present study. Decorin is also abundant in the skin [4], and its localization in skeletal muscles and vasculature coincides with that of OG presented here. Moreover, decorin localization in chick aortic arch vessels shows high homology to OG localization in rabbit aortas [18]. OG antibodies have been found to react with chondrocytes, osteocytes, and osteoblasts in rat forelimbs [14], but no other histological structure was examined in that study. The study of Shanahan [15] however, showed a localization of OG mRNA in arteries similar to that presented here. Taken together, all these results show a coincident localization of structurally related SLRPs in tissues. We can hypothesize then that structurally related SLRPs have similar physiological roles in those tissues. The functions of these SLRPs are most probably related to protein–protein interactions, more precisely to collagen fibrillogenesis. This assumption arrives from several biochemical, *in vitro*, and *in vivo* studies [4, 5, 9–11, 19–21]. These studies suggest that the function of SLRPs such as decorin, lumican, proteoglycan-Lb, and OG is to keep the spatial order of collagen fibers. This molecular function of SLRPs may determine physical properties of tissues such as elasticity and tensile strength. Indeed, a deficit in the lumican gene causes skin laxity and fragility in mutant mice [11]. In addition decorin, a dermatan sulfate proteoglycan with a high sequence homology to OG, has been demonstrated to interact with collagen type III in order to stabilize the large helical configuration of the aortic arch arteries [18]. In our study, abundance of OG was restricted to anatomical structures where tension and elastic forces play a crucial role, like uterus, intestine, muscles, lung, ducts, and vessels. In this regard, the unique distribution of OG in the pulmonary vasculature, which must adapt to significant physiological changes in flood blow, has special relevance. Therefore, it is tempting to speculate that OG is implicated in the regulation of elastic properties of organs.

Acknowledgements

We thank Professor Dr. Jutta Schaper and Dr. Sawa Kostin for their invaluable advice and suggestions in the morphological evaluations. We thank also Kerstin Wiczorek, and Claudia Ullmann for technical assistance, and Geard Staemmler for his help in editing pictures and text.

References

1. Bentz H, Nathan RM, Rosen DM, Armstrong RM, Thompson AY, Segarini PR, Mathews MC, Dasch JR, Piez KA, Seyedin SM: Purification and characterization of a unique osteoinductive factor from bovine bone. J Biol Chem 264: 20805–20810, 1989
2. Funderburgh JL, Corpuz LM, Roth MR, Funderburgh ML, Tasheva ES, Conrad GW: Mimecan, the 25-kDa corneal keratan sulfate proteoglycan, is a product of the gene producing osteoglycin. J Biol Chem 272: 28089–28095, 1997
3. Blochberger TC, Vergnes J-P, Hempel J, Hassells JR: cDNA to chick Lumican (corneal keratan sulfate proteoglycan) reveals homology to the small interstitial proteoglycan gene family and expression in muscle and intestine. J Biol Chem 267: 347–352, 1992
4. Danielson KG, Baribault H, Holmes DF, Graham H, Kadler KE, Iozzo RV: Targeted disruption of decorin leads to abnormal collagen fibril morphology and skin fragility. J Cell Biol 136: 729–743, 1997
5. Rada JA, Cornuet PK, Hassell JR: Regulation of corneal collagen fibrillogenesis *in vitro* by corneal proteoglycan (lumican and decorin) core proteins. Exp Eye Res 56: 635–648, 1993
6. Yamaguchi Y, Mann DM, Ruoslahti E: Negative regulation of transforming growth factor-β by the proteoglycan decorin. Nature 346: 281–284, 1990
7. Krusius T, Finne J, Margolis RK, Margolis RU: Identification of an O-glycosidic mannose-linked sialylated tetrasaccharide and keratan sulfate oligosaccharides in the chondroitin sulfate proteoglycan of brain. J Biol Chem 261: 8237–8242, 1986
8. Funderburgh JL, Caterson B, Conrad GW: Distribution of proteoglycans antigenically related to corneal keratan sulfate proteoglycan. J Biol Chem 262: 11634–11640, 1987
9. Scott JE: Proteoglycan-fibrillar collagen interactions. Biochem J 252: 313–323, 1988
10. Vogel KG, Paulsson M, Heinegard D: Specific inhibition of type I and type II collagen fibrillogenesis by the small proteoglycans of tendon. Biochem J 223: 587–597, 1984

11. Chakravarti S, Magnuson T, Lass JH, Jepsen KJ, LaMantia C, Carroll H: Lumican regulates collagen fibril assembly: Skin fragility and corneal opacity in the absence of lumican. J Cell Biol 14: 1277–1286, 1998
12. Corpuz LM, Funderburgh JL, Funderburgh ML, Bottomley GS, Prakash S, Conrad GW: Molecular cloning and tissue distribution of keratocan. J Biol Chem 271: 9759–9763, 1996
13. Tasheva ES, Corpuz LM, Funderburgh JL, Conrad GW: Differential splicing and alternative polyadenylation generate multiple mimecan mRNA transcripts J Biol Chem 272: 32551–32556, 1997
14. Dash JR, Pace DR, Avis PD, Bentz H, Chu S: Characterization of monoclonal antibodies recognizing bovine bone osteoglycin. Conn Tissue Res 30: 11–21, 1993
15. Shanahan CM, Cary NRB, Osbourn JK, Weissberg PL: Identification of osteoglycin as a component of the vascular matrix. Differential expression by vascular smooth muscle cells during neointima formation and in atherosclerotic plaques. Arterioscl Thromb Vascular Biol 17: 2437–2447, 1997
16. Chomczynski P, Sacchi N: Single-step method of RNA isolation by acid guanidinium thiocyanate-phenol-chloroform extraction. Anal Biochem 162: 156–159, 1987
17. Sambrook J, Fritsch EF, Maniatis T: Molecular Cloning: A Laboratory Manual. Cold Spring Harbor Laboratory Press, New York, 1989
18. Thieszen SL, Rosenquist TH: Expression of collagens and decorin during aortic arch artery development: Implications for matrix pattern formation. Matrix Biol 14: 573–582, 1994
19. Scott JE, Orford CR: Dermatan sulphate-rich proteoglycan associates with rat tail-tendon collagen at the d band in the gap region. Biochem J 197: 213–216, 1981
20. Hedbom E, Heinegard D: Binding of fibromodulin and decorin to separate sites on fibrillar collagens. J Biol Chem 268: 27307–27312, 1993.
21. Schonherr E, Hausser H, Beavan L, Kresse H: Decorin-type I collagen interaction. J Biol Chem 270: 8877–8883, 1995

Molecular and Cellular Biochemistry **246:** 13–17, 2003.

Effects of midazolam on the contraction and relaxation of segments of thoracic aorta stripped of endothelium and stimulated by adrenaline – experimental study in rabbits

Antônio Aires Ferreira Rodrigues Borges and Otoni Moreira Gomes

Fundação Cardiovascular São Francisco de Assis/ServCor, EMESCAN, Santa Mônica, Belo Horizonte, Brasil

Abstract

To evaluate the effects of midazolam on the angiokinesis of segments of rabbits' thoracic aorta stripped of endothelium and stimulated by adrenaline.

Two groups of aortic rings removed from albinic rabbits anesthetized with thiopental were used (Group I – 6 animals; Group II – 12 animals), stripped of endothelium, studied in an organ chamber, perfused by Krebs-Henseleit solution. The groups were stimulated by adrenaline, recording the maximum contraction and dT/dt at 12″, 36″, 60″ and 120″. When the plateau phase was reached, the vessel was washed with perfusion solution, recording relaxation at 2′, 4′ and 6′. When the base values were reached, Group I underwent a new adrenergic stimulus; and Group II was stimulated with midazolam and then with adrenaline, and the same values were recorded. T test was applied as a statistical analysis when two variables were studied. When studying more than two variables the Anova test was used, supplemented by the Tuckey test.

Group I did not show any significant difference between the two stimuli. Group II – the midazolam significantly reduced the maximum contraction induced by adrenaline (83.01 ± 4.11%) ($p < 0.01$). The dT/dt was reduced at 12″ (57.06 ± 8.47%), and also at 36″ (70.59 ± 5.26%). There was no significance at 60″ and 120″ ($p < 0.01$).

The relaxation increased significantly at all measurements – at 2′-adrenaline 39.31 ± 9.60%; adrenaline/midazolam: 44.06 ± 9.62% ($p < 0.05$). At 4′-adrenaline: 53.08 ± 8.3%; adrenaline/midazolam: 61.68 ± 8.50% ($p < 0.01$). At 6′-adrenaline: 76.26 ± 5.45%; adrenaline/midazolam: 84.20 ± 7.96% ($p < 0.01$).

Midazolam significantly reduced the maximum contraction obtained by the adrenergic stimulus as well as the dT/dt in the initial phases of contraction. The relaxation speed also increased. (Mol Cell Biochem **246**: 13–17, 2003)

Key words: arterial contraction, arterial relaxation, experimental cardiology, midazolam

Introduction

Midazolam is an imidazole benzodiazepine which differs from other benzodiazepines because its imidazole ring is fused with the benzodiazepine and it is unique in that, at a pH lower than 4, it is soluble in aqueous solutions [1], which makes it extremely easy to handle and administer. Its use has therefore been growing, covering all the way from the use in seriously ill patients in ICUs [2], an auxiliary in subsidiary examinations [3], treatment in epilepsy [4], dental interventions [5] anesthesia inducer [6] and even in pediatrics [7]. However, there began to appear reports of probable adverse effects, such as an increase of arterial PCO_2 [8], reduction of the cerebral flow associated with the reduction of the rate of the use of oxygen and an increase in cerebral vascular resistance [9]. With regard to the cardiovascular system, papers

Address for offprints: O. Moreira Gomes, Fundação Cardiovascular São Francisco de Assis/ServCor, EMESCAN, Rua José do Patrocínio, 522, Santa Mônica 31525-160, Belo Horizonte, MG / Brasil (E-mail: servicor@servicor.com.br)

have shown a reduction in coronary flow in human beings [10], and also studies on isolated rat hearts have shown myocardial depression and a reduced coronary flow [9]. It was also found that midazolam provokes a slight drop in heart output in humans with coronary disease, associated to a sharp blood pressure drop, both systolic and diastolic. More serious reactions, including respiratory seizures and deaths, have been reported associated with the use of midazolam [7, 11].

However, in spite of the ever wider use of midazolam, especially in seriously ill patients, and although all the authors agree that one of its most important cardiovascular effects is to reduce peripheral vascular resistance with its resulting drop in blood pressure, the literature examined shows very few studies on the actions of this substance on the kinetics of blood vessels. Therefore, it is the objective of this work to research its interference on the contraction and relaxation of segments of the chest aorta of rabbits, stripped of the endothelium and stimulated by adrenaline.

Materials and methods

With approval from the Ethics Committee of the School Hospital of the Santa Casa de Misericódia of Vitória, State of Espírito Santo, 18 albino rabbits were studied, without gender distinction, divided into two groups (Group I – 6 animals; Group II – 12 animals). They were anesthetized with thiopental 170 mg intra-peritoneally. A segment of the chest aorta 0.5 cm long was removed, carefully cleaned in a Petri dish with Krebs-Henseleit solution with the following composition in millimole per liter: NaCl – 126, NaHCO – 25, KH2PO4 – 1.2, KCl – 4.8, MgSO4 – 1.2, CaCl2 – 2.5, Glycose – 11.0, and enough de-ionized water to complete 1000 ml. Once cleaned, the endothelium was damaged by running a straight tweezers tip along the segment turning it against the Petri dish 3 times clockwise and 3 times anti-clockwise. Just after the ring was put in a vertical position, fixed in two triangles of stainless steel thread, one of which was fixed and the other linked to the GRASS Force Displacement Transducer (model FT-03, Grass Instrument Company, Quink, MA, USA), nr. 914475, and placed in an organ bath (Cleveland Clinic Foundation, OH, USA) model, with 25 ml Kreb's solution, with carbogenic bubbles, at 37.5°C. After a period of stabilization, the damage to the endothelium was confirmed because 10 micromoles of acetylcholine were unable to inhibit the contraction provoked by 0.05 micromoles of adrenaline.

In order to observe and register the tensions of the ring, a Besse III biomonitor, model A9E, manufactured by Bioengenharia de Sistemas e Equipamentos BH, was used. The ring then rested for 40′, and was washed with a perfusion solution at 13′, 26′ and 40′. The vessels were then submitted to successive stimuli of 736 micromoles of KCl, increasing its tension gradually until reaching the most effective contraction. The vessels that did not reach a contraction of 1.5 g were excluded from the experiment.

At the end of this phase, both groups were stimulated with 0.05 micromoles of adrenaline, registering all the contractions and, once the plateau phase was reached, the vessel was washed twice with perfusion solution, registering its relaxation until pre-stimulus values. Subsequently, Group I was stimulated again with 0.05 micromoles of adrenaline, Group II was stimulated with 2 micromoles of midazolam, followed by a new stimulus with 0.05 micromoles of adrenaline. The phases of contraction were also registered, until pre-stimulus values were reached, when they were washed with perfusion solution and relaxation was registered until pre-stimulus values were reached.

In both groups, the maximum concentration obtained was studied (absolute and percentage values) at dT/dt at 12″, 36″, 60″ and 120″ (in absolute and percentage values of the stimulus with adrenaline/midazolam, as a result of the purely adrenergic stimulus), as well as a relaxation rate at 2′, 4′ and 6′ (in percentage values relative to maximum contraction obtained by each ring).

In this study, the curve obtained by the stimulus of the vessel segment and its relaxation was divided into 4 phases: a – rapid increment phase; b – slow increment phase; c – plateau phase; d – relaxation phase. In Fig. 1 there is an example of this curve and its division.

The statistical study was carried out using the Student test, when using two variables, and the Anova supplemented by the Tuckey test, when studying more than 2 variables, considering significance level of < 0.05.

Results

In Group I there were not significant differences in any of the values studied.

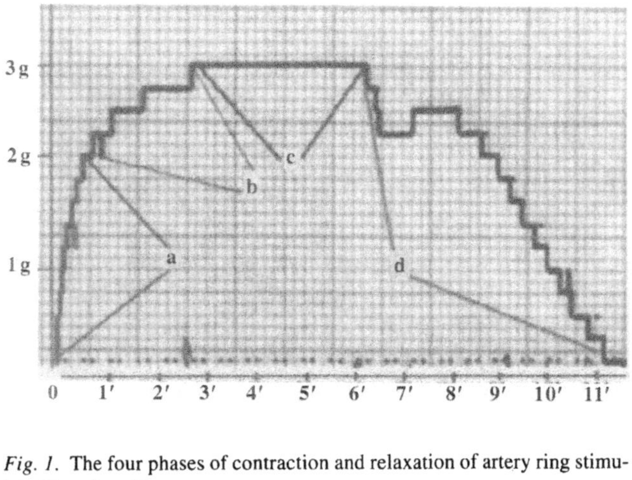

Fig. 1. The four phases of contraction and relaxation of artery ring stimulated by adrenaline.

In Group II it was found that midazolam significantly reduced the maximum contraction, with the following values: maximum contraction reached by adrenergic stimulus – 2.75 ± 0.58 g, stimulus by adrenaline/midazolam = 2.25 ± 0.54 g ($p < 0.01$). The percentage values are shown in graph 1.

In Group II, with regard to dT/dt, the stimulus by adrenaline/midazolam significantly reduces the values reached at 12′ and 36′. In terms of percentage for adrenergic stimulus, the values obtained were: At 12″ – 57.06 ± 8.47% ($p < 0.01$); at 36″ – 70.59 ± 5.26% ($p < 0.1$); at 60″ – 78.62 ± 5.26% (p = NS); at 120″ – 84.20 ± 7.96% (p = NS). The absolute values are shown in graph II.

With regard to the speed of relaxation, measured at 2′, 4′ and 6′ in percentage values of maximum contraction obtained by each ring, the adrenaline/midazolam association increased significantly, as shown in graph III.

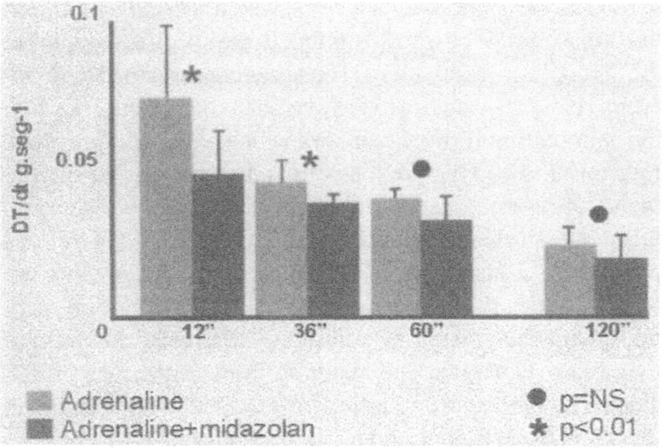

Fig. 3. dT/dt variation in stimulus by adrenaline and adrenaline + midazolan.

Discussion

This work chose to use artery segments stripped of endothelium, since in many papers, among which those of Vanhoute, Rubany and Miller *et al.* [12], as well as of Furchgott [13], Angus, Cocks and Sato [14], it was shown that the whole endothelium interferes strongly with the vasomotion of the vessel and can in this way alter the results of the action of drugs on the smooth muscle. Also, Nilsson [15] quotes that the presence of whole endothelium strongly reduces the response to stimuli by agonists.

When evaluating the results obtained in this work, the importance of the action of the deposits of calcium of the sarcoplasmic reticulum must be considered in the dynamics of vascular smooth muscle. The most important of these are the IICR (Induced IP3 Calcium Release) stimulated by the IP3 and inhibited by the CICR (Calcium Induced Calcium Release) released by the calcium and the caffeine and inhibited by the ryanodine. The outstanding work of Walsh [16] demonstrates the mechanics of this phenomena, especially in the case of stimulation of the smooth muscle by adrenergic agonists. Therefore, when stimulated by adrenergic agonist, the receptor of the membrane processes the activation of the beta-isoenzymes of C phospholipase via the G protein found in the cellular membrane. This leads to the hydrolysis of bi-phosphate 4.5 phosphatidylinositol, which gives rise to two

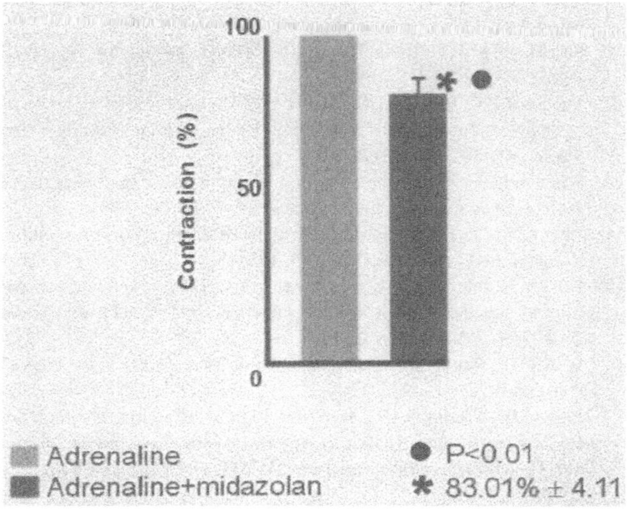

Fig. 2. Maximum contraction obtained by stimulus by adrenaline and adrenaline + midazolam.

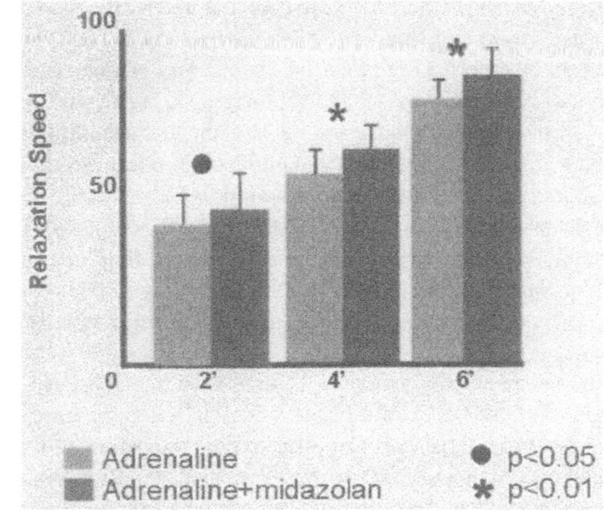

Fig. 4. Speed of relaxation obtained by stimulus of adrenaline and adrenaline + midazolam.

second messengers: diacylglycerol, which will activate C protein kinase and IP3, which will link to the specific receptor of the sarcoplasmic reticulum. The latter controls the IICR calcium, releasing calcium to the cytosol, which, coupling with the calmodulin, forms a complex which activates free chain myosin kinase. This provokes the phosphorylation of filaments of myosin, leading to the first rapid and wide amplitude contraction (1st phase of contraction of rapid increment phase). Next, also by a signal issued by the adrenaline, unspecific ionic channels are activated, as well as chloride channels, with the closing of potassium channels, the cellular membrane is depolarized resulting in the opening of the L channels and the strong inflow of extra-cellular calcium to the cytosol, bringing about the second phase of contraction (slow), not as wide as the first, but longer lasting (2nd phase or slow increment). The increase of cytosolic calcium, leading to the plateau phase (third phase) which, according to Jaggar, Wellman and Heppener *et al.* [17], will lead to a discharge of 'lightning' of calcium from the CICR deposit. This activate calcium dependent potassium channels existing in the cell membrane, leading to its hyper-polarization, closing the L channels and reducing the level of cytosolic calcium, beginning the fourth phase (relaxation). There is a simultaneous stimulation of a myosin phosphatase through the action of GMP, which, as well as dephosphorlating the myosin, reduces the sensitivity of the contractile elements to calcium, at the same time as its inflow ceases, and its extrusion occurs through calcium pumps of the cell membrane, when there is also the activation of the phospholambam collecting cytosolic calcium from the sarcoplasmic reticulum.

During this work, we concluded that midazolam reduces the maximum contraction obtained, and also interferes in the rapid increment phase and the slow increment phase, not interfering in the plateau phase, and increasing the speed of relaxation. These three facts are corroborated by the work of Karaki, Osaki and Hori *et al.* [18], who also quote that midazolam has a blocking effect on the calcium channels. The work of Shiraishi, Ohashi and Kanmura *et al.* [19], studying the action of noradrenaline and midazolam on the mesenterial artery of rabbits, concluded that midazolam interferes in the first and second phases of contraction, which is confirmed by the work of Yamaguchi, Kanmura and Yoshimura [20]. Cheng, Mazzeo and Bosnjak *et al.* [21] also conclude that midazolam has a relaxing effect, in their study of the action of adrenaline and midazolam on smooth muscles of the airways.

In this paper, analyzing the dT/dt, it was found that at 12″ and 36″ there was a significant reduction when midazolam was used ($p < 0.01$), with no difference at the plateau phase, which means that midazolam interfered in the rapid increment phase (release of calcium from the sarcoplasmic reticulum) as well as in the slow increment phase (inflow of calcium form the extracellular medium, through membrane channels). Also, analyzing its action on relaxation, it was found that there was a significant increase of the speed of relaxation throughout the process ($p < 0.01$).

It may therefore be conclude that midazolam reduces the maximum contraction obtained by adrenergic stimulus, as well as that it acted on the rapid and slow increment phases, reducing both of them. It did not, however, act on the plateau phase, and increased the speed of relaxation during all the process.

References

1. Fragen RJ, Gahl F, Caldwell N: A water soluble benzodiazepine, RO-21398, for induction of anesthesia. Anesthesiology 49: 41–43, 1978
2. Shafer A: Complications of sedation with midazolam in the Intensive Care Unit: A comparison with other sedative regimens. Crit Care Med 26: 947–956, 1998
3. Al-Khudairi D, Whitwam JG, McCloy W: Midazolam and diazepam for gastrocopy. Anaesthesia 37: 1001–1006, 1982
4. Anley DF Jr, Pozo M: Treatment of status epilepticus with midazolam in the critic care setting. Int J Clin Pract 1: 34–35, 2000
5. Aun C, Flynn PJ, Richar DSJ, Major E: A comparison of midazolam and diazepam for intravenous sedation in dentistry. Anaesthesia 39: 589–593, 1984
6. Baber R, Hobest A, Munro I, Purcell G, Binstead R: Midazolam as an intravenous induction agent for general anahestesy: A clinical trial. Anaesth Intens Care 10: 29–35, 1982
7. Blumer JL: Clinical pharmacology of midazolam in infants and children. Clin Pharmacol 35: 37–47, 1998
8. Bergreen L, Erikson I, Mollhenolt P, Sunzel M: Changes in respiratory pattern after repeated doses of diazepam and midazolam in healthy subjects. Acta Anaesthiol Scand 3: 667–678, 1987
9. Knudsen L, Cold GE, Holgard HO, Johansen UT, Jensen S: The effects of midazolam on cerebral blood flow and oxygen consumption. Anaesthesia 45: 1016–1019, 1990
10. Marty J, Nitemberg A, Blanchet F, Zouioueche S, Desmonts JM: Effects of midazolam on the coronary circulation in patients with coronary artery disease. Anaesthesiology 14: 206–300, 1986
11. Rao RB, Ely SF, Hoffman RS: Deaths related to liposuction. New Engl J Med 13: 1471–1475, 1999
12. Vanhoute PM, Rubany GM, Miller VM, Houston DS: Modulation of vascular smooth muscle contraction by the endothelium. Ann Rev Physiol 48: 307–320, 1986
13. Furchgott RF: Role of endothelium in responses of vascular smooth muscles. Ann Rev Pharmacol Toxicol 24: 175–179, 1984
14. Angus JA, Cocks TM, Sato K: The alpha adrenoreceptors on endothelial cells. Fed Proc 45: 2355–2359, 1986
15. Nilsson H: Interactions between membrane potential and intracellular calcium concentration in vascular smooth muscle. Acta Physiol Scand 1664: 559–566, 198?
16. Walsh MP: Regulation of vascular smooth muscle tone. Can J Physiol Pharmacol 72: 919–935, 1994
17. Jaggar JH, Wellman GC, Heppner TJ *et al.*: Calcium channel, ryanodine receptors and calcium activated potassium channels: A functional unit for regulating arterial tone. Acta Physiol Scand 164: 577–585, 1998
18. Karaki H, Osaki H, Hori M, Saito MM, Amano KI, Myamoto S, Nakazawa H, Won KJ, Sato K: Calcium movements, distribution and function in smooth muscle. Pharmacol Rev 49: 157–230, 1997

19. Shiraishi Y, Kanmura Y, Yamagushi S, Yoshimura N, Itoh T: Possible mechanisms underlying the midazolam induced relaxation of the nor adrenaline contraction in rabbit mesenteric resistance artery. Br J Pharmacol 121: 1155–1163, 1997
20. Yamaguchi S, Kanmura Y, Yoshimura N: Effects of midazolam in smooth muscle of the rabbit mesenteric artery. Anesth Anal 84: 199–205, 1997
21. Cheng EY, Mazzeo AJ, Bosnjak ZJ, Coon RL, Kampine JP: Direct relaxant effects of intravenous anesthesics on airway smooth muscles. Anesth Anal 83: 162–168, 1996

Molecular and Cellular Biochemistry **246:** 19–24, 2003.

Improved pulmonary artery buffering function during phenylephrine-induced pulmonary hypertension

Daniel Bia,[1] Juan C. Grignola,[1] Ricardo L. Armentano[1,2] and Fernando F. Ginés[1]

[1]*Departamento de Fisiología, Facultad de Medicina, Universidad de la República, Montevideo, Uruguay;* [2]*Universidad Favaloro, Buenos Aires, Argentina*

Abstract

The goal of this study was to determine the *in vivo* pulmonary arterial buffering function (BF) during acute and moderate pulmonary hypertension achieved by phenylephrine-induced smooth muscle activation.

Pulmonary pressure (Konigsberg P7) and diameter (sonomicrometry) were measured in nine anesthetized sheep. Transit pulmonary arterial hypertension was induced by mechanical occlusion of the pulmonary artery (HP) and by phenylephrine infusion (5 μg/kg/min) (PHE). A viscoelastic Kelvin-Voigt model was used. By increasing the values of the viscous modulus, the pressure-diameter hysteresis area was reduced to a minimum in order to obtain the purely elastic pressure-diameter relationship. The elastic index (E) was calculated as the first derivative of the exponential model of the purely elastic pressure-diameter relationship at the mean pressure point.

Systolic, diastolic, mean and pulse pressures were similar during HP and PHE, but significantly higher with regard to control steady state. In HP, E and arterial diameter (both its minimum and maximum values) increased significantly. In contrast, when pulmonary hypertension was induced by VSM activation, E was maintained concomitantly with pulmonary artery vasoconstriction.

Pulmonary hypertension produced by occlusion of the pulmonary artery increases elasticity. Smooth muscle activation may offset the deleterious effect of pulmonary hypertension on arterial wall elasticity by reducing E and impeding arterial dilatation and collagen recruitment, maintaining BF during pulmonary hypertension. (Mol Cell Biochem **246**: 19–24, 2003)

Key words: buffering function, elasticity, pulmonary artery, sheep, vascular smooth muscle

Introduction

It is known that the large arteries have two distinct but interrelated major functions: (a) to be low resistance blood distribution conduits to deliver an adequate supply of blood to the peripheral organs, named conduit function, and (b) to smooth out the pressure and flow pulsatility due to intermittent ventricular ejection ('Windkessel' effect), in order to transform it into an almost continuous arteriolar-capillary flow and pressure, named buffering function (BF) [1–3]. The BF exerts a protective cardiovascular action. The efficiency of BF depends on viscoelastic properties of arterial walls and the 'geometric' characteristics of the arteries including their diameter, cross-section and length. The principal alteration in BF is due to stiffening of the arterial wall.

Today there is a growing interest in studying the large arteries BF, because its accurate characterization could contribute to a better understanding of both arterial and ventricular mechanics and the ventricle-arterial coupling alterations accompanying cardiovascular diseases [4, 5].

Among the arterial wall mechanical properties, elasticity (E) plays the principal role in BF [6]. The function of the elas-

Address for offprints: D. Bia Santana, Departamento de Fisiología, Facultad de Medicina, Universidad de la República, General Flores 2125, 11800 Montevideo, Uruguay, South America (E-mail: dbia@fmed.edu.uy)

tin and collagen fibers is to maintain a steady tension to hold the wall against the transmural pressure present in the vessel. The vascular smooth muscle (VSM) contribution to the elastic behavior of large arteries is a controversial topic [7].

The active tension developed by the VSM has not often been believed to be an important factor in the mechanical properties of the blood vessel, although it alters E as well as the viscosity of the vessel wall [3, 8, 9].

Barra *et al.* showed in experiments performed in descending thoracic aorta of conscious dogs that muscle activation by phenylephrine increased the elastic modulus when this was plotted as a function of strain (isometric analysis) and reduced E when the plot was performed as a function of pressure or stress (isobaric analysis) [8]. While the former analysis is, in fact, a comparison of two different materials (VSM with a very low E versus the collagen fibers whose high E determines the mechanical properties of a largely distended artery), the latter one appears to be the most appropriate approach to use in functional terms.

Arterial viscoelasticity under smooth muscle activation has been extensively studied in the systemic circulation [6, 8, 9].

In the pulmonary circulation, Cox [10, 11], working with extra and intralobar canine pulmonary artery circumferential rings, and Bevan *et al.* [12], working with an isolated rabbit sympathetic nerve-pulmonary artery preparation, demonstrated that the activation of VSM at specific values of transmural pressure resulted in a reduction of values the elastic modulus. Evans *et al.* [13] found similar results studying active and passive mechanics of rabbit pulmonary artery segments rings. However, to our knowledge, no study assessing the effect of the *in vivo* smooth muscle activation on pulmonary artery BF has been reported.

From a clinical point of view, the pulmonary circulation has created great interest among pediatricians because of the frequency and significance of pulmonary vascular changes in congenital heart disease, and among adult cardiologists dealing with acute and chronic pulmonary hypertension (i.e. thromboembolic disease, chronic obstructive pulmonary disease, congestive heart failure) [2].

The goal of this study was to determine the *in vivo* pulmonary arterial BF during acute and moderate pulmonary hypertension induced by phenylephrine smooth muscle activation.

Materials and methods

Instrumentation and surgery

Nine Merino sheep, weighting 26 ± 4.5 kg were anesthetized with intravenous injection of pentobarbital (35 mg/kg). The animals were tracheotomized and respiration was maintained with a positive pressure respirator (Dragger SIMV Polyred 201, Spain), using an inspired oxygen fraction of 40%. Arterial oxygen and carbon dioxide partial pressures were monitored. Respiratory rate and tidal volume were adjusted to maintain pCO_2 at 35–45 mm Hg and pH at 7.35–7.4. Arterial pO_2 always exceeded 80 mm Hg. The main pulmonary artery, and its mains branches (left and right pulmonary arteries), were exposed through a left thoracotomy at the fourth intercostal space. The right safenous vein was catheterized with a polyvinyl chloride fluid-filled catheter (2.8 mm o.d.) for the administration of saline solution and pharmacological agents. A pressure microtransducer (model P7, Konigsberg Instruments, Inc., Pasadena, CA, USA) was inserted in the PA through a stab wound. To measure external pulmonary diameter, a pair of ultrasonic crystals (5 MHz, 3 mm diameter) were sutured to the adventicia of the main pulmonary artery 2–3 mm distal to the site where the microtransducer was placed to avoid interference of the radio frequency signal between crystals. The transit time of the ultrasonic signal (1580 msec) was converted into distance through a sonomicrometer (Triton Technology Inc. San Diego, CA, USA) and observed on the screen of an oscilloscope (model 465B, Tektronix) to confirm optimal positioning of the dimension gauges. This methodology, which has been used by us in previous studies [8, 9], allows accurate and reproducible measurements because of the high frequency and the linearity of response of both dimension gauges and the pressure transducer. A cuff occluder was implanted around the left pulmonary artery.

Experimental protocol

After surgical instrumentation, pulmonary pressure and external diameter were recorded and stored under control steady state (CTL), during the occlusion of the left pulmonary artery (HP) and during intravenous infusion of phenylephrine (PHE, 5 μg/kg/min, Sigma, St. Louis, MO, USA). Ten min were allowed to elapse between these interventions to return to basal values.

Because the variations in pressure and diameter were extremely rapid, the response of the pulmonary artery wall to the occlusion manoeuver during HP only reflects the intrinsic passive elastic properties of the vessel wall. During PHE infusion the instantaneous pressure and diameter recordings were monitored until stabilization was achieved, in fact, we waited 15–20 min to ensure a steady state under PHE infusion. To activate the VSM, we chose phenylephrine, which is a sympathomimetic drug with a direct α_1-receptor stimulant effect and minimal central action, mimicking the actions of the nervous sympathetic system [14].

All experimental procedures were approved by the Animal Care and Use Committee and were conducted in accordance with the National Institutes of Health Guidelines for the care and use of laboratory animals [15].

Data collection

Pulmonary pressure and diameter were monitored in real time and digitized every 5 msec, using a burst sampling with only 50 μsec delay between signals and a 75 Hz low-pass filter. Under steady-state conditions, approximately 20 consecutive beats during control, HP and PHE conditions were sampled and analyzed in a IBM PC. A specific software, developed in our laboratory, was used to obtain the pressure and diameter signals and the elastic parameters [16].

Three main components of the arterial wall were taken into account: elastin, collagen fibers and VSM, each of whom influence the whole elastic behavior of the arterial wall. A two-component model was used (Kelvin-Voigt Viscoelastic model). Accordingly, total pressure developed by the wall to resist stretching can be separated into an elastic and a viscous pressure component [17]:

$$P_{total} = P_{elastic} + P_{viscous} \quad (1)$$
$$P_{elastic} = P_{total} - P_{viscous} \quad (2)$$

As viscous pressure is proportional to the first derivative of the artery diameter, we can develop the following equation:

$$P_{elastic} = P_{total} - \eta_p \frac{dD}{dt} \quad (3)$$

where η_p is the arterial wall viscous index. To separate the purely elastic wall properties, one must subtract the viscous pressure from the total pulmonary pressure, finding the optimal value through the criterion of disappearance of the hysteresis loop. Increasing values of η_p were given by analyzing by visual inspection the reduction of the hysteresis loop area. When the area reached a minimum (considered to be the value that maintained the clockwise course of the loop), an exponential relationship was fitted to the following equation:

$$P = \alpha \cdot e^{\beta \cdot D} \quad (4)$$

E was assessed by using a linear elastic theory and assuming an isotropic homogeneous elastic material for the pulmonary wall. E was calculated as the slope of the pressure-diameter curve at the mean diastolic pressure [8, 9]:

$$E = \left.\frac{dP}{dD}\right|_{\text{mean pressure}} \quad (5)$$

Statistical analysis

All measurements and calculated values are expressed as mean ± S.D. The presence of significant differences was assessed using ANOVA for repeated measures followed by a *t*-test. A p value of 0.05 was considered significant.

Results

Figure 1 shows pulmonary diameter and pressure waveforms during each experimental condition in a typical sheep. Arterial diameter shows a time lag due to viscous effects. Note arterial vasoconstriction during phenylephrine infusion. In contrast, during pulmonary occlusion arterial diameter showed the same trend than pulmonary pressure.

Table 1 shows the hemodynamic parameters (mean ± S.D.), during CTL, HP and PHE experimental conditions. During HP and PHE systolic, diastolic, mean and pulse pressures, were similar but significantly higher than CTL. Sys-

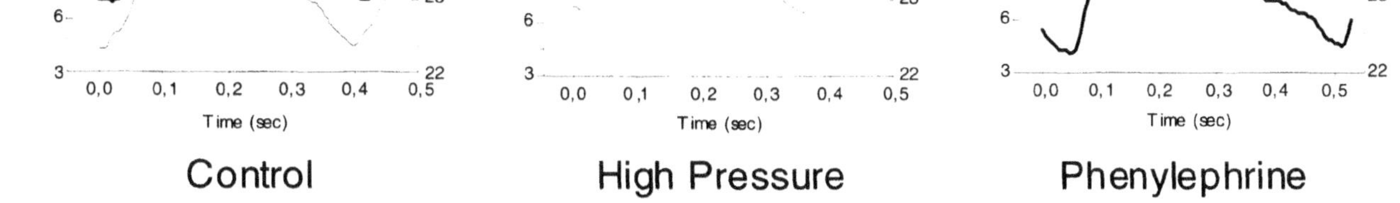

Fig. 1. Pulmonary diameter (thick line) and pressure (thin line) waveforms during control steady state, high pressure (HP) and vascular smooth muscle activation (PHE), in a typical sheep.

Table 1. Basal hemodynamic parameters during control condition, pulmonary occlusion and smooth muscle activation

	Control	Occlusion	Activation
SAP (mm Hg)	16.8 ± 5.92	27.99 ± 8.21*	26.21 ± 6.18*
DAP (mm Hg)	9.16 ± 4.95	13.28 ± 6.02*	11.45 ± 5.11#
MAP (mm Hg)	11.7 ± 5.22	18.18 ± 6.67*	16.37 ± 5.26**
PP (mm Hg)	7.64 ± 1.86	14.71 ± 3.11*	14.76 ± 3.3*
SD (mm)	24.75 ± 3.71	25.76 ± 3.76*	24.72 ± 3.71§
DD (mm)	22.99 ± 3.36	24.1 ± 3.59*	22.03 ± 3.33§**
SD-DD (mm)	1.75 ± 0.94	1.66 ± 1.1	2.69 ± 0.96§**
HR (bpm)	114.6 ± 23.4	114 ± 21.6	96.6 ± 24§**
E (mm Hg/mm)	4.41 ± 2.87	8.76 ± 4.13#	5.02 ± 2.77§

Mean values (mean ± S.D.): n = 9. DD and SD – pulmonary diastolic and systolic diameters respectively; SD-DD – systolic-diastolic difference of pulmonary diameter; HR – heart rate; SAP, DAP and MAP – systolic, diastolic and mean pulmonary pressures, respectively; PP – pulse pressure, E – elastic index. *$p < 0.001$ with respect to control; **$p < 0.01$ with respect to control; #$p < 0.05$ with respect to control and §$p < 0.05$ with respect to occlusion.

tolic diameter during HP increased by 4% ($p < 0.001$) with regard to CTL while it did not change during PHE. Diastolic diameter increased significantly during HP and decreased when VSM was activated by PHE infusion. Despite pulse pressure was similar during both HP and PHE, only in HP the systolic-diastolic difference of pulmonary diameter was maintained. Finally, heart rate was not changed in HP condition and decreased significantly in PHE.

The purely elastic pressure-diameter relationship was successfully fitted to an exponential function, with a correlation coefficient of 0.95 ± 0.04. Elasticity, calculated as the slope at the mean diastolic pressure point, showed a significant in-

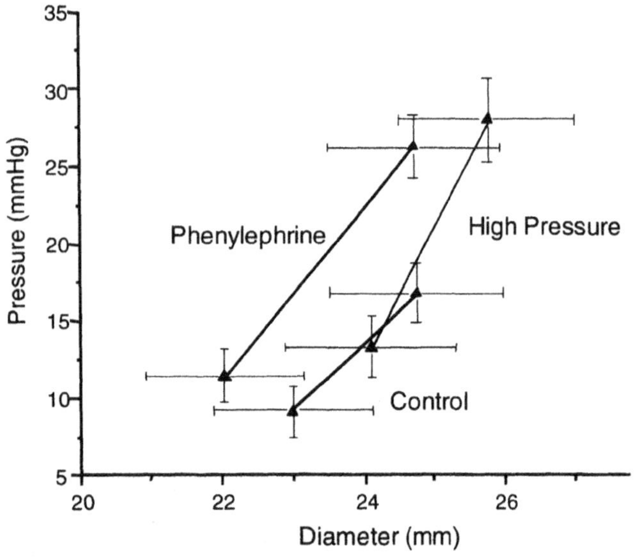

Fig. 2. Plot showing systolic-diastolic values of pulmonary pressure and diameter of all animals (n = 9) (mean ± S.E.M.).

crease (100%, $p < 0.05$) during HP, while under PHE administration E did not change. Figure 2 shows averaged (mean ± S.E.M.) systolic-diastolic values of arterial pressure and diameter during each experimental condition. Note that the activation of VSM by phenylephrine provokes isobaric contraction in diameter and a decreased slope (lesser stiffness) with regard to HP, shifting the pressure-diameter relationship toward the left.

Discussion

The goal of this study was to assess the effect of smooth muscle activation induced by phenyleprine on the *in vivo* pulmonary arterial BF during acute and moderate pulmonary hypertension and its possible beneficial effect on the pulmonary circulation. To this end, the VSM was activated by using phenylephrine, which could modify the wall elasticity of the pulmonary artery by a direct effect on the wall and/or by wall distention secondary to the intravascular pressure increase generated by pulmonary vasoconstriction.

To determine the pressure-diameter relationship in the *in vivo* preparation the ultrasonic dimension technique and a miniature pressure gauge (Konigsberg P7) were used [8, 9, 16, 17]. These techniques have been used in our laboratory [8, 9, 16, 17] allowing accurate and reproducible measurements over a long time due to the high frequency responses of the dimension gauge and the pressure transducer and the linearity of their responses. The 200 Hz sampling rate used in the digitalization of the data is at least two times higher than the highest frequency components in the pressure and diameter spectra, thus allowing signal reconstruction without distortion.

An original procedure of hysteresis elimination was used to determine the purely elastic behavior, based on a previous work where complete assessment of aortic wall mechanics was performed [9, 18].

The mathematical model used to characterize the elastic behavior of the arterial wall has been previously validated in several arteries as well as under different conditions [9, 18] and can be easily extrapolated to non-invasive measurements. From the exponential model, the arterial elasticity can be calculated at each level of pressure to obtain the elastic behavior of the vessel.

We increased arterial pressure mechanically by use of a cuff occluder around the left pulmonary artery. Therefore, we could compare the elastic behavior of the pulmonary artery wall with and without VSM activation at the same level of arterial pressure. Reaching similar increased values of arterial pressure during both HP and PHE, only in HP E was significantly increased, with a concomitant increase in systolic and diastolic values of pulmonary diameter (Fig. 1). Our data show that, under PHE administration, it could be possible to increase pulmonary arterial pressure without distending the arterial wall in anesthetized sheep. Furthermore, the diastolic diameter was significantly reduced with regard to CTL. The isobaric contraction of the diameter could counteract the increased E observed during HP. The reduction of E observed under VSM activation in relation to HP, could be regarded as a balance between the augmentation of peripheral resistance and the reduction of pulmonary stiffness resulting from phenylephrine-induced pulmonary vasoconstriction, which maintains unchanged arterial impedance or limits its variations. Our results agree with previous publications on systemic arteries which concluded that VSM activation decreased arterial wall stiffness when the elastic modulus was plotted against transmural pressure [7, 8, 9].

Taking into account the characterization of aortic wall behavior by Armentano *et al.* [9], during VSM activation the isobaric contraction diameter could reduce the collagen participation with regard to HP, avoiding the collagen fibers recruitment which makes the vessel wall stiffer; in other words, VSM activation transfers the vessel wall stress from collagen to the VSM. Thus, while elastin and collagen passively contribute to stiffness of the vessel, the degree of smooth muscle activation appears to actively modulate BF, independently of arterial pressure effects.

Cox [10, 11] demonstrated in pulmonary artery circumferential rings that the activation of VSM shifted the stress-strain curve upward and to the left. Under conditions of constant blood vessel radius, when E was plotted as a function of transmural pressure, activation produced consistent decreases in values of E among all intra and extralobar arteries sites. This is in agreement with the results presented in this work.

The systemic circulation and the pulmonary circulation receive the same blood flow from the heart with the same periodicity. However, there are many structural and physiological differences between both vascular beds [2]: the pulmonary bed has higher pulmonary arterial distensibility, less wall thickness, no well- developed arterioles, lower peripheral reflection coefficient, absence of increasing stiffness between central and peripheral sites, shorter path lengths along vascular segments, not the same asymmetry shown by the systemic circulation between the upper and lower part of the body, more homogeneous vascular resistance distribution between arteries, capillaries and veins (46, 35 and 20% respectively), tapering to its point of bifurcation, lower mean and pulse arterial pressure. All these differences impede extrapolation of the mechanical properties of systemic wall arteries to the pulmonary artery. In addition, pulmonary arteries with muscular vessel walls are mostly extra-alveolar and involved in active control of pulmonary vascular resistance by mechanisms such as nervous, humoral or gaseous (hypoxic pulmonary hypertension) control. In contrast, pulmonary capillaries are intimately associated with the alveolus, so resistance of these vessels is in high degree influenced passively by alveolar pressure and volume [19, 20].

Pulmonary arteries possess significant amounts of vascular smooth muscle, which is innervated by postganglionic sympathetic fibers. The stimulation of them by their sympathetic nerve supply affects predominantly the compliance or distensibility of the pulmonary artery, with small or negligible effects on pulmonary vascular resistance [19].

Hence, sympathetic stimulation or vasoactive substances like norepinephrine, nitric oxide and endothelins were thought to modulate the VSM tone of the larger arteries, keeping 'Windkessel' function during pulmonary hypertension [19, 21].

In relation with this, Fitch *et al.* [22] demonstrated that acute inhibition of nitric oxide synthase by L-NAME increased aortic stiffness (measured by pulse wave velocity) independent of the concomitant increase in blood pressure, indicating that nitric oxide plays a role in modulating aortic compliance.

The decreased heart rate during PHE could be explained by an increased vagal output secondary to phenylephrine-induced systemic hypertension. It is well known that the ventricular pulsatile power (wasted in arterial pulsations) is markedly altered by changes in heart rate. However, Milnor *et al.* [23] demonstrated that during a heart rate of around 100 beats/min the pulsatile component of right ventricular external power dissipated in the lungs is approximately equal.

Conclusion

The present study demonstrates that acute and moderate pulmonary hypertension modifies pulmonary artery elasticity. The reduction of pulmonary wall stiffness at high levels of pressure, mediated by VSM activation, could be considered as a beneficial contribution to circulation. The elastic behavior of active muscle makes the vessel wall more compliant and appears to actively modulate vascular compliance. When pulmonary hypertension is produced without VSM activation, the vessel wall expands, recruiting more noncompliant collagen and increasing beat to beat energy dissipation, losing storing potential energy. In summary, activation of the VSM (which has a very low E), maintains the efficiency of Windkessel function during pulmonary hypertension.

Acknowledgements

We thank Dr. Alberto Crottogini for revising the manuscript. This work was partially supported by research grants from Universidad de la República (Uruguay)-Universidad Favaloro (Argentina) cooperation agreement and Universidad de la República (CSIC).

References

1. O'Rourke MF: Arterial Function in Health and Disease. Churchill Livingstone, New York, 1982
2. Nichols WW, O'Rourke MF: McDonald's Blood Flow in Arteries: Theoretical, Experimental and Clinical Principles, 3rd edn. Edward Arnold, London, 1990, pp 125–142 and 381–397
3. O'Rourke MF: Second workshop on structure and function of large arteries. Part I. Mechanical principles in arterial disease. Hypertension 26: 2–9, 1995
4. Simon A, O'Rourke MF, Levenson J: Arterial distensibility and its effects on wave reflection and cardiac loading in cardiovascular disease. Coron Artery Dis 2: 1111–1120, 1991
5. O'Rourke MF, Avolio AP, Nichols WW: Left ventricular-systemic arterial coupling in humans and strategies to improve coupling in disease states. In: F.C.P. Yin (ed). Ventricular Vascular Coupling: Clinical, Physiological and Engineering Aspects. Springer-Verlag, New York, 1987, pp 1–19
6. Milnor WK: Hemodynamics. Williams & Wilkins, Baltimore, 1982, pp 56–96
7. Dobrin PB, Rovick AA: Influence of vascular smooth muscle on contractile mechanics and elasticity of arteries. Am J Physiol 217: 1644–1652, 1969
8. Barra JG, Armentano RL, Levenson J, Cabrera Fischer EI, Pichel RH, Simon A: Assessment of smooth muscle contribution to descending thoracic aortic elastic mechanics in conscious dogs. Circ Res 73: 1040–1050, 1993
9. Armentano RL, Barra JG, Levenson J, Simon A, Pichel RH: Arterial wall mechanics in conscious dogs: assessment of viscous, inertial, and elastic moduli to characterize the aortic wall behavior. Circ Res 76: 468–478, 1995
10. Cox RH: Viscoelastic properties of canine pulmonary arteries. Am J Physiol 246: H90–H96, 1984
11. Cox RH: Comparison of mechanical and chemical properties of extra- and intralobar canine pulmonary arteries. Am J Physiol 242: 245–253, 1982
12. Bevan JA, Johnson RC, Verity MA: Changes in elasticity of pulmonary reflexogenic area with sympathetic activity. Am J Physiol 206: 36–42, 1964
13. Evans JN, Little B, Doyle JM: Mechanical properties of isolated pulmonary arterial vessels. Physiologist 23: (abstr) 76, 1980
14. Hoffman BB, Lefkowitz RJ: Alfa-adrenergic agonists, sympathomimetic drugs and adrenergic receptors agonists. In: J.G. Hardman, L.E. Limbird, P.B. Molinoff, R.W. Ruddon, A. Goodman Gilman (eds). The Pharmacological Basis of Therapeutics, 9th edn. McGraw-Hill Interamericana, 1996, pp 211–264
15. Guide for the Care and Use of Laboratory Animals. National Institutes of Health (NIH Publication N° 86–23, revised 1985
16. Bia D, Gamero L, Grignola JC, Rodríguez M, Vallarino M, Núñez L, Armentano RL, Ginés F: Pulmonary artery buffering function characterization. J Mol Cell Cardiol 33: (abstr) A13, 2001
17. Armentano RL, Cabrera Fischer EI: Biomecánica Arterial: Fundamentos para su abordaje en la clínica médica. Editorial Akadia, Buenos Aires, 1994, p 91
18. Armentano RL, Megnien JL, Simon A, Bellenfant F, Barra JG, Levenson J: Effects of hypertension on viscoelasticity of carotid and femoral arteries in humans. Hypertension 26: 48–54, 1995
19. Barnes PJ, Liu SHF: Regulation of pulmonary vascular tone. Pharmacol Rev 47: 87–131, 1995
20. Lumb AB: The pulmonary circulation. In: A.B. Lumb (ed). Nunn's Applied Respiratory Physiology, 5th edn. Butterworth-Heinemann, Oxford, 2000, pp 138–162
21. Veyssier-Belot C, Cacoub P: Role of endothelial and smooth muscle cells in the physiopathology and treatment management of pulmonary hypertension. Cardiovasc Res 44: 274–282, 1999
22. Fitch RM, Vergona R, Sullivan ME, Wang Y: Nitric oxide synthase inhibition increases aortic stiffness measured by pulse wave velocity in rats. Cardiovasc Res 51: 351–358, 2001
23. Milnor WR, Bergel DH, Baragainer JD: Hydraulic power associated with pulmonary blood flow and its relation to heart rate. Circ Res 19: 467–480, 1966

Molecular and Cellular Biochemistry **246**: 25–30, 2003.

Correlation between the extent of coronary atherosclerosis and lipid profile

Janusz Tarchalski,[1] Przemysław Guzik[2] and Henryk Wysocki[2]

[1]*Department of Cardiology, Municipal Hospital in Kalisz, Kalisz;* [2]*Department of Cardiology – Intensive Therapy, Karol Marcinkowski's University School of Medical Sciences, Poznań, Poland*

Abstract

Increased concentration of low density lipoprotein (LDL) cholesterol or decreased level of high density lipoprotein (HDL) cholesterol are important risk factors for coronary atherosclerosis. However, an independent association of triglycerides (TG) with atherosclerosis is uncertain.

The aim of this prospective study was to evaluate the relationship between serum lipid levels and the extent of coronary atherosclerosis in patients with suspected coronary artery disease (CAD) and no previous myocardial infarction who were not treated with lipids lowering therapy or low-lipid diet.

The study was conducted in 141 patients (53.6 ± 7.8 years old; 32 female) who underwent a routine coronary angiography for CAD diagnosis. A modified angiographic Gensini Score (GS) was used to reflect the extent of coronary atherosclerosis. Fasting serum lipid concentrations were determined using cholesterol esterase/peroxidase (CHOD/PAP) enzymatic method for total cholesterol and its fractions and lipase glycerol kinase (GPO/PAP) enzymatic method TG evaluation. The association of Gensini Score with variables characterising lipid profile was analysed with the use of Pearson correlation (r co-efficient; p value).

GS was positively correlated with total cholesterol ($r = 0.404$; $p < 0.001$), LDL cholesterol ($r = 0.484$; $p < 0.001$) and TG ($r = 0.235$; $p = 0.005$). There was a negative correlation between Gensini Score and HDL cholesterol ($r = -0.396$; $p < 0.001$).

In angina pectoris patients with no previous myocardial infarction, the extent of coronary atherosclerosis is positively correlated with pro-atherogenic lipids, i.e. total cholesterol, LDL cholesterol and TG and negatively correlated with antiatherogenic HDL cholesterol. (Mol Cell Biochem **246**: 25–30, 2003)

Key words: coronary artery atherosclerosis, Gensini Score, cholesterol, LDL, HDL, triglycerides

Introduction

Atherosclerosis is responsible for most cases of coronary artery disease (CAD). One of the major predisposing factors to atherosclerosis is an abnormal lipoprotein metabolism and it may be present in over 70% of patients with premature CAD [1]. High concentration of low density lipoprotein (LDL) cholesterol and lipoprotein (a) or low levels of high density lipoprotein (HDL) cholesterol are able to promote atheroma formation and are recognised as particularly important risk factors for atherosclerosis and CAD [1–7]. Randomised trials showed that LDL reduction decreases mortality and coronary events both in a primary and in a secondary prevention [8, 9]. It was found that the reduction in non-fatal myocardial infarction and CAD death was strongly correlated with the small incremented serum HDL-cholesterol level [10]. An independent association of triglycerides (TG) with atherosclerosis and/or CAD remains uncertain. Some studies suggest that TG level does not influence the CAD risk whereas others provide some proofs that it does [11–13].

CAD extent may be evaluated in various ways, for instance by the number of diseased coronary arteries or by CAD score according to Gensini [14, 15]. Gensini Score (GS) is a point scale and it is based on the number of coronary artery segments with stenosis, segment localisation (in respect to its

Address for offprints: P. Guzik, Department of Cardiology – Intensive Therapy, Karol Marcinkowski's University School of Medical Sciences, Przybyszewskiego 49, 60-355 Poznań, Poland (E-mail: pguzik@ptkardio.pl)

significance in myocardial blood supply) and the degree of its lumen narrowing.

Ten years ago it was suggested that there is an inverse association between HDL level and the number of diseased coronary vessels [16]. However, some other studies manifested that there is no association of CAD severity with lipid concentration [17, 18]. This conflicting data persuaded us to perform a prospective study for evaluation of the relationship between serum lipid levels and the extent of atherosclerosis within coronary arteries in patients with suspected CAD and no previous myocardial infarction, and who were not treated with lipids lowering therapy before entering the study.

Materials and methods

Patients

The study inclusion criteria assumed that no patient with suspected CAD had a history of previous myocardial infarction and no one was treated with lipids lowering drugs or low-lipid diet. Routine coronary angiography for CAD diagnosis was performed in 141 consecutive patients (33–71 years old, 109 male and 32 female) who fulfilled inclusion criteria. All patients gave informed consent for participation in the study.

Coronary angiography

Catheterization of coronary arteries was performed with the use of Judkins approach. Selective left and right coronary arteriograms were taken in all patients (Multiskop, Siemens, Germany) and all stenotic segments were evaluated digitally (Ancor, Siemens, Germany). A modification of the Gensini Score was used to reflect the extent of coronary atherosclerosis. Calculation of GS is based on the evaluation of the number of stenotic CA segments, the degree of their lumen stenosis and the localisation of stenotic changes. Each GS score is expressed as a percent of maximal possible score for the individual coronary anatomy in each patient (to compensate for differences in maximal scores due to variations in coronary anatomy). Gensini Score equals 0 if there is no stenotic change within analysed CA segment whereas the maximal GS equals 100% for theoretical total occlusion of all segments.

Blood chemistry

Fasting venous blood was collected before cardiac catheterization from all patients. Total cholesterol, HDL cholesterol and TG were measured in fresh serum samples. Total cholesterol and HDL cholesterol were measured using cholesterol esterase/peroxidase (CHOD/PAP) enzymatic method (respectively Liquick Cor-CHOL 60/120, PZ CORMAY, Poland and CORMAY HLD, PZ CORMAY, Poland). Triglycerides were measured using lipase glycerol kinase (GPO/PAP) enzymatic method (Liquick Cor-TG 60/120, PZ CORMAY, Poland). LDL was calculated by the Friedewald formula. The value of LDL was not calculated for two patients with TG level exceeding 400 mg/dl (respectively 471 and 486 mg/dl).

Statistical analysis

According to Kolmogorov–Smirnov test, all continuous variables had normal distribution. Therefore, association of Gensini Score with variables characterising lipid profile was analysed with the use of Pearson correlation with presentation of r co-efficient and p value. Single variable linear regression was used for better graphical presentation of the relationship between GS and lipid concentration. The differences between continuous data were analysed using Student t-test. Qualitative data is presented as the number of cases and a percentage of all patients. Quantitative data is shown as the mean and standard deviation (S.D.) with minimal and maximal data range. Only p values < 0.05 were considered as significant. All statistical analyses and graphs were done using Prism 3.0 (GraphaPad, USA).

Results

Mean age of patients was 53.6 ± 7.8 years (33–71 years), their average BMI was 27.7 ± 3.6 kg/m^2 (18–44 kg/m^2) and mean left ventricle ejection fraction was 66.8 ± 8.7% (23–81%). Fifteen patients had diabetes mellitus (10.6%), 75 patients had hypertension (53.2%) and 75 were active or current smokers (53.2%). The mean total cholesterol, HDL, LDL and TG and Gensini Score are summarized in Table 1.

Gensini Score was positively correlated with total cholesterol ($r = 0.394$; $p < 0.001$), LDL cholesterol ($r = 0.460$; $p < 0.001$) and TG ($r = 0.235$; $p = 0.005$). There was negative cor-

Table 1. The mean GS, HDL, LDL, TG

	Mean	S.D.	Minimum	Maximum
Total cholesterol	257.6	44.6	391	181
LDL (mg/dl)	186.9	41.6	118	330
HDL (mg/dl)	41.7	8.6	24	75
TG (mg/dl)	195.1	67.1	78	483
Gensini score (%)	17.3	14.1	0	56

GS – Genisni Score; LDL – low density lipoprotein; HDL – high density lipoprotein; TG – triglycerides.

relation between GS and HDL cholesterol ($r = -0.396$; $p < 0.001$). All correlation samples are shown in Fig. 1.

Analysis of association of clinical characteristics with lipids revealed only two significant differences. HDL level was significantly lower in male than in female patients (40.73 ± 8.20 vs. 44.81 ± 9.35 mg/dl; $p = 0.018$) whereas TG level was significantly higher in smokers than in non-smoking patients (208.07 ± 72.65 vs. 180.41 ± 57.21 mg/dl; $p = 0.014$).

Discussion

We observed that the severity of coronary atherosclerosis is significantly related to total cholesterol, LDL, and HDL and triglyceride concentration, i.e. lipids associated with the risk of atherosclerotic changes.

The observed positive association of the extent of coronary atherosclerosis with total cholesterol seems to be caused by cholesterol present in LDL particles. In our study LDL cholesterol was not measured directly but evaluated indirectly (calculated according to Friedewald formula). Nevertheless, there is a positive association of calculated LDL cholesterol with Gensini Score. Circulating LDL can enter cells via apo B/E receptors or through an unregulated scavenger receptor [19]. The latest mechanism, present in smooth muscle cells and macrophages, can result in excess accumulation of intracellular cholesterol and the formation of foam cells. The unregulated scavenger receptors are able to uptake both native and modified LDL particles. The most common form of LDL modification is its oxidation, which can occur in any of the cells within the artery, including the endothelial cells, T lymphocytes, smooth muscle cells and macrophages [19–22]. Foam cells can rupture leading to the release of oxy LDL, proteolytic enzymes and toxic oxygen derivatives that can altogether damage the vessel wall [20, 23]. Oxidized LDL particles can promote atherosclerotic changes by several mechanisms, for instance acting as a monocyte chemoattractant, inducing endothelial cells dysfunction and apoptosis, reducing nitric oxide release and endothelium-dependent vasodilatation [20, 24–26]. Oxy LDL may increase platelet

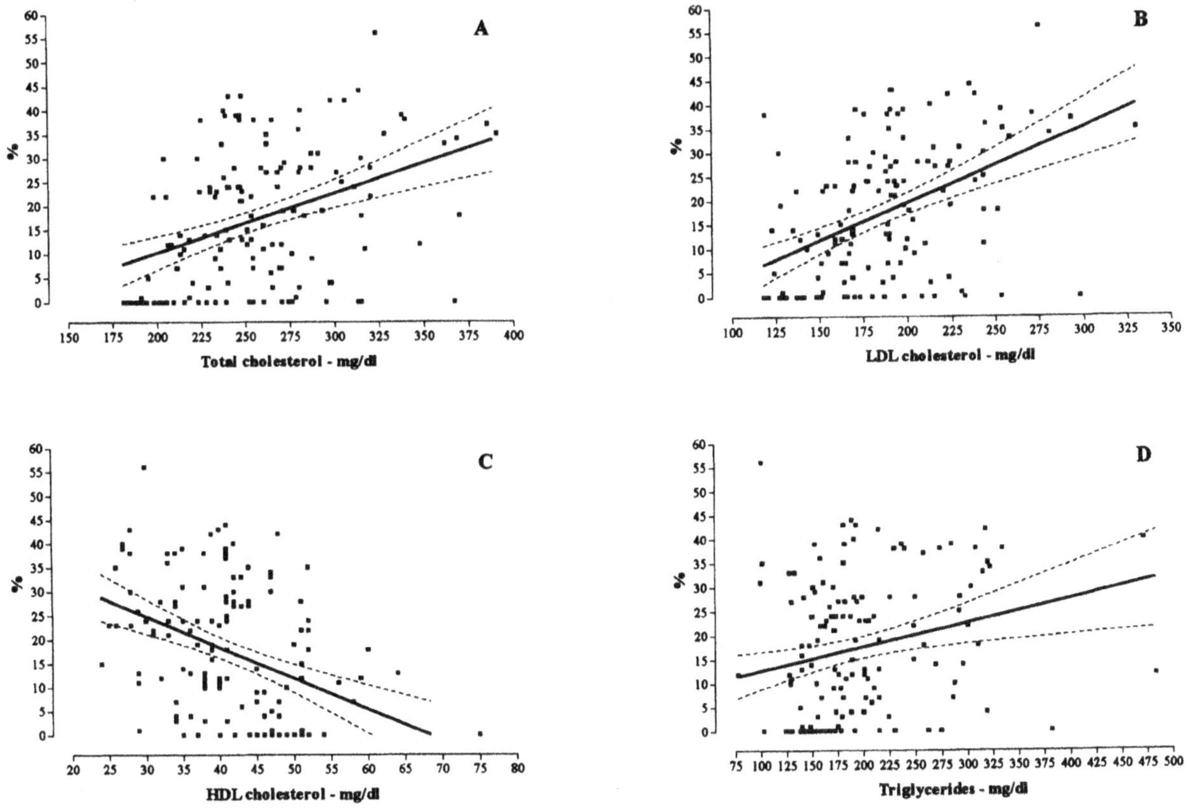

Fig. 1. Correlations between lipid profile and the extent of coronary artery atherosclerosis, i.e. Gensini Score. Correlation and regression line for the association of GS with total cholesterol ($r^2 = 0.1549$; $p < 0.0001$) is shown on between panel A, with LDL cholesterol ($r^2 = 0.2117$; $p < 0.0001$) on panel B, with HDL cholesterol ($r^2 = 0.1568$; $p < 0.0001$) on panel C and with triglycerides ($r^2 = 0.0552$; $p = 0.0050$) on panel D. GS – Genisni Score; LDL – low density lipoprotein; HDL – high density lipoprotein.

aggregation and thromboxane release what causes vasoconstriction and thrombus formation [27, 28]. Cytokines released from activated platelets and injured endothelial cells can stimulate smooth muscle proliferation and the formation of atherosclerotic plaque [29]. Thus, the increased LDL cholesterol is related to the development of atherosclerotic changes within arteries.

The inverse association of GS with HDL cholesterol observed in our study can be explained by antiatherogenic properties of HDL particles. HDL is engaged in reverse cholesterol transport from cells and atherosclerotic plaques into the liver or to other tissues [30]. The uptake of cellular cholesterol by HDL is controlled by apolipoprotein (apo) A-I and cholesterol efflux regulatory protein. Apo A-I serves as a signal transduction protein to mobilize cholesterol esters from intracellular pools and also activates lecithin-cholesterol acyltransferase dependent cholesterol esterification. Cholesterol efflux regulatory protein promotes the transfer of intracellular cholesterol to the cell membrane and surface [31]. Further, cholesterol is removed from HDL particles by direct liver uptake or it is transferred to apo B-containing lipoproteins, i.e. very low density lipoprotein (VLDL), intermediate density lipoproteins (IDL) and LDL [30, 31]. In addition to reverse cholesterol transport, HDL has a variety of vascular actions that can counteract atherogenesis. HDL contains paraoxonase, an enzyme that protects lipoproteins from oxidative modification and it is able to destroy oxidized lipids present in oxy-LDL and to hydrolyze lipid peroxides in human atherosclerotic lesions. It was observed that lower serum paraoxonase is linked to more severe coronary artery disease [32–34]. HDL is thought to maintain endothelial function and a low blood viscosity through a permissive action on red cell deformability [27, 35, 36]. It is speculated that HDL may downregulate thrombin generation and in this way limit its influence on atherosclerosis [37].

There is a weak but positive correlation between triglyceride concentration and GS in our patients. However, according to literature it remains uncertain if there is an independent association of TG with coronary atherosclerosis [11–13]. Lipoproteins rich in triglycerides, i.e. VLDL and IDL (lipoprotein remnants) have been identified in human atherosclerotic plaques [38]. Moreover, hypertriglyceridemia is associated with other abnormalities predisposing to the development of atherosclerosis. Hypertriglyceridemia is usually accompanied by low HDL level and thus it is suspected that the reduction of HDL rather than the elevation in TG is responsible for atherogenesis [38, 39]. It was even demonstrated that enrichment of HDL with TG enhances the clearance of HDL apo A-I from the circulation [40]. Nevertheless, increased TG level reflects the accumulation of triglyceride-rich lipoproteins, which are also transporting cholesterol and are atherogenic. In hypertriglyceridemia, VLDL is enriched in apo E what can lead to a conformational change in the VLDL particle and promotes its binding to the macrophage scavenger receptor similarly to that seen with oxidized LDL [38]. Patients with elevated TG concentration have increased blood viscosity and coagulability that may contribute to endothelial dysfunction and tissue ischemia [27, 38, 41]. In patients with lipoprotein lipase deficiency, triglyceride-enriched lipoproteins have a prolonged circulation time that allows for more interactions with the endothelium [38]. Additionally, insulin resistance is frequently present in hypertriglyceridemic patients and lipoprotein lipase activity is decreased in this situation. Insulin resistance syndrome is associated with the presence of other atherogenic risk factors, i.e. abdominal obesity, hypertension, diabetes mellitus and impaired fibrinolytic capacity [42].

It is noteworthy that increased level of LDL and TG and decreased concentration of HDL is accompanied by an abnormal fibrinolysis, which can be observed in patients with CAD [27, 37, 38]. It was observed that TG concentration and VLDL level are positively correlated with plasminogen activator inhibitor type 1 (PAI-1) antigen, PAI-1 activity, and tissue plasminogen activator (t-PA) antigen, and negatively correlated with t-PA activity. Lowering of plasma triglycerides in hypertriglyceridemic subjects was associated with a significant reduction in PAI-1 antigen levels [27, 43]. LDL is able to inhibit the release of t-PA and it can stimulate PAI-1 release from endothelial cells whereas HDL may suppress PAI-1 activity [27, 44].

Limitation of the study

The performed study has some limitations. In our study LDL was calculated and not measured directly, so its value may be just an approximation of the real LDL level but this approach is widely accepted in clinical practice. Moreover, we did not analyze the association of apo-lipoproteins concentration and lipoprotein (a) that is believed to be particularly atherogenic [2, 6]. Additionally, there is a problem of interactions between lipids and sex and smoking behavior. Gensisni Score is affected by both sex and smoking behaviour in our patients (data not shown). Further, coronary angiography is not the best way for the evaluation of the severity of coronary atherosclerosis. However, this is actually a 'gold standard' method and Gensini Score is used as an indicator of the extent of coronary artherosclerosis in several studies [15, 17, 18, 45, 46].

Conclusions

In patients with angina pectoris and no previous myocardial infarction, the extent of CA atherosclerosis is positively correlated with pro-atherogenic lipids, i.e. total cholesterol, LDL cholesterol and TG. It is also and negatively correlated with antiatherogenic HDL cholesterol.

References

1. Genest JJ Jr, Martin-Munley SS, McNamara JR, Ordovas JM, Jenner J, Myers RH, Silberman SR, Wilson PW, Salem DN, Schaefer EJ: Familial lipoprotein disorders in patients with premature coronary artery disease. Circulation 85: 2025–2033, 1992
2. Stein JH, Rosenson RS: Lipoprotein Lp(a) excess and coronary heart disease Arch Intern Med 157: 1170–1176, 1997
3. Executive Summary of the Third Report of the National Cholesterol Education Program (NCEP) Expert Panel on Detection, Evaluation, and Treatment of High Blood Cholesterol in Adults (Adult Treatment Panel III). JAMA 285: 2486–2497, 2001
4. LaRosa JC, Hunninghake D, Bush D, Criqui MH, Getz GS, Gotto AM Jr, Grundy SM, Rakita L, Robertson RM, Weisfeldt ML: The cholesterol facts: A summary of the evidence relating dietary fats, serum cholesterol, and coronary heart disease: A joint statement by the American Heart Association and the National, Heart, Lung, and Blood Institute. Circulation 81: 1721–1733, 1990
5. Grundy SM, Goodman DS, Rifkind BM, Cleeman JI: The place of HDL in cholesterol management. A perspective from the National Cholesterol Education Program. Arch Intern Med 149: 505–510, 1989
6. Rosenson RS: Beyond low-density lipoprotein cholesterol. A perspective on low high-density lipoprotein disorders and Lp(a) lipoprotein excess. Arch Intern Med 156: 1278–1284, 1996
7. Kinosian B, Glick H, Garland G: Cholesterol and coronary heart disease: Predicting risks by levels and ratios. Ann Intern Med 121: 641–647, 1994
8. Shepherd J, Cobbe SM, Ford I, Isles CG, Lorimer AR, MacFarlane PW, McKillop JH, Packard CJ: Prevention of coronary heart disease with pravastatin in men with hypercholesterolemia. West of Scotland Coronary Prevention Study Group. N Engl J Med 333: 1301–1307, 1995
9. Sacks FM, Pfeffer MA, Moye LA, Rouleau JL, Rutherford JD, Cole TG, Brown L, Warnica JW, Arnold JM, Wun CC, Davis BR, Braunwald E: The effect of pravastatin on coronary events after myocardial infarction in patients with average cholesterol levels. Cholesterol and Recurrent Events Trial investigators. N Engl J Med 335: 1001–1009, 1996
10. Robins SJ, Collins D, Wittes JT, Papademetriou V, Deedwania PC, Schaefer EJ, McNamara JR, Kashyap ML, Hershman JM, Wexler LF, Rubins HB: Relation of gemfibrozil treatment and lipid levels with major coronary events: VA-HIT: A randomized controlled trial. JAMA 285: 1585–1591, 2001
11. Avins AL, Neuhaus, JM: Do triglycerides provide meaningful information about heart disease risk? Arch Intern Med 160: 1937–1944, 2000
12. Austin MA, Hokanson JE, Edwards KL: Hypertriglyceridemia as a cardiovascular risk factor. Am J Cardiol 81: 7B–12B, 1998
13. Ballantyne CM, Grundy SM, Oberman A, Kreisberg RA, Havel RJ, Frost PH, Haffner SM: Hyperlipidemia: Diagnostic and therapeutic perspectives. J Clin Endocrinol Metab 85: 2089–2112, 2000
14. Marwick T, D'Hondt A-M, Baudhuin T, Willemart B, Wijns W, Detry J-M: Optimal use of dobutamine stress for the detection and evaluation of coronary artery disease: Combination with echocardiography or scintigraphy, or both? J Am Coll Cardiol 22: 159–167, 1993
15. Hoffmeister A, Rothenbacher D, Bazner U, Frohlich M, Brenner H, Hombach V, Koenig W: Role of novel markers of inflammation in patients with stable coronary heart disease. Am J Cardiol 87: 262–266, 2001
16. Romm PA, Green CE, Reagan K, Rackley CE: Relation of serum lipoprotein cholesterol levels to presence and severity of angiographic coronary artery disease. Am J Cardiol 67: 479–83, 1991
17. Lekakis JP, Papamichael CM, Cimponeriu AT, Stamatelopoulos KS, Papaioannou TG, Kanakakis J, Alevizaki MK, Papapanagiotou A, Kalofoutis AT, Stamatelopoulos SF: Atherosclerotic changes of extracoronary arteries are associated with the extent of coronary atherosclerosis. Am J Cardiol 85: 949–952, 2000
18. Cerne D, Kaplan-Pavlovcic S, Kranjec I, Jurgens G: Mildly elevated serum creatinine concentration correlates with the extent of coronary atherosclerosis. Ren Fail 22: 799–808, 2000
19. Brown MS, Goldstein JL: A receptor-mediated pathway for cholesterol homeostasis. Science 232: 34–47, 1986
20. Steinberg D, Parthasarathy S, Carew TE, Khoo JC, Witztum JL: Beyond cholesterol. Modifications of low-density lipoprotein that increase its atherogenicity. N Engl J Med 320: 915 – 924, 1989
21. Hiltunen TP, Luoma JS, Nikkari T, Yla-Herttuala S: Expression of LDL receptor, VLDL receptor, LDL receptor-related protein, and scavenger receptor in rabbit atherosclerotic lesions. Marked induction of scavenger receptor and VLDL receptor expression during lesion development. Circulation 97: 1079–1086, 1998
22. Klouche M, Rose-John S, Schmiedt W, Bhakdi S: Enzymatically degraded, nonoxidized LDL induces human vascular smooth muscle cell activation, foam cell transformation, and proliferation. Circulation 101: 1799–1805, 2000
23. Dimmeler S, Haendeler J, Galle J, Zeiher AM: Oxidized low-density lipoprotein induces apoptosis of human endothelial cells by activation of CPP32-like proteases. A mechanistic clue to the 'response to injury' hypothesis. Circulation 95: 1760–1763, 1997
24. Li D, Mehta JL: Antisense to LOX-1 inhibits oxidized LDL-mediated upregulation of monocyte chemoattractant protein-1 and monocyte adhesion to human coronary artery endothelial cells. Circulation 101: 2889–2895, 2000
25. Mathew V, Cannan CR, Miller VM, Barber DA, Hasdai D, Schwartz RS, Holmes DR Jr, Lerman A: Enhanced endothelin-mediated coronary vasoconstriction and attenuated basal nitric oxide activity in experimental hypercholesterolemia. Circulation 96: 1930–1936, 1997
26. Anderson TJ, Meredith IT, Charbonneau F, Yeung AC, Frei B, Selwyn AP, Ganz P: Endothelium-dependent coronary vasomotion relates to the susceptibility of LDL to oxidation in humans. Circulation 93: 1647–1650, 1996
27. Rosenson RS, Lowe GD; Effects of lipids and lipoproteins on thrombosis and rheology. Atherosclerosis 140: 271–80, 1998
28. Chen LY, Mehta P, Mehta JL: Oxidized LDL decreases L-arginine uptake and nitric oxide synthase protein expression in human platelets: Relevance of the effect of oxidized LDL on platelet function. Circulation 93: 1740–1746, 1996
29. Ross R: The pathogenesis of atherosclerosis. A perspective for the 1990s. Nature 362: 801–809, 1993
30. Tall AR: Plasma high density lipoproteins: Metabolism and relationship to atherogenesis. J Clin Invest 86: 379–384, 1990
31. Marcil M, Brooks-Wilson A, Clee SM, Roomp K, Zhang LH, Yu L, Collins JA, van Dam M, Molhuizen HO, Loubster O, Ouellette BF, Sensen CW, Fichter K, Mott S, Denis M, Boucher B, Pimstone S, Genest J Jr, Kastelein JJ, Hayden MR: Mutations in the ABC1 gene in familial HDL deficiency with defective cholesterol efflux. Lancet 354: 1341–1346, 1999
32. Aviram M, Hardak E, Vaya J, Mahmood S, Milo S, Hoffman A, Billicke S, Draganov D, Rosenblat M: Human serum paraoxonases (PON1) Q and R selectively decrease lipid peroxides in human coronary and carotid atherosclerotic lesions: PON1 esterase and peroxidase-like activities. Circulation 101: 2510–2517, 2000

33. James RW, Leviev I, Righetti A: Smoking is associated with reduced serum paraoxonase activity and concentration in patients with coronary artery disease. Circulation 101: 2252–2257, 2000
34. Shih DM, Gu L, Xia YR, Navab M, Li WF, Hama S, Castellani LW, Furlong CE, Costa LG, Fogelman AM, Lusis AJ: Mice lacking serum paraoxonase are susceptible to organophosphate toxicity and atherosclerosis. Nature 394: 284–287, 1998
35. Kuhn FE, Mohler ER, Satler LF, Reagan K, Lu DY, Rackley CE: Effects of high-density lipoprotein on acetylcholine-induced coronary vasoreactivity. Am J Cardiol 68: 1425–1430, 1991
36. Epand RM, Stafford A, Leon B, Lock PE, Tytler EM, Segrest JP, Anantharamaiah GM: HDL and apolipoprotein A-1 protect erythrocytes against the generation of procoagulant activity. Arterioscler Thromb 14: 1775–1783, 1994
37. Griffin JH, Kojima K, Banka CL, Curtiss LK, Fernandez JA: High-density lipoprotein enhancement of anticoagulant activities of plasma protein S and activated protein C. J Clin Invest 103: 219–227, 1999
38. Rosenson RS: Approach to the patient with hypertriglyceridemia-I. In: B.D. Rose (ed). UpToDate 2001; Version 9.3. UpToDate Inc., Wellesley, MA, USA, http://www.uptodate.com
39. Garber AM, Avins AL: Triglyceride concentration and coronary heart disease. Not yet proved of value as a screening test. BMJ 309: 2–3, 1994
40. Lamarche B, Uffelman KD, Carpentier A, Cohn JS, Steiner G, Barrett PH, Lewis GF: Triglyceride enrichment of HDL enhances *in vivo* metabolic clearance of HDL apo A-I in healthy men. J Clin Invest 103: 1191–1199, 1999
41. Rosenson RS, Shott S, Lu L, Tangney CC: Hypertriglyceridemia and other factors associated with plasma viscosity. Am J Med 110: 488–492, 2001
42. Reaven GM: Role of insulin resistance in human disease (syndrome X): An expanded definition. Annu Rev Med 44: 121–131, 1993
43. Mussoni L, Mannucci L, Sirtori M, Camera M, Maderna P, Sironi L, Tremoli E: Hypertriglyceridemia and regulation of fibrinolytic activity. Arterioscler Thromb 12: 19–27, 1992
44. Levin EG, Miles LA, Fless GM, Scanu AM, Baynham P, Curtiss LK, Plow EF: Lipoproteins inhibit the secretion of tissue plasminogen activator from human endothelial cells. Arterioscler Thromb 14: 438–442, 1994
45. Oishi Y, Wakatsuki T, Nishikado A, Oki T, Ito S: Circulating adhesion molecules and severity of coronary atherosclerosis. Coron Artery Dis 11: 77–81, 2000
46. Inoue T, Matsunaga R, Morooka S, Uehara Y: Serum N-acetyl-beta-D-gulucosaminidase activity increases in association with insulin resistance in patients with coronary artery disease. Atherosclerosis 149: 117–122, 2000

Molecular and Cellular Biochemistry **246**: 31–38, 2003.

Regulation of cytosolic phospholipase A_2, cyclooxygenase-1 and -2 expression by PMA, TNFα, LPS and M-CSF in human monocytes and macrophages

Yan J. Jiang,[1] Biao Lu,[2] Patrick C. Choy[2] and Grant M. Hatch[2,3]

Departments of [1]Pathology; [2]Biochemistry and Medical Genetics; [3]Pharmacology and Therapeutics and Center for Research and Treatment of Atherosclerosis, Faculty of Medicine, University of Manitoba, Winnipeg, Canada

Abstract

Cytosolic phospholipases A_2 (cPLA$_2$) and cyclooxygenases-1 and -2 (COX-1 and -2) play a pivotal role in the metabolism of arachidonic acid (AA) and in eicosanoid production. The coordinate regulation and expression of these enzymes is not well defined. In this study, the effect of phorbol 12-myristate 13-acetate (PMA), tumor necrosis factor α (TNFα), lipopolysaccharide (LPS) and macrophage-colony stimulating factor (M-CSF) on AA release and prostaglandin E_2 (PGE$_2$) production and the expression of cPLA$_2$ and COX-1 and -2 were investigated in U937 human pre-monocytic cells and fully differentiated macrophages. Treatment of U937 cells with PMA or macrophages with LPS increased AA release and PGE$_2$ production. Incubation of U937 cells or macrophages for 8 h with all stimuli elevated cPLA$_2$ expression. In contrast, cPLA$_2$ expression was reduced upon further incubation of U937 cells or macrophages for 24 h with all stimuli indicating a bi-phasic expression pattern of this enzyme. PMA induced COX-1 expression in U937 cells whereas LPS induced COX-2 expression in macrophages. Although TNFα and M-CSF induced a significant amount of AA release in both cell models, they failed to induce a comparable production of PGE$_2$ since they were unable to induce the coordinate expression of the downstream key enzymes, COX-1 or COX-2. The results suggest that the enhancement of AA release in both U937 cells and macrophages may be caused by both increased cPLA$_2$ activity and elevated cPLA$_2$ protein expression. In addition, PMA stimulates PGE$_2$ production via up-regulation of COX-1, and likely COX-2, expression in U937 cells whereas LPS stimulates PGE$_2$ production via induction of COX-2 expression in macrophages. (Mol Cell Biochem **246**: 31–38, 2003)

Key words: cPLA$_2$, COX-1, COX-2, PMA, LPS, TNFα, U937, macrophages

Abbreviations: AA – arachidonic acid; cPLA$_2$ – cytosolic phospholipase A_2; COX-1/2 – cyclooxygenase 1/2; PG – prostaglandin; PGE$_2$ – prostaglandin E_2; PMA – phorbol 12-myristate 13-acetate; TNFα – tumor necrosis factor α; LPS – lipopolysaccharide; M-CSF – macrophage-colony stimulating factor

Introduction

Human monocytes are known to metabolize arachidonic acid (AA) and to release prostaglandins upon stimulation [1, 2]. When monocytes undergo maturation and differentiation into macrophages, this property changes. For example, monocyte-derived-macrophages are more readily responsive to lipopolysaccharide (LPS) treatment and release a large amount of prostaglandin E_2 (PGE$_2$) [3]. PGE$_2$ is a vasodilator which has been implicated in signal transduction [4] and the maintenance of vascular integrity [5, and references therein]. Hence, studies on regulation of PGE$_2$ biosynthesis are important for

Address for offprints: G.M. Hatch, Department of Pharmacology and Therapeutics, Faculty of Medicine, University of Manitoba, A307 Chown Building, 753 McDermot Avenue, Winnipeg, Manitoba, Canada, R3E 0W3 (E-mail: hatchgm@ms.umanitoba.ca)

our understanding of chronic inflammatory diseases such as atherosclerosis.

The biosynthesis of PGE_2 is regulated by the bio-availability of AA and the activity of cyclooxygenase (COX) enzyme. Among several groups of phospholipase A_2, the high molecular weight cytosolic form ($cPLA_2$) is responsible for increased AA release in U937 cells [6]. There are two isoforms of cyclooxygenase (COX-1 and -2) which are encoded by different genes [7]. Generally, COX-1 is constitutively expressed in most tissues and appears to be involved in cellular homeostasis [8]. In contrast, COX-2 is highly inducible by reagents including phorbol esters, growth factors, cytokines, and plays a central role in inflammatory and mitogenesis [9, 10]. Functional coupling of $cPLA_2$ to COX-1 and/or COX-2 in AA metabolism and prostaglandin production is beginning to emerge [7]. However, few reports have addressed the coordinate regulation of these three enzymes in the cyclooxygenase pathway during inflammatory stimulation.

In this study, the effects of inflammatory stimuli (PMA, TNFα, LPS) and growth factor stimuli (M-CSF) on AA release and PGE_2 production were examined in human premonocytic U937 cells and fully differentiated macrophages. Since $cPLA_2$, COX-1 and -2 play important roles in the release of AA and PGE_2 production, the mechanisms underlying AA metabolism via the COX pathway were investigated by the simultaneous examination of the expression of these three key enzymes in response to stimuli. The expression pattern of these enzymes during early (8 h) and prolonged (24 h) stimulation were examined. Our results demonstrate that PMA stimulates AA release and subsequent PGE_2 production via up-regulation of $cPLA_2$, COX-1, and likely COX-2, expression in undifferentiated U937 cells. In contrast, LPS preferentially stimulates AA release and PGE_2 production via induction of $cPLA_2$ and COX-2 expression in macrophages.

Materials and methods

Materials

RPMI-1640 medium, phosphate-buffered saline (PBS), phorbol 13-myristate 12-acetate (PMA), bovine serum albumin (BSA), and other chemicals were obtained from Sigma Chemical Company (St. Louis, MO, USA). Polyclonal anti-COX-2 antibodies, monoclonal anti-$cPLA_2$ antibody, and polyclonal anti-α-actin antibodies were purchased form Santa Cruz Biotechnology, Inc. (Santa Cruz, CA, USA). Monoclonal anti-COX-1 antibody and PGE_2 enzyme immunoassay (EIA) assay kit were purchased form Cayman Chemical Co. (Ann Arbor, MI, USA). Calcium ionophore Ionomycin was obtained from BioMol Inc. (Plymouth Meeting, PA, USA). [5,6,8,9,11,12,14,15-^{3}H]-arachidonic acid (209 mCi/mmol) was obtained from Amersham Corp. (Quebec, Canada). All lipid standards were obtained from Serdary Research Laboratory (London, Ontario, Canada). Thin layer chromatograph plates (silica gel G) were the products of Fisher Scientific.

Cell culture

Human pre-monocytic cell line U937 (CRL-1593.2) was obtained from the American Type Culture Collection (ATCC) (Rockville, MD., USA). In the undifferentiated cell model, U937 cells were cultured in suspension in RPMI-1640 medium containing 10% fetal bovine serum, 100 units/ml of penicillin G, 10 μg/ml of streptomycin, and 0.25 μg/ml amphotericin B. Cells were seeded at a density of 7×10^5/ml in 60-mm Petri dishes and were made quiescent by incubation with FBS-free medium containing 0.1% BSA (basal medium) before the treatment of the appropriate reagents. U937 cells were differentiated into macrophages by incubation with PMA (160 nM, dissolved in basal medium) for 48 h. At the end of the incubation, macrophages became firmly attached to the bottom of the Petri dish. Undifferentiated floating cells and dead cells were washed away with three rinses of warm PBS. The attached macrophages were used in subsequent experiments.

[^{3}H]-arachidonic acid radio-labeling

U937 cells or macrophages were radio-labeled in flasks or Petri dishes respectively, for 12 h in basal medium containing 1.0 μCi/ml [^{3}H] AA. Cells were then rinsed three times with basal medium to remove any non-specific binding of AA to the cell surface prior to agonist stimulation [11]. Subsequently, cells were incubated in basal medium containing either PMA (160 nM, U937 cell model), TNFα (10 ng/ml), LPS (100 ng/ml) or M-CSF (10 μg/ml) for the indicated periods of time.

Measurement of AA-release

The AA released from the cells was determined as described previously [11, 12]. Briefly, following the treatment with reagents (described above) or vehicle, cells were stimulated with 5 μM calcium ionophore ionomycin for 15 min. The medium was collected and centrifuged at 1,000 xg for 8 min at 40°C, to remove the suspension cells. The recovered medium was acidified with 50 μl of glacial acetic acid. A 0.8-ml aliquot was resolved by thin layer chromatography (TLC) in a solvent system consisting of hexane/diethyl ether/acetic acid (70:30:1, by volume). The fatty acid fraction was visualized by iodine vapor, and its radioactivity was determined by liquid scintillation counting.

Measurement of PGE_2 release

Following the treatment of U937 cells with PMA, TNFα, LPS or M-CSF or macrophages with TNFα, LPS or M-CSF for 8 or 24 h, respectively, the medium was collected and PGE_2 in the medium was determined. The PGE_2 level was determined by enzyme immunoassay (EIA). Briefly, 50 μl of PGE_2 standard or samples were added to the pre-coated mouse anti-rabbit IgG macrotitre plates. Subsequently, 50 μl PGE_2 tracer and 50 μl monoclonal antiserum of PGE_2 were added into each well and the mixture was incubated for 18 h at 4°C. After incubation, the content in each well was removed and the wells were washed 12 times with PBS buffer containing 0.05% Tween-20. An aliquot of 200 μl Ellman's reagent was added into each well, and the mixture was incubated for 90 min at room temperature with occasional shaking, until the development of a yellow color solution. The optical density of solution was determined by a multi-well spectrophotometer (ELISA reader) at 412 nm. PGE_2 concentration in each well was calculated from a standard curve.

Electrophoresis and Western blot analyses

Cell lysate was prepared by treating cells on ice for 20 min with lysis buffer (15 nM Tris, pH 8.0, 0.5% Tween X-100, 1 nM DTT, 0.5 nM EDTA and 1 mM phenylmethysulfonyl fluoride). Subsequently, the supernatant was obtained by centrifuging the crude lysates at 4,000 g for 8 min to remove cell debris. The protein concentration of the supernatant was determined by the protein assay kit (Bio-Rad) using BSA as a reference. Samples containing 30–50 μg of protein were subjected to sodium dodecylsulfate/8.0% polyacrylamide gel electrophoresis (SDS-PAGE) with pre-stained protein markers as molecule weight references. Protein fractions in the gel were transferred to the Immobilon-P transfer membrane (Millipore). The membrane was incubated with the specific antibodies for $cPLA_2$ (1:200–1:500), COX-1 (1:1000), COX-2 (1:500), or α-actin (1:2000) overnight at 4°C. Subsequently, the membrane was washed and incubated with peroxidase labeled with secondary antibodies (1:2000–4000) for 30 min at room temperature. Protein bands on the membrane were visualized by chemiluminescence detection system (Amersham). The relative intensities of protein bands on film were acquired using Bio-Rad MAZ MultiImaging System (Bio-Rad), and the digital data were analyzed by the software Quantity One (Bio-Rad).

Statistical analysis

The data were analyzed with a two-tailed Students *t*-test. In all cases, the level of significance was defined as $p < 0.05$ or $p < 0.01$. Results are presented as the mean ± S.D.

Results

PMA, TNFα, LPS and M-CSF enhance AA release in U937 pre-monocytes and macrophages

AA is released from cell membrane phospholipids via the action of a family of phospholipase A_2, mainly $cPLA_2$ [6]. The effect of stimuli on AA release was examined in undifferentiated U937 pre-monocytes and in differentiated macrophages. As shown in Fig. 1A, treatment of U937 cells for 8 h with PMA, LPS and TNFα enhanced AA release. Prolonged incubation of cells with PMA, LPS or TNFα for up to 24 h resulted in a higher amount of AA release. PMA had the maximal effect (40% increase at 8 h and 150% increase at 24 h compared to control). Treatment of U937 cells for 8 h with M-CSF did not affect AA release. In contrast, M-CSF enhanced AA release after 24 h treatment. The PMA-treated

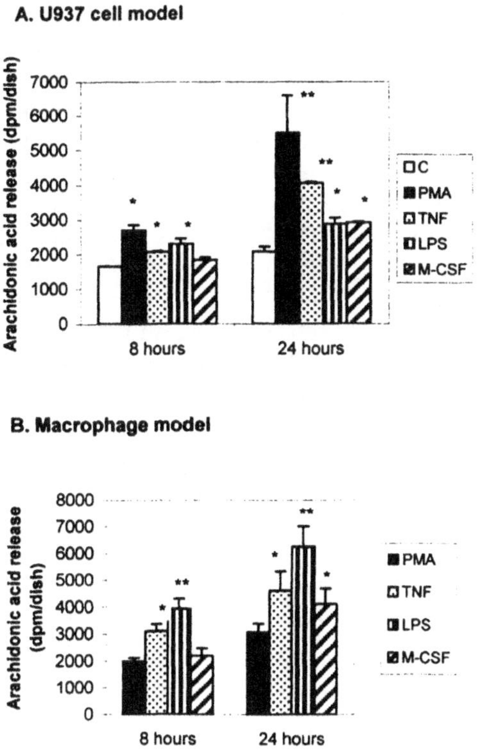

Fig. 1. The effect of stimuli on arachidonic acid (AA) release in U937 cells and macrophages. U937 cells or macrophages were incubated for 12 h with 1.0 μCi/ml [^{3}H] AA, washed and subsequently U937 cells were incubated in the absence (C) or presence of PMA (160 nM) or TNFα (10 ng/ml) or LPS (100 ng/ml) or M-CSF (10 μg/ml) for 8 or 24 h. PMA-differentiated macrophages (PMA) were treated with the same amount of TNFα or LPS or M-CSF for the same periods of time as in U937 cells. The AA released from cells were determined as described in 'Materials and methods'. (A) U937 cells. (B) Macrophages. Results presented are means ± S.D. of 3 independent experiments. *$p < 0.05$; **$p < 0.01$.

U937 cells were not fully differentiated into macrophages since they did not adhere to the culture dish, an indication of full differentiation. Incubation of U937 cells with PMA for 48 h caused full differentiation into macrophages and adherence to culture dishes.

In fully differentiated macrophages, treatment with TNFα and LPS for 8 h resulted in an increase in AA release compared to controls (Fig. 1B). After prolonged (24 h) incubation, TNFα and LPS induced a further enhancement of AA release from macrophages. Treatment of macrophages for 8 h with M-CSF did not affect AA release. In contrast, treatment of macrophages with M-CSF for 24 h enhanced AA release. These results demonstrated that PMA, TNFα, LPS and M-CSF were capable of inducing AA release in U937 cells and macrophages.

PMA, TNFα, LPS and M-CSF differentially affect PGE_2 production in U937 cells and macrophages

PGE_2 is an important prostanoid produced via the cyclooxygeanse pathway. The availability of AA and its subsequent metabolism by cyclooxygeanses are responsible for PGE_2 biosynthesis. Hence, the effect of stimuli on PGE_2 production was examined in U937 cells and macrophages. As shown in Fig. 2, the effects of stimuli on PGE_2 production were different from that of AA release. The basal level of PGE_2 production in U937 cells was low compared to macrophages (Fig. 2). An 8 h challenge of U937 cells with stimuli had little effect on PGE_2 production. In contrast, PMA treatment of U937 cells for 24 h resulted in a 6.4-fold ($p < 0.05$) stimulation of PGE_2 production (Fig. 2A). TNFα, LPS and M-CSF did not affect PGE_2 production in U937 cells. Treatment of macrophages with LPS stimulated PGE_2 production at 8 and 24 h compared to control (Fig. 2B). In contrast, TNFα and M-CSF did not stimulate PGE_2 production in macrophages.

Differential expression of $cPLA_2$, COX-1 and COX-2 by PMA, TNFα, LPS and M-CSF in U937 cells and macrophages

The mechanism of the stimuli-mediated alterations in AA release and PGE_2 production was examined. The effects of PMA, TNFα, LPS and M-CSF on AA release and PGE_2 production in U937 cells and macrophages could result from altered enzyme activities and/or protein mass. Hence, the expression of $cPLA_2$, COX-2 and COX-1 enzyme proteins were determined. PMA, TNFα, LPS and M-CSF treatment of U937 cells resulted in a bi-phasic expression of $cPLA_2$ protein (Fig. 3). During the early course (8 h) of treatment, $cPLA_2$ protein was up-regulated 1.8–3.4 fold ($p < 0.05$) by all stimuli in U937 cells. Prolonged treatment (24 h) of U937 cells with the same stimuli attenuated $cPLA_2$ expression. The attenuation of $cPLA_2$ expression by TNFα was the greatest presumably due to the pro-apoptotic action of this compound. $cPLA_2$ expression was higher (1.8–1.9-fold, $p < 0.05$) in macrophages than in U937 cells (Fig. 4). Early (8 h) treatment of macrophages with TNFα, LPS or M-CSF did not affect $cPLA_2$ expression (Fig. 4). Prolonged treatment (24 h) of macrophages with all stimuli attenuated $cPLA_2$ expression indicating a biphasic expression of $cPLA_2$ protein.

U937 cells expressed COX-1 protein and treatment with TNFα, LPS or M-CSF for 8 or 24 h did not affect COX-1 expression compared to control (Fig. 5). In contrast, PMA treatment for 8 or 24 h enhanced expression of COX-1 protein compared to control (Fig. 5). Macrophages expressed COX-1 protein at a level 2.5–3.5 fold higher than in U937 cells (Fig. 6). Treatment of macrophages with TNFα, LPS or M-CSF did not enhance expression of COX-1 protein com-

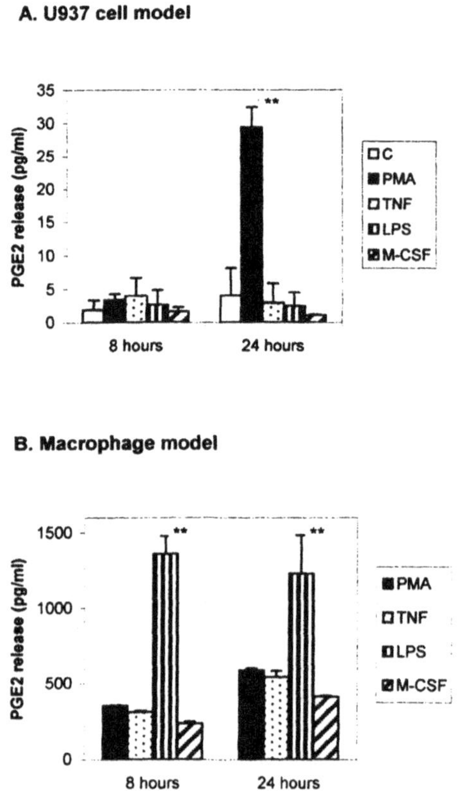

Fig. 2. The effect of stimuli on PGE_2 production in U937 cells and macrophages. U937 cells were incubated in the absence (C) or presence of PMA (160 nM) or TNFα (10 ng/ml) or LPS (100 ng/ml) or M-CSF (10 μg/ml) for 8 or 24 h. PMA-differentiated macrophages (PMA) were treated with the same amount of TNFα or LPS or M-CSF for the same periods of time as in U937 cells. PGE_2 released into the medium was measured by EIA, as described in 'Materials and methods.' (A) U937 cells. (B) Macrophages. Results presented are means ± S.D. of 3 independent experiments. **$p < 0.01$.

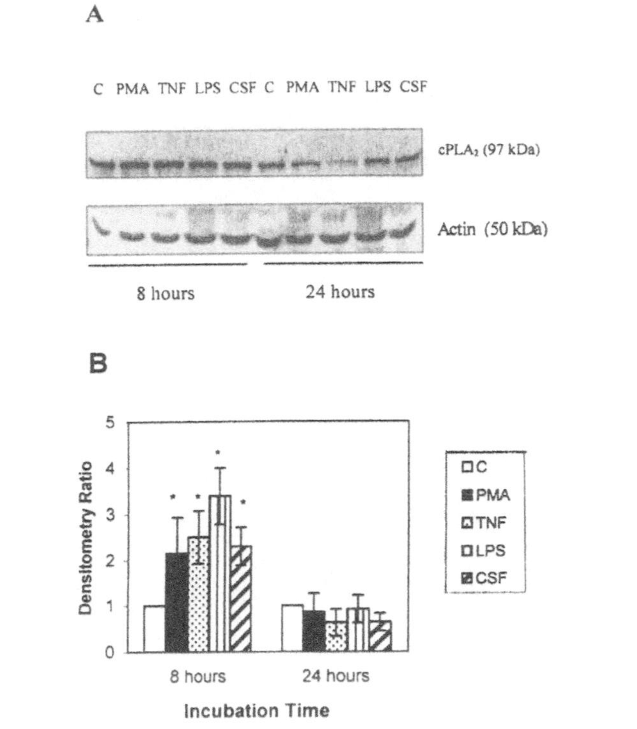

Fig. 3. The effect of stimuli on $cPLA_2$ protein expression in U937 cells. Cells were treated with stimuli as described in Fig. 2. Following treatment, the cell lysate was subjected to Western blot analysis, using specific antibodies for $cPLA_2$ and α-actin, as described in 'Materials and methods'. (A) A representative blot is depicted. (B) Densitometry ratio (target protein $cPLA_2$ over α-actin) plot of A. Results presented are means ± S.D. of 3 independent experiments. C – untreated U937 cells. $^*p < 0.05$.

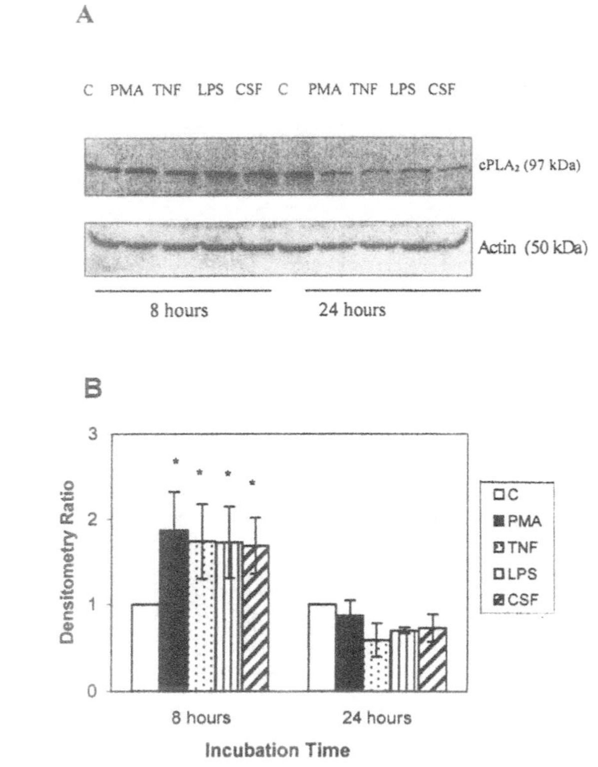

Fig. 4. The effect of stimuli on $cPLA_2$ protein expression in macrophages. Cells were treated with stimuli as described in Fig. 2. Following treatment, the cell lysate was subjected to Western blot analysis, using specific antibodies for $cPLA_2$ and α-actin. (A) A representative blot is depicted. (B) Densitometry ratio (target protein $cPLA_2$ over α-actin) plot of A. Results presented are means ± S.D. of 3 independent experiments. C – untreated U937 cells; PMA – U937 cells induced by PMA into macrophages. $^*p < 0.05$.

pared to control (Fig. 6). COX-2 was not expressed in U937 cells (Fig. 7A). COX-2 was expressed in macrophages and treatment of macrophages for 8 or 24 h with LPS up-regulated COX-2 expression 3.3–4 fold ($p < 0.05$) compared to control (Fig. 7). TNFα and M-CSF did not affect COX-2 expression in macrophages. These data were consistent with an LPS-mediated elevation in PGE_2 production in macrophages.

In summary, the data indicate that PMA-treatment of U937 cells enhances AA release and PGE_2 production via up-regulation of $cPLA_2$, COX-1, and likely COX-2, protein expression. In contrast, short term LPS-treatment of macrophages enhances AA release and PGE_2 production via up-regulation of $cPLA_2$ and COX-2 protein expression.

Discussion

U937 cells and monocyte-derived macrophages represent different stages of cell differentiation, and these two types of cells are distinct in morphology and function. The phorbol ester PMA is a tumor promoting agent commonly used as an inducer for U937 pre-monocyte differentiation into macrophages. Under our experimental conditions (8 or 24 h incubation with PMA), it is unlikely that full differentiation of U937 cells to macrophages occurred. This is supported by the fact that none of the cells were found to attach to the culture dish, an indication of full differentiation. In contrast, prolonged incubation of U937 cells with PMA for 48 h caused cells to attach to the culture dish. PMA is a potent protein kinase C activator and regulates COX-2 expression via the mitogen-activated protein kinase (MAPK) signalling pathway [16]. In the present study, PMA induced a significant increase in AA release and PGE_2 production in U937 cells. The peak of AA release and PGE_2 production occurred 24 h post-PMA addition. This dramatic increase in AA release in the late stage of treatment could not be explained solely by changes in $cPLA_2$ protein mass since 24 h post-incubation, $cPLA_2$ protein levels in PMA-treated cells dropped. Previously, Rehfeldt *et al.* reported that PMA activated $cPLA_2$ by inducing enzyme translocation to the membrane fraction in U937 cells [13]. Hence, the increase in AA release seen at 24 h post-PMA treatment

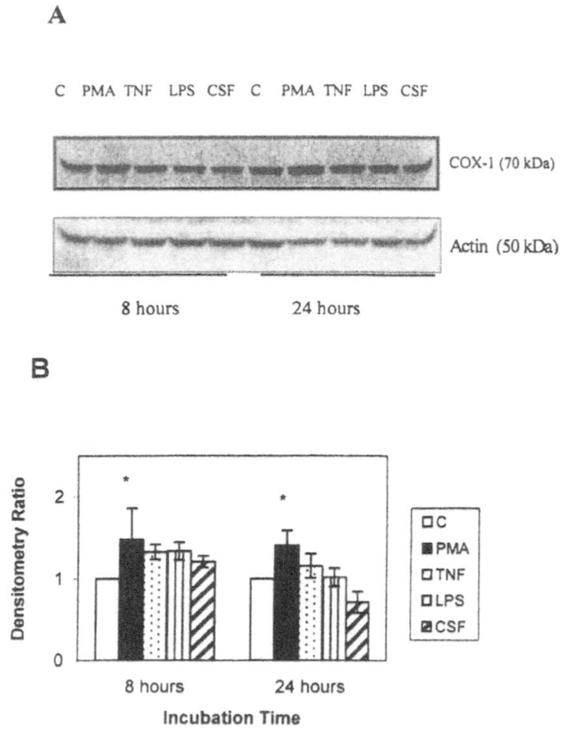

Fig. 5. The effect of stimuli on COX-1 protein expression in U937 cells. Cells were treated with stimuli as described in Fig. 2. Following treatment, the cell lysate was subjected to Western blot analysis, using specific antibodies for COX-1 and α-actin. (A) A representative blot is depicted. (B) Densitometry ratio (target protein COX-1 over α-actin) plot of A. Results presented are means ± S.D. of 3 independent experiments. C – untreated U937 cells. $*p < 0.05$.

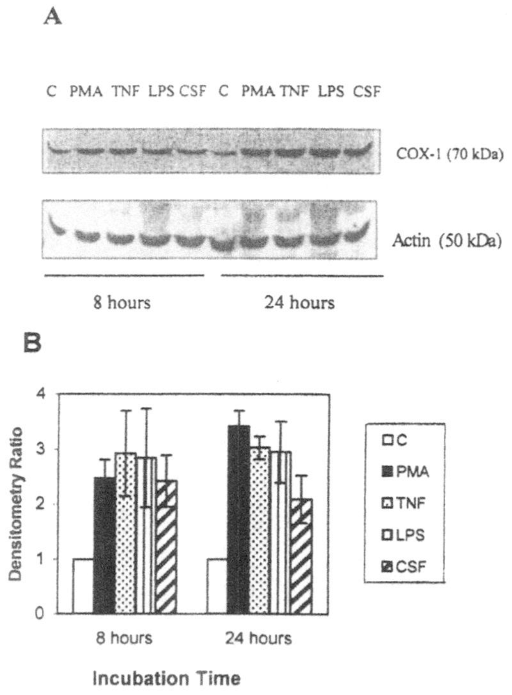

Fig. 6. The effect of stimuli on COX-1 protein expression in macrophages. Cells were treated with stimuli and the expression levels of COX-1 and *a*-actin were determined by Western blot analysis, as described in Fig. 2. (A) A representative blot is depicted. (B) Densitometry ratio (target protein COX-1 over α-actin) plot of A. Results presented are means ± S.D. of 3 independent experiments. C – untreated U937 cells; PMA – U937 cells induced by PMA into macrophages.

was likely due to an increase in enzyme activation. Alternatively, the induction of $cPLA_2$ protein in the early stage (8 h) may trigger signaling that leads to the expression of secretory PLA_2 and COX-2 which may be responsible for the late-phase PGE_2 synthesis [14]. Since early (8 h) treatment with PMA increased AA release and up-regulation of $cPLA_2$ and COX-1 expression in U937 cells, the elevated PGE_2 production likely resulted from a combined action of both COX-1 and COX-2.

In U937 cells, a significant amount of AA release but not PGE_2 production was observed after LPS treatment. Undifferentiated U937 cells expressed COX-1 but not COX-2. The lack of PGE_2 production in LPS-treated U937 cells might result from the inability of this compound to induce COX-1 protein expression. LPS is a bacteria-derived endotoxin. In addition to the MAPK-mediated signaling pathway, LPS may mediate COX-2 transcription via NFκB or p38/RK/Mpk2 pathways [15, 16]. Macrophages readily express CD14, a distinct cell surface marker absent in monocytes and may function as the receptor for the complex of LPS with LPS-binding protein [17]. The data from the present study indicated that the LPS-mediated enhancement of PGE_2 production in macrophages was contributed by elevated COX-2 protein expression. Similar to the action of PMA, LPS may enhance AA release in macrophages via directly increasing the enzyme activity of $cPLA_2$ or secretory PLA_2. Prolonged treatment (24 h) of macrophages with LPS attenuated $cPLA_2$ protein, indicating that both the bioavailability of AA as well as COX-1 and/or COX-2 expression were equally important in regulating PGE_2 biosynthesis in these cells.

TNFα failed to induce PGE_2 production in macrophages or U937 cells, in spite of enhancing $cPLA_2$ protein expression and subsequent stimulation of AA release. TNFα has been shown to induce COX-2 mRNA expression in mouse osteoblasts [18] but not in human monocytes [19]. Our results were consistent with the report on human monocytes [19] and suggest that TNFα has limited influence on PGE_2 production in U937 cells. Hence, PGE_2 production likely requires coordinated induction and functional coupling of $cPLA_2$, COX-1 and/or COX-2.

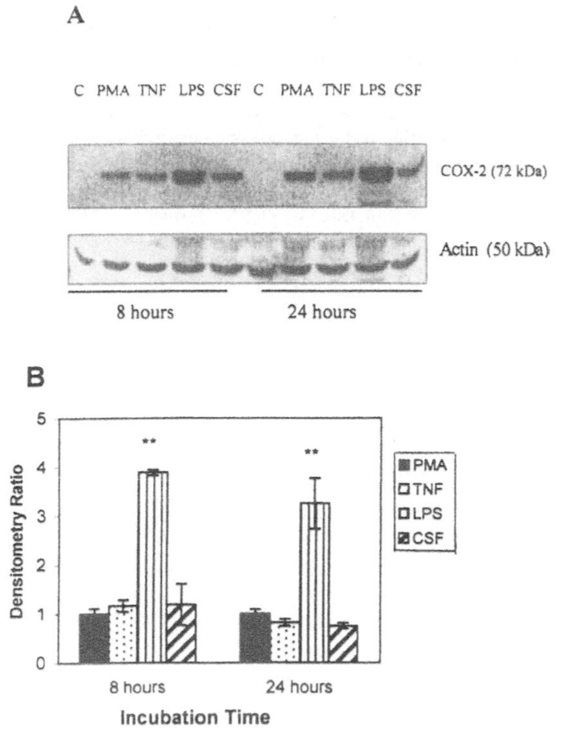

Fig. 7. The effect of stimuli on COX-2 protein expression in macrophages. Cells were treated with stimuli and the expression levels of COX-2 and α-actin were determined. (A) A representative blot is depicted. (B) Densitometry ratio (target protein $cPLA_2$ over α-actin) plot of A. Results presented are means ± S.D. of 3 independent experiments. C – untreated U937 cells; PMA – U937 cells induced by PMA into macrophages. $^{**}p < 0.01$.

M-CSF had little capacity to release PGE_2 in U937 cells or macrophages, similar to the results obtained in bone marrow macrophages [20]. The cellular response to growth factor is generally regarded to be mediated by sequential activation of receptor tyrosine kinase, Src, Ras, and one or more of MAPK pathways [7]. However, activation of the COX-2 promoter by serum and platelet-derived growth factor was inhibited by dominant negative Ras, MEKK-1 and Raf1 [21]. Thus, the lack of COX-2 induction by M-CSF may be due to the inability of this compound to activate any of the Ras, MEKK-1 or Raf-1 pathways.

In summary, the enhancement of AA release in both U937 cells and macrophages may result from a combination of increased $cPLA_2$ activity and its $cPLA_2$ protein expression. PMA stimulated PGE_2 production via up-regulation of COX-1, and likely COX-2, expression in U937 cells, whereas LPS elicits a similar effect in macrophages via induction of COX-2 expression. The complex regulation of AA release and PGE_2 production in U937 cells and in macrophages clearly depends upon the coordinated work of key enzymes in the cyclooxygenase pathway, $cPLA_2$, COX-2, COX-1, and probably other enzymes, such as secretory PLA_2 and PGE_2 synthase.

Acknowledgements

This work was supported by a grant from the Heart and Stroke Foundation of Manitoba and the Canadian Institutes of Health Research (CIHR). G.M.H. is a CIHR Scientist.

References

1. Sanduja SK, Mehta K, Xu XM, Hsu S M, Sanduja R, Wu KK: Differentiation associated expression of prostaglandin H and thromboxane A syntheses in monocytoid leukemia cell lines. Blood 78: 3178–3185, 1991
2. Schenkein HA: LPS-stimulated release of prostaglanin E2 and thromboxane B2 from the U937 cells. Cell Immunol 102: 307–314, 1986
3. Barrios-Rodiles M, Tiraloche G, Chadee K: Lipopolysaccharide modulates cyclooxygenase-2 transcriptionally and posttranscriptionally in human macrophages independently from endogenous IL-1 beta and TNF-alpha. J Immunol 163: 963–969, 1999
4. Hajjar DP, Pomerantz KB: Signal transduction in atherosclerosis: Integration of cytokines and the eicosanoids network. FASEB J 6: 2933–2941, 1992
5. Janssen-Timmen U, Tomic I, Specht E, Beilecke U, Habenicht AJ: The arachidonic acid cascade, eicosanoids and signal transduction. Ann NY Acad Sci 733: 325–334, 1994
6. Withnall MT, Pennington A, Wiseman D: Characterisation of cytosolic phospholipase A_2 as mediator of the enhanced arachidonic acid release from dimethyl sulphoxide differentiated U937 cells. Biochem Pharmacol 50: 1893–1895, 1995
7. Smith WL, DeWitt DL, Garavito RM: Cyclooxygenases: Structural, cellular, and molecular biology. Ann Rev Biochem 69: 145–182, 2000
8. Dubois RN, Abramson SB, Crofford L, Gupta RA, Simon LS, Van De Putte LB, Lipsky PE: Cyclooxygenase in biology and disease. FASEB J 12: 1063–1073, 1998
9. Rimarachin JA: Regulation of cyclooxygenase-2 expression in aortic smooth muscle cells. Artierioscler Thromb 14: 1021–1031, 1994
10. Smith WL: Prostanoids biosynthesis and the mechanism of actions. Am J Physiol 263: F181–F191, 1994
11. Wong JT, Tran K, Pierce GN, Chan AC, O, K, Choy PC: Lysophosphatidylcholine stimulates the release of arachidonic acid in human endothelial cells. J Biol Chem 273: 6830–6836, 1998
12. Tran K, Wong JT, Lee E, Chan AC, Choy PC: Vitamin E potentiates arachidonate release and phospholipase A_2 activity in rat heart myoblastic cells. Biochem J 319: 385–391, 1996
13. Rehfeldt W, Hass R, Goppelt-Struebe M: Characterization of phospholipase A_2 in monocytic cell lines. Functional and biochemical aspects of membrane association. Biochem J 276: 631–636, 1991
14. Fujishima H, Sanchez Mejia RO, Bingham CO III, Lam BK, Sapirstein A, Bonventre JV, Austen KF, Arm JP: Cytosolic phospholipase A_2 is essential for both the immediate and the delayed phases of eicosanoid generation in mouse bone marrow-derived mast cells. Proc Natl Acad Sci USA 91: 3228–3232, 1999
15. Zhang FX, Kirschning CJ, Mancinelli R, Xu XP, Jin Y, Faure E, Mantovani A, Rothe M, Muzio M, Arditi M: Bacterial lipopolysaccharide activates nuclear factor-kappaB through interleukin-1 signaling mediators in cultured human dermal endothelial cells and mononuclear phagocytes. J Biol Chem 274: 7611–7614, 1999
16. Dean JL, Brook M, Clark AR, Saklatvala J: p38 Mitogen-activated protein kinase regulates cyclooxygenase-2 mRNA stability and transcription in lipopolysaccharide-treated human monocytes. J Biol Chem 274: 264–269, 1999

17. Wright SD, Ramos RA, Tobias PS, Ulevitch RJ, Mathison JC: CD 14, a receptor for complexes of lipopolysaccharide (LPS) and LPS binding protein. Science 249: 1431–1433, 1990
18. Murakami M, Kuwata H, Amakasu Y, Shimbara S, Nakatani Y, Atsumi G, Kudo I: Prostaglandin E2 amplifies cytosolic phospholipase A_2- and cyclooxygenase-2-dependent delayed prostaglandin E2 generation in mouse osteoblastic cells. Enhancement by secretory phospholipase A_2. J Biol Chem 272: 19891–19897, 1997
19. Glaser KB, Lock YW: Regulation of prostaglandin H synthase 2 expression in human monocytes by the marine natural products manoalide and scalaradial. Novel effects independent of inhibition of lipid mediator production. Biochem Pharmacol 50: 913–922, 1995
20. Shibata Y, Bjorkman DR, Schmidt M, Oghiso Y, Volkman A: Macrophage colony-stimulating factor-induced bone marrow macrophages do not synthesize or release prostaglandin E2. Blood 84: 3316–3323, 1994
21. Xie W, Herschman HR: Transcriptional regulation of prostaglandin synthase 2 gene expression by platelet-derived growth factor and serum. J Biol Chem 271: 31742–31748, 1996

Molecular and Cellular Biochemistry **246:** 39–44, 2003.

FR167653 suppresses the progression of experimental autoimmune myocarditis

Seitaro Maruyama,[1] Kiminori Kato,[1] Makoto Kodama,[1] Yuji Okura,[1] Satoru Hirono,[1] Koichi Fuse,[1] Haruo Hanawa,[1] Osamu Nakagawa,[1] Mikio Nakazawa,[2] Takashi Miida,[3] Eisin Yaoita,[4] Tadashi Yamamoto,[4] Ikuo Inoue[5] and Yoshifusa Aizawa[1]

[1]*First Department of Internal Medicine;* [2]*Department of Medical Technology;* [3]*Department of Laboratory Medicine;* [4]*Department of Pathology, Institute of Nephrology, Niigata University School of Medicine, Niigata;* [5]*The Fourth Department of Internal Medicine, Saitama Medical School, Saitama, Japan*

Abstract

Experimental autoimmune myocarditis (EAM) induced in rats by injection of cardiac myosin is an animal model of human myocarditis and post-myocarditis dilated cardiomyopathy. It has been reported that proinflammatory cytokines play crucial roles in the induction of EAM and in the progression of myocardial injury in this disease. FR167653 (1-[7-(4-fluorophenyl)-1,2,3,4-tetrahydro-8-(4-pyridyl) pyrazolo [5,1-c] [1,2,4] triazin-2-yl]-2-phenylethanedione sulfate monohydrate) as been reported to suppress tumor necrosis factor-alpha (TNF-α). We hypothesized that FR167653 would suppress the progression of EAM if TNF-α and/or interleukin-1 beta (IL-1β) were the culprit cytokines in EAM. To investigate the effects of FR167653 in EAM, FR167653 was given to rats for 4 weeks, immediately after they had been immunized with cardiac myosin. The ratio of heart weight to body weight and the area of inflammatory lesions were less in the FR167653 groups than in the control rats. FR167653 reduced serum sialic acid levels significantly. The control group showed a deterioration in cardiac function. The FR167653 groups had significantly better hemodynamic parameters, including improved left ventricular end-diastolic pressure, central venous pressure, aortic pressure, and positive and negative left ventricular pressure derivatives. mRNA expression of IL-1β in the heart was significantly lower in rats given FR167653. However, mRNA of TNF-α was not detected in any groups. Our results suggest that FR167653 suppresses the development of myocarditis by suppression of IL-1β. (Mol Cell Biochem **246**: 39–44, 2003)

Key words: experimental autoimmune myocarditis (EAM), FR167653, IL-1β, TNF-α, ribonuclease protection assay

Introduction

Giant cell myocarditis is a very severe inflammatory myocardial disease of unknown origin and is frequently fatal [1, 2]. Experimental autoimmune myocarditis (EAM), produced in mice and rats by challenging the animals with cardiac myosin together with complete Freund's adjuvant, serves as an animal model for understanding the pathophysiology of human giant cell myocarditis [3, 4]. In rats, EAM is evoked 2 weeks after myosin injection into the footpads, peaks at about day 19, gradually subsides between days 25 and 40, and is followed by dilated cardiomyopathy [5]. According to several studies, the inflammatory process of EAM is induced by a cellular immune response rather than by a humoral immune reaction, because the disease is adoptively transferable into naive rats by T cells, not by antibodies [6–8]. Proinflammatory cytokines, such as IL-1β and TNF-α, are exclusively detected only during the maximum inflammatory phase (day19) [9]. These cytokines probably play an important role in the induction of EAM [10]. Subsequently, Th2 type cytokines, interleukin-10, and transforming growth factor-β are expressed in the recovery phase [9].

Address for offprints: S. Maruyama, First Department of Internal Medicine, Niigata University School of Medicine, 1-754 Asahimachi, Niigata, Zip #951-8510, Japan (E-mail: maruyama@med.niigata-u.ac.jp)

FR167653 (1-[7-(4-fluorophenyl)-1,2,3,4-tetrahydro-8-(4-pyridyl) pyrazolo [5-1-c] [1,2,4] triazin-2-yl]-2-phenylethanedione sulfate monohydrate) is a low-molecular-weight compound that obstructs the production of interleukin-1 alpha (IL-1α), IL-1β, and TNF-α [11–13]. The cytokine-inhibitory function of FR167653 has been demonstrated in human monocytes stimulated with lipopolysaccharide and in human lymphocytes stimulated with phytohemagglutinin-M. FR167653 has been proposed to inhibit the mitogen-activated protein kinase, which is activated by various extracellular stimuli [14]. We further hypothesized that FR167653 would suppress EAM by affecting the cytokine network.

Materials and methods

Animals

Male Lewis rats were purchased from Charles River Japan Inc. (Atsugi, Japan) and maintained under specific pathogen-free conditions at the Facilities for Comparative Medicine and Animal Experimentation, Niigata University School of Medicine, Japan.

Immunization

Antigen was prepared from the ventricular muscle of porcine hearts by a procedure previously described [4, 15]. The cardiac myosin was emulsified with an equal volume of complete Freund's adjuvant supplemented with *Mycobacterium tuberculosis* H37Ra (Difco, Detroit, MI, USA) at a concentration of 10 mg/mL. On day 0, each rat received a subcutaneous injection of 0.125 mL of the emulsified solution in each rear footpad.

Treatment

First treatment

Treatment with FR167653 began immediately after myosin administration. FR167653 in dimethylsulfoxide (DMSO) solution was given at a low dose of 0.1 mg/kg/h (group 4L) to 6 rats with EAM, and at a high dose of 0.32 mg/kg/h (group 4H) to 8 rats with EAM via an osmotic pump. The 5 control rats (group 4C) with EAM were treated with DMSO solution alone (4.6 mg/kg/h). These three groups were used for hemodynamic and pathologic study.

Second treatment

To confirm the effects of FR167653 on the cytokine network in EAM, rats treated with 0.32 mg/kg/h FR167653 (group 3H; n = 9) and controls (group 3C; n = 10) were killed at day 21 of the first experiment. Levels of mRNA of IL-1β and TNF-α have been reported to peak on day 21 in EAM [9].

Hemodynamic study

On day 28, rats were anesthetized for hemodynamic study. Anesthesia was induced with 2% halothane, and the concentration was then reduced to 0.5% to minimize its hemodynamic effects. Mean blood pressure (mean BP) was recorded through a catheter introduced into the right femoral artery. Central venous pressure (CVP) was recorded through a catheter introduced into the superior vena cava from the right jugular vein. These polyethylene catheters (PE-50) were connected to Statham P50 pressure transducers (Statham P50; Gould Statham, Oxnard, CA, USA) and a carrier amplifier (Nihon-Kohden, Tokyo, Japan). A catheter-tip transducer (Millar SPR 249; Millar Inst., Houston, TX, USA) was inserted into the left ventricle from the right common carotid artery to measure peak left ventricular pressure (LVP) and left ventricular end-diastolic pressure (LVEDP). The rates of intraventricular pressure rise and decline (dP/dt max and dP/dt min) were measured with a differential amplifier (NEC San-ei, Tokyo, Japan). Heart rate was calculated from the electrocardiogram. All hemodynamic parameters were recorded on a thermostylus recorder (NEC San-ei, Tokyo, Japan). The hemodynamic parameters were measured after a stabilizing period of 20 min [16].

Histopathology

After the hemodynamic measurements had been completed, the rats' hearts were removed. The cardiac ventricles were separated from the atria and weighed (heart weight, HW). Thereafter, the ventricles were transversely sliced, fixed in 10% formalin, embedded in paraffin, and stained. The extent of inflammatory cell infiltration and of myocardial necrosis was estimated after hematoxylin and eosin staining. The extent of fibrosis was estimated after Mallory-Azan staining. The area of residual intact cardiac muscle was analyzed by computerized densitometry with an optical scanner (Hewlett Packard; San Jose, CA, USA) on a Power Macintosh 9500/132 (Apple Computer; Cupertino, CA, USA) and an NIH Image 1.62 software program (NIH; Bethesda, MD, USA). The percentage of intact area in a transversely sliced section of the heart was presented.

Ribonuclease protection assay

Ribonuclease protection assay was carried out as described previously [17–19]. ^{32}P-labeled cRNA probes were synthesized by *in vitro* transcription, using linearized plasmids inserted with rat IL-1β (182 bp), TNF-α (147 bp), and glyceraldehyde-3-phosphate dehydrogenase (GAPDH) (114 bp) [20]. Ten micrograms of total RNA was hybridized with the cRNA at 45°C overnight. The unhybridized probes were digested with ribonuclease A (1.2 mg/mL) and ribonuclease

T1 (120 units/mL) for 60 min at 30°C. The ribonucleases were digested with proteinase K (0.45 mg/mL) at 37°C for 30 min. After phenol/chloroform extraction, the hybridized probes were precipitated with ethanol and heat-denatured for electrophoresis on 6% polyacrylamide gels. The dried gels were exposed to X-ray film (Fuji Photo Film Co.; Kanagawa, Japan) at –70°C to detect mRNA expression.

Measurement of density of mRNA bands on autoradiograms

To quantify the levels of mRNA expression in each sample, the autoradiograms were analyzed by computerized densitometry with an optical scanner (Hewlett Packard; San Jose, CA, USA) on a Power Macintosh 9500/132 (Apple Computer; Cupertino, CA, USA) and an NIH Image 1.62 software program (NIH; Bethesda, MD, USA). The utility of this method for quantification of RNA detected on autoradiograms has been confirmed by Masters [21]. Each specific mRNA expression was presented as a ratio to GAPDH mRNA expression in the same sample.

Measurement of serum albumin and sialic acid levels

We planned to measure levels of inflammation-mediated proteins, such as C-reactive protein (CRP), in this study. However, we failed to estimated rat CRP by routine methods. However, we measured the serum levels of sialic acid and albumin. Serum albumin was measured by the BCG method [22]. The serum concentration of albumin is considered to reflect the nutritional and inflammatory state. Serum sialic acid was quantified by neuraminidase/N-acetylneuraminic acid-aldose rate assay (enzymatic method) [23]. Sialic acid is considered to be a more direct inflammatory marker than serum albumin [24].

Statistical analysis

Data are presented as means ± S.E. Statistical analyses were conducted by using Student's *t*-test, one-way ANOVA, and Bonferroni test, where appropriate. Difference was considered significant if $p < 0.05$.

Results

General findings

Body weight did not differ significantly among groups 4C, 4L, and 4H. Heart weight was significantly lowest in group 4H ($p < 0.01$). The heart-to-body-weight ratio was significantly higher in group 4H than in group 4C ($p < 0.001$, Table 1).

Rats (3C) given FR167653 had significantly lower sialic acid levels than controls at 3 weeks (Table 2). Rats given FR167653 (3H) had significantly lower serum albumin levels than did control rats at 3 weeks [9].

Hemodynamic findings

In both treatment groups, LVP, mean BP, dP/dt max and dP/dt min were significantly higher than in group 4C. LVEDP and CVP were significantly lower. Rats given FR167653 had significantly better hemodynamics than controls (Table 3). Heart rate was not affected by FR167653 treatment. Every hemodyanamic parameters except HR was significantly better in group 4H than in group 4L, revealing a dose-dependent effect.

Pathological findings and quantification of mRNA

Pathological examination revealed that the intact area of cardiac muscle was significantly broader in the FR167653 groups than in the control group (4C: 57.2 ± 8.3%, 4L: 82.3 ± 2.1%, 4H: 85.2 ± 3.7%; $p < 0.01$, Fig. 1).

Expression of mRNAs of IL-1β or TNF-α was not detected at 4 weeks (Data were not shown). At 3 weeks, however, IL-1β mRNA expression was decreased by treatment with FR167653 (Fig. 2A). Bands of TNF-α were not detected in groups 3C and 3H (Fig. 2B).

Table 1. General findings

	4 C (n = 5)	4 L (n = 6)	4 H (n = 8)	
BW (g)	312.00 ± 8.50	304.17 ± 8.00	321.88 ± 7.20	NS
HW (g)	1.22 ± 0.06	1.07 ± 0.08	0.98 ± 0.04	*
HW/BW (mg/g)	3.89 ± 0.14	3.51 ± 0.22	3.04 ± 0.11	

BW – body weight; HW – heart weight. 4l: FR167653 0.1 mg/kg/h for FR167653; 4H: 0.32 mg/kg/h of FR 167653. The ratio of HW to BW (mg/g) is calculated by HW/BW × 1000. Values are mean ± S.E. NS: not significant. *p = 0.007 between group 4C and group 4H.

Table 2. General findings

	3 C (n = 10)	3 H (n = 9)	
Albumin (g/dl)	1.56 ± 0.04	1.48 ± 0.02	NS
Sialic acid (mg/dl)	87.90 ± 2.30	74.10 ± 2.50	**

BW – body weight; HW – heart weight. 3C: Control, 3H: 0.32 mg/kg/h of FR167653. Values are mean ± S.E. NS: not significant. **p = 0.0008 between group 3C and group 3H.

Table 3. Haemodynamic findings

	4C (n = 5)	4L (n = 6)	4H (n = 8)	
CVP (mmHg)	2.1 ± 0.4	1.3 ± 0.7	–0.1 ± 0.5	*
BP (mmHg)	68.6 ± 3.1	72.9 ± 1.5	79.4 ± 2.2	*, **
LVP (mmHg)	74.3 ± 6.4	86.1 ± 3.0	90.6 ± 4.0	*
EDP (mmHg)	14.9 ± 1.0	16.3 ± 2.0	9.8 ± 1.3	*, **
dP/dt max (mmHg/sec)	4464 ± 407	5347 ± 485	6690 ± 461	*
dP/dt min (mmHg/sec)	4720 ± 578	5640 ± 574	7340 ± 564	*
HR (beats/min)	343 ± 6	336 ± 8	328 ± 7	NS

4C: control group; 4L: low dose, 0.1 mg/kg/h of FR167653; 4H: high dose, 0.32 mg/kg/h of FR167653; CVP: central vein pressure; BP: blood pressure; LVP: left ventricular pressure; EDP: end-diastolic pressure; HR: heart rate; NS: not significant. *$p < 0.05$ between group 4C and group 4H; **$p < 0.05$ between group 4L and group 4H.

Discussion

FR167653, a suppressor of cytokine production, partly prevented the progression of rat autoimmune myocarditis. FR-167653 treatment suppressed expression of IL-1β mRNA in hearts of rats with EAM and resulted in decreased serum sialic acid levels on day 21 of treatment. More prominent effects of FR167653 were seen in cardiac histopathology and hemodynamic parameters at the 28th day. Although FR167653 administration could not prevent the induction of EAM, left ventricular remodeling during myocarditis was suppressed in rats treated with FR167653. Consequently, left ventricular function was preserved in those rats.

Effective strategies for treating myocarditis have not yet been established [25]. The disease has various causes, including viral infection, allergic or autoimmune reaction, and exposure to toxins. A significant number of patients die during the active phase of the disease from congestive heart failure and ventricular arrhythmias [26]. If patients can survive the active phase, the myocardial inflammation subsides spontaneously [27]. The morbidity of patients who recover from myocarditis depends on the degree of ventricular remodeling that occurs during and after the myocarditis [28]. Accordingly, there are three therapeutic targets in myocarditis. The first target is to determine the cause, which unfortunately

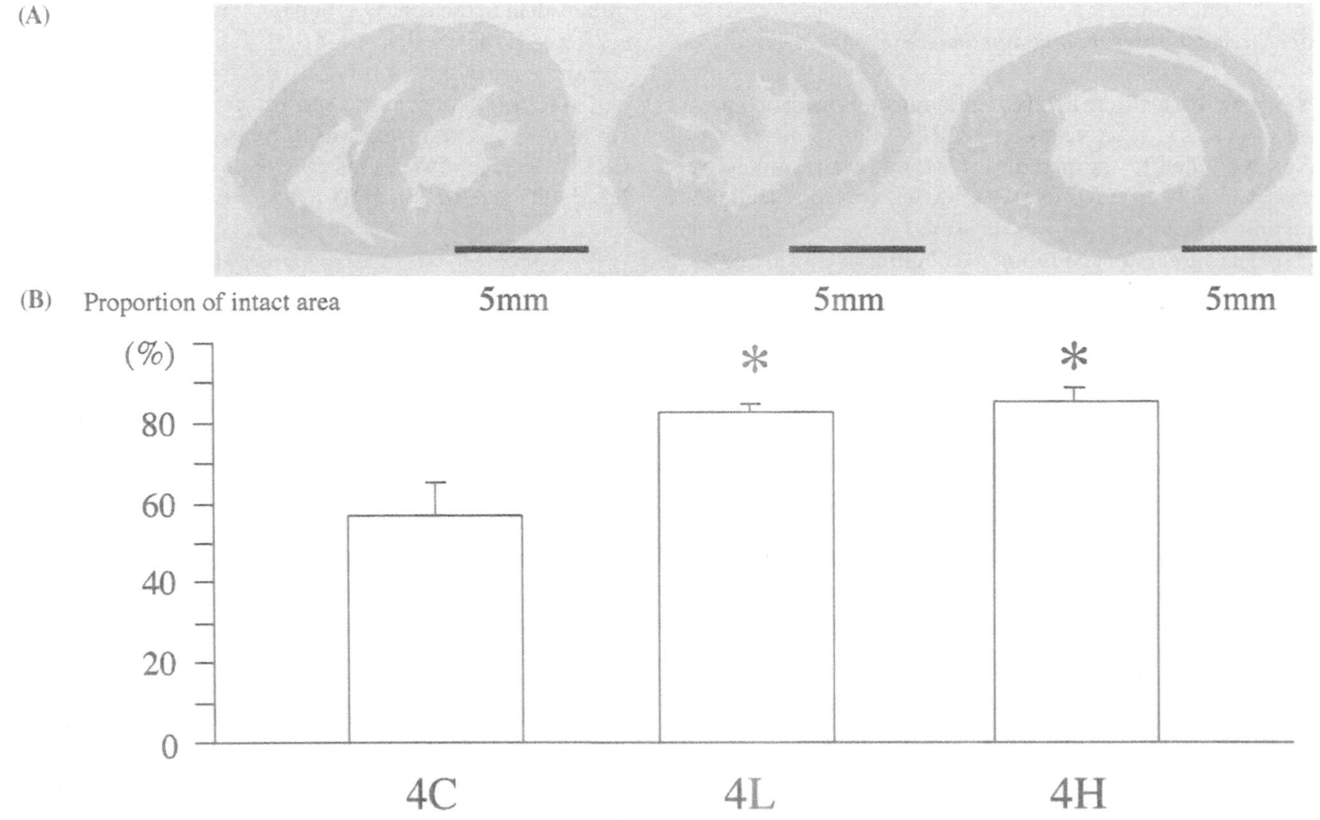

Fig. 1. Pathology and area of residual intact myocardium at 4 weeks. (A) (Left) section from a control rat (group 4C). (Center) section from a rat treated with low dose of FR167653 (0.1 mg/kg/h; group 4L). (Right) section from a rat treated with high dose of FR167653 (0.32 mg/kg/h; group 4H). All sections were stained by Azan-Mallory stain. (B) This graph represents proportion of intact area among groups 4C, 4L, and 4H. 4C: Control, 4L: Low dose (0.1 mg/kg/h), 4H: High dose (0.32 mg/kg/h). *$p < 0.05$ vs. Control.

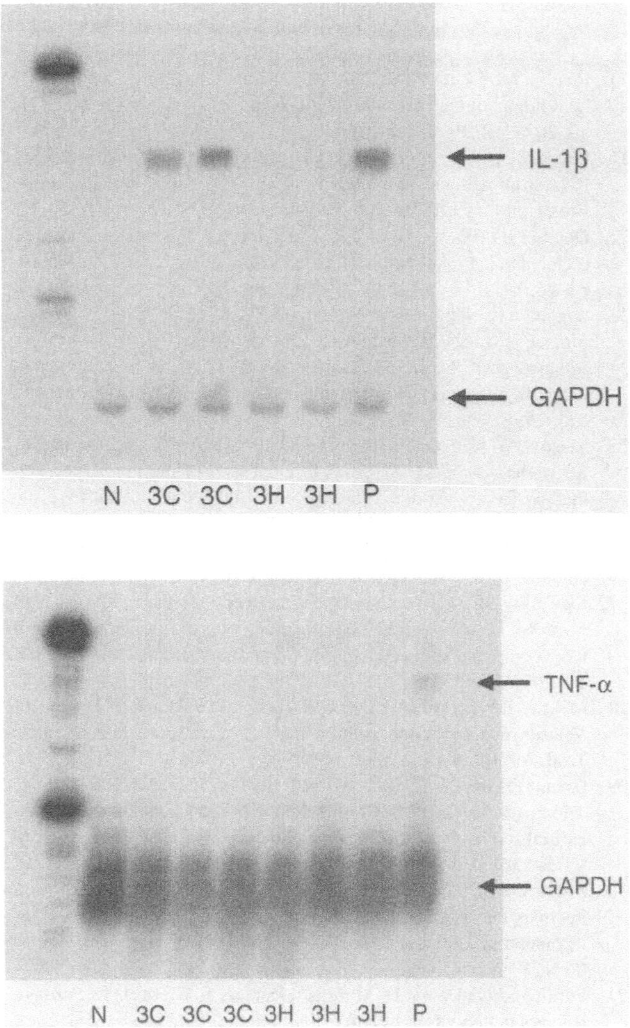

Fig. 2. Representative blots from the ribonuclease protection assay, depicting the effects of FR167653 on the mRNAs for IL 1β (A) and TNF-α (B) at 3 weeks. P: mRNA from spleen cells cultured with lipopolysaccharide (positive control); N: mRNA from the heart of a normal rat (negative control); 3C: mRNA from the heart of a control rat (myocarditis without FR167653 (0.32 mg/kg/h) treatment); 3H: mRNA from the heart of a rat with myocarditis treated with FR167653 (0.32 mg/kg/h).

cannot be satisfactorily confirmed in every patient [26]. The second target is to minimize myocardial functional suppression and subsequent hemodynamic derangement during the active phase. Cytokines are considered to play important roles in myocardial suppression [29–31]. The third target is to prevent ventricular remodeling. Our results suggest that FR167653 can act effectively on the second and third therapeutic targets in myocarditis.

EAM is an animal model for human giant cell myocarditis. Observational evidence supports the hypothesis that autoimmune myocardial injuries are involved in the pathogenesis of human giant cell myocarditis [2]. Immunosuppressive agents, such as cyclosporine and tacrolimus, have been demonstrated to effectively prevent the induction of EAM [32, 33], and the former agent is efficacious in some human cases [2]. However, powerful immunosuppressants are harmful in treating viral myocarditis [25]. We therefore have to strictly distinguish between viral myocarditis and giant cell myocarditis before we use immunosuppressants, and this is impossible. FR167653 is a selective cytokine blocker that interferes with the production of IL-1β and TNF-α. IL-1β and TNF-α are pro-inflammatory cytokines commonly acting in the pathogenesis of both viral and autoimmune myocarditis. Therefore, we may be able to use selective cytokine blockers in treating patients with myocarditis of unknown etiology. Powerful immunosuppressants frequently have adverse systemic effects. Selective cytokine blockers may have advantages in this regard, although we did not evaluate side effects in this study.

The precise mechanisms of the cytokine-blocking action of FR167653 are still uncertain. Some reports have suggested that FR167653 interferes with signal transduction through mitogen-activated protein kinase in lymphocytes. Details of the actions of FR167653 on cardiomyocytes as well as on lymphocytes need to be clarified. FR167653 had beneficial effects on EAM but its effects were rather weak. To treat myocarditis after the onset of disease, we will need to use a stronger cytokine blocker.

Acknowledgement

We thank Fujisawa Pharmaceutical Co. Ltd., for generously providing the FR167653.

References

1. Davidoff R, Palacios I, Southern J, Fallon JT, Newell J, Dec GW: Giant cell vs. lymphocytic myocarditis. A comparison of their clinical features and long-term outcomes. Circulation 83: 953–961, 1991
2. Cooper LT Jr, Berry GJ, Shabetai R: Idiopathic giant-cell myocarditis – natural history and treatment. Multicenter Giant Cell Myocarditis Study Group Investigators. N Engl J Med 336: 1860–1866, 1997
3. Neu N, Rose NR, Beisel KW, Herskowitz A, Gurri-Glass G, Craig SW: Cardiac myosin induces myocarditis in genetically predisposed mice. J Immunol 139: 3630–3636, 1987
4. Kodama M, Matsumoto Y, Fujiwara M, Masani F, Izumi T, Shibata A: A novel experimental model of giant cell myocarditis induced in rats by immunization with cardiac myosin fraction. Clin Immunol Immunopathol 57: 250–262, 1990
5. Kodama M, Hanawa H, Saeki M, Hosono H, Inomata T, Suzuki K, Shibata A: Rat dilated cardiomyopathy after autoimmune giant cell myocarditis. Circ Res 75: 278–284, 1994
6. Smith SC, Allen PM: Myosin-induced acute myocarditis is a T cell-mediated disease. J Immunol 147: 2141–2147, 1991

7. Kodama M, Matsumoto Y, Fujiwara M: *In vivo* lymphocyte-mediated myocardial injuries demonstrated by adoptive transfer of experimental autoimmune myocarditis. Circulation 85: 1918–1926, 1992
8. Okura Y, Takeda K, Honda S, Hanawa H, Watanabe H, Kodama M, Izumi T, Aizawa Y, Seki S, Abo T: Recombinant murine interleukin-12 facilitates induction of cardiac myosin-specific type 1 helper T cells in rats. Circ Res 82: 1035–1042, 1998
9. Okura Y, Yamamoto T, Goto S, Inomata T, Hirono S, Hanawa H, Feng L, Wilson CB, Kihara I, Izumi T, Shibata A, Aizawa Y, Seki S, Abo T: Characterization of cytokine and iNOS mRNA expression *in situ* during the course of experimental autoimmune myocarditis in rats. J Mol Cell Cardiol 29: 491–502, 1997
10. Smith SC, Allen PM: Neutralization of endogenous tumor necrosis factor ameliorates the severity of myosin-induced myocarditis. Circ Res 70: 856–863, 1992
11. Yamamoto N, Sakai F, Yamazaki H, Nakahara K, Okuhara M: Effect of FR167653, a cytokine suppressive agent, on endotoxin-induced disseminated intravascular coagulation. Eur J Pharmacol 314: 137–142, 1996
12. Aiba M, Takeyoshi I, Ohwada S, Kobayashi J, Iwanami K, Sunose Y, Kawashima Y, Mastumoto K, Muramoto M, Morishita Y: Optimal end point of FR167653 administration and expression of interleukin-8 messenger RNA on extended liver resection with ischemia in dogs. J Am Coll Surg 191: 251–258., 2000
13. Otani Y, Takeyoshi I, Koibuchi Y, Matsumoto K, Muramoto M, Morishita Y: The effect of FR167653 on pulmonary ischemia-reperfusion injury in rats. J Heart Lung Transplant 19: 377–383, 2000
14. Ohashi N, Matsumori A, Furukawa Y, Ono K, Okada M, Iwasaki A, Miyamoto T, Nakano A, Sasayama S: Role of p38 mitogen-activated protein kinase in neointimal hyperplasia after vascular injury. Arterioscler Thromb Vasc Biol 20: 2521–2526, 2000
15. Inomata T, Hanawa H, Miyanishi T, Yajima E, Nakayama S, Maita T, Kodama M, Izumi T, Shibata A, Abo T: Localization of porcine cardiac myosin epitopes that induce experimental autoimmune myocarditis. Circ Res 76: 726–733, 1995
16. Hirono S, Islam MO, Nakazawa M, Yoshida Y, Kodama M, Shibata A, Izumi T, Imai S: Expression of inducible nitric oxide synthase in rat experimental autoimmune myocarditis with special reference to changes in cardiac hemodynamics. Circ Res 80: 11–20, 1997
17. Feng L, Tang WW, Loskutoff DJ, Wilson CB: Dysfunction of glomerular fibrinolysis in experimental antiglomerular basement membrane antibody glomerulonephritis. J Am Soc Nephrol 3: 1753–1764, 1993
18. Sambrook JF: Molecular Cloning. In: A Laboratory Manual. Cold Spring Harbor Lab., New York, 1989, pp 71–78
19. Yamamoto T, Sasaki S, Fushimi K, Ishibashi K, Yaoita E, Kawasaki K, Fujinaka H, Marumo F, Kihara I: Expression of AQP family in rat kidneys during development and maturation. Am J Physiol 272: F198–F204, 1997
20. Ponassi M, Jacques TS, Ciani L, Constant CF: Expression of the rat homologue of the *Drosophila* fat tumour suppressor gene. Mech Dev 80: 207–212, 1999
21. Masters DB, Griggs CT, Berde CB: High sensitivity quantification of RNA from gels and autoradiograms with affordable optical scanning. Biotechniques 12: 902–906, 908–911, 1992
22. Doumas BT, Watson WA, Biggs HG: Albumin standards and the measurement of serum albumin with bromcresol green. Clin Chim Acta 31: 87–96, 1971
23. Sugahara K, Sugimoto K, Nomura O, Usui T: Enzymatic assay of serum sialic acid. Clin Chim Acta 108: 493–498, 1980
24. Sillanaukee P, Ponnio M, Jaaskelainen IP: Occurrence of sialic acids in healthy humans and different disorders. Eur J Clin Invest 29: 413–425, 1999
25. Mason JW, O'Connell JB, Herskowitz A, Rose NR, McManus BM, Billingham ME, Moon TE: A clinical trial of immunosuppressive therapy for myocarditis. The Myocarditis Treatment Trial Investigators. N Engl J Med 333: 269–275, 1995
26. Kodama M, Oda H, Okabe M, Aizawa Y: Early and long-term mortality of the clinical subtypes of myocarditis. Jpn Circ J 65: 961–964, 2001
27. McCarthy RE 3rd, Boehmer JP, Hruban RH, Hutchins GM, Kasper EK, Hare JM, Baughman KL: Long-term outcome of fulminant myocarditis as compared with acute (nonfulminant) myocarditis. N Engl J Med 342: 690–695, 2000
28. Mendes LA, Picard MH, Dec GW, Hartz VL, Palacios IF, Davidoff R: Ventricular remodeling in active myocarditis. Myocarditis Treatment Trial. Am Heart J 138: 303–308, 1999
29. Bryant D, Becker L, Richardson J, Shelton J, Franco F, Peshock R, Thompson M, Giroir B: Cardiac failure in transgenic mice with myocardial expression of tumor necrosis factor-alpha. Circulation 97: 1375–1381, 1998
30. Siwik DA, Chang DL, Colucci WS: Interleukin-1beta and tumor necrosis factor-alpha decrease collagen synthesis and increase matrix metalloproteinase activity in cardiac fibroblasts *in vitro*. Circ Res 86: 1259–1265, 2000
31. Ferdinandy P, Danial H, Ambrus I, Rothery RA, Schulz R: Peroxynitrite is a major contributor to cytokine-induced myocardial contractile failure. Circ Res 87: 241–247, 2000
32. Zhang S, Kodama M, Hanawa H, Izumi T, Shibata A, Masani F: Effects of cyclosporine, prednisolone and aspirin on rat autoimmune giant cell myocarditis. J Am Coll Cardiol 21: 1254–1260, 1993
33. Hanawa H, Kodama M, Zhang S, Izumi T, Shibata A: An immunosuppressant compound, FK-506, prevents the progression of autoimmune myocarditis in rats. Clin Immunol Immunopathol 62: 321–326, 1992

Molecular and Cellular Biochemistry **246**: 45–50, 2003.

The effect of early treatment by cerivastatin on the serum level of C-reactive protein, interleukin-6, and interleukin-8 in the patients with unstable angina and non-Q-wave myocardial infarction

Petr Ostadal, David Alan, Petr Hajek, David Horak, Jiri Vejvoda, Jiri Trefanec, Martin Mates and Jan Vojacek
Department of Internal Medicine, University Hospital Motol and 2nd Medical Faculty, Charles University, Prague, Czech Republic

Abstract

The aim of our study was to evaluate whether a single dose of cerivastatin at the time of admission of patients with unstable angina pectoris (UAP) or non-Q-wave myocardial infarction (NQMI) can influence the serum level of C-reactive protein (CRP), interleukin-6 (IL-6) and interleukin-8 (IL-8) 24 h later. Forty-four patients with rest chest pain and subendocardial ischemia on ECG were randomized to receive cerivastatin 0.3 mg at the time of admission (group C+) to standard therapy or to remain just on standard therapy (group C–). Blood samples for determination of troponin I (TI), CRP, IL-6 and IL-8 were collected at admission (entry level) and 24 h later (final level). Patients with non-physiological baseline levels of TI, as well as patients with progression to Q wave MI were excluded. All baseline, clinical and demographic data and final values of TI were comparable in the two groups. In patients treated with cerivastatin (group C+, $n = 13$) we observed decrease in the CRP level (–6.73 ± 3.93 mg/L); on the other hand, in group C– ($n = 17$) the CRP level increased (+7.92 ± 2.77 mg/L, $p = 0.004$). Similar differences were observed also in IL-6: in group C+ the level was significantly reduced as compared with the increase in group C– (–0.76 ± 0.52 vs. 4.58 ± 1.49 ng/L, $p = 0.005$). The level of IL-8 was not affected. Our results suggest that early treatment with cerivastatin can decrease the serum level of CRP and IL-6 in patients with UAP/NQMI; this might positively influence their prognosis. Nevertheless, further studies are needed to support this hypothesis. (Mol Cell Biochem **246**: 45–50, 2003)

Key words: cerivastatin, inflammation, C-reactive protein, interleukin, unstable angina, myocardial infarction

Introduction

During the past decade inflammation has been repeatedly suggested as a major factor in the pathogenesis of atherosclerosis and all its consequences, including ischemic heart disease. It was found that 'classical' risk factors, such as hypertension, hypercholesterolemia, diabetes, and smoking could not explain the development of coronary stenosis in all patients suffering from ischemic heart disease. Intensive study of the pathogenesis of coronary plaque development and plaque rupture led to the hypothesis of involvement of inflammatory factors in this process. The presence of T-lymphocytes and monocytes/macrophages was repeatedly reported directly in the plaque; elevated levels of the acute phase proteins (C-reactive protein, serum amyloid A, fibrinogen), cytokines (interleukin 1, interleukin 6, interleukin 8, tumor necrosis factor), and adhesive molecules (ICAM-1) correlate with the worse prognosis in patients suffering from the ischemic heart disease [1–5]. Furthermore, the increased level of C-reactive protein became widely accepted as a risk

Address for offprints: P. Ostadal, Department of Internal Medicine, University Hospital Motol, V uvalu 84, 150 18 Prague 5, Czech Republic (E-mail: ostadal@yahoo.com)

factor of the higher incidence of non-fatal and fatal coronary events in both chronic ischemic heart disease and acute coronary syndromes [6–9]. Activation of the immune reaction in the acute forms of ischemic heart disease has probably two origins: (i) pathological events occurring in the culprit artery wall, responsible for the plaque rupture with all its consequences and (ii) myocardial necrosis that activates mechanisms leading to removal of the necrotic mass and to its replacement by scar. Whereas activation of the plaque inflammation (as listed above) is a marker of plaque instability, elevation of inflammatory factors from a second source correlates with the extent of necrosis.

Statins, 3-hydroxy-3-methylglutaryl coenzyme A (HMG-CoA) reductase inhibitors, lower the cholesterol level by decreasing the production of low-density lipoproteins (LDL) and up-regulating the expression of LDL-receptor [10]. These drugs are widely used in patients with hypercholesterolemia in primary and secondary prevention of coronary artery disease because of their efficacy in reducing cardiovascular morbidity and mortality [11–13]. Surprisingly, it has been observed that statin therapy improves prognosis also in patients with normal or low cholesterol level [14]. Study of the non-lipid activities of statins revealed that a possible beneficial effect is mediated by reduction of the parameters of inflammation, namely CRP; this effect seems to be independent of the cholesterol level [1, 15]. Mechanisms by which statins inhibit inflammation are not fully understood; it has been reported that statins suppress production of monocyte chemotactic protein-1 (MCP-1) [16] and matrix metalloproteinases (MMPs) [5, 17, 18]; statins also decrease macrophage expression of soluble ICAM-1 and secretion of IL-1, IL-6, TNF-α [5, 19–24].

During the past few years several clinical trials have been designed to investigate the effect of statins in unstable angina (UAP) and non-Q wave myocardial infarction (NQMI) [25]. Although these studies were designed for early treatment, patients were randomized up to 48–96 h after admission. There are no data available about the effect of treatment as early as at the time of admission. Furthermore, almost all clinical trials with statins have been designed as longer-term trials focused in the majority of cases on clinical outcomes or lipid parameters; almost nothing has been reported about the effect of statins as early as 24 h after administration. The aim of our study was, therefore, to investigate the effect of cerivastatin at admission on the C-reactive protein, IL-6, and IL-8 levels 24 h later in patients with UAP or NQMI. To our best knowledge, this is the first clinical report showing the effect of an early statin administration at the time of admission and, in addition, this is the first report describing the immediate effect of cerivastatin on the serum level of CRP, IL-6, and IL-8.

Materials and methods

Patients

The study protocol was approved by the institutional ethics review committee and written informed consent was obtained from all participating patients. We enrolled patients with UAP or NQMI, presented at the Coronary Care Unit with ischemic chest pain at rest in the absence of extracardiac cause and with either ST segment depression of ≥ 0.1 mV or T wave inversion in two or more contiguous leads on 12 lead ECG at admission. Exclusion criteria were (i) elevated entry level of troponin I, (ii) evolution of Q waves, (iii) statin therapy in the pre-hospitalization period, and (iv) ST segment elevation on ECG. All patients received standard therapy according to the Guidelines for management of unstable angina and non-Q wave myocardial infarction [26], including coronary angiography during hospitalization.

Study design

Patients were randomized to receive standard therapy plus cerivastatin 0.3 mg (Cholstat 0.3, Laboratoire Fournier, France) immediately after admission (group C+) or just standard therapy (group C–). Blood samples for the determination of troponin I, CRP, IL-6, and IL-8 were taken at admission (entry level) before cerivastatin administration and 24 h after admission (final level).

Laboratory assays

Troponin I was assayed by an automated monoclonal antibody solid phase sandwich-type enzyme immunoassay on the IMMULITE instrument (DPC, Los Angeles, CA, USA). Monoclonal antibody, other reagents, and calibrators were obtained from the same company (DPC, Los Angeles, CA, USA). CRP was estimated by a turbidimetry technique using CRP-HS set (WAKO, Japan) on the ADVIA 1650 analyzer (Bayer Diagnostic Division, Tarrytown, USA). IL-6 and IL-8 were measured by an enzyme immunoassay based on the chemiluminiscence using an IMMULITE instrument (DPC, Los Angeles, CA, USA). Monoclonal antibodies, other reagents, and calibrators were obtained from the same company (DPC, Los Angeles, CA, USA).

Statistical analysis

The values are expressed as means ± S.E.M. Statistical comparison of the groups was performed by one-way ANOVA test and t-test for baseline characteristics and by non-parametric Mann-Whitney U test for troponin I, CRP, IL-6, and IL-8. $p < 0.05$ was considered to be statistically significant.

Results

A total of 44 patients were randomized, 14 of them were excluded because of positive entry level of troponin I. Group C+ included 13 patients treated by cerivastatin, group C– comprised 17 patients. We did not observe any adverse effect of cerivastatin treatment. There were no differences between the groups in demographic characteristics, medical history, and in concomitant therapy (Table 1). Moreover, there were no differences in the entry levels of troponin I, CRP, IL-6, IL-8 and in the final level of troponin I between group C+ and group C– (Table 2).

Cerivastatin treatment (group C+) induced a reduction in CRP level by 6.73 ± 3.93 mg/L, whereas in group C– the level of CRP increased by 7.92 ± 2.77 mg/L, p = 0.004. The final CRP level was, therefore, significantly lower in group C+, compared to group C– (2.59 ± 1.03 vs. 11.45 ± 3.17 mg/L, p = 0.025) (Fig. 1).

Whereas cerivastatin treatment (group C+) decreased the IL-6 level by 0.76 ± 0.52 ng/L, in group C– the IL-6 level increased by 4.58 ± 1.49 ng/L, p = 0.005. The final value of IL-6 was 2.5 ± 0.05 ng/L in group C+ and 7.57 ± 1.7 ng/L in group C–, p = 0.015 (Fig. 2).

Table 1. Baseline characteristics

	Group C+ (n = 13)	Group C– (n = 17)	p
Demographic data			
Age	59.2 ± 7.2	59.8 ± 8.4	ns
Male/female	11/2	14/3	ns
Medical history			
Hypertension (%)	54	59	ns
Diabetes (%)	15	18	ns
Hyperlipoproteinemia (%)	38	41	ns
Ischemic heart disease (%)	85	82	ns
Concomitant medication			
Aspirin (%)	100	100	ns
Beta-blocker (%)	85	88	ns
Heparin/LMWH (%)	100	100	ns
ACE-inhibitors (%)	23	24	ns
Diuretics (%)	15	18	ns
Nitrates (%)	54	59	ns

Data are expressed as mean ± S.E.M., absolute values, or percentage of occurrence. LMWH – low molecular weight heparin; ACE – angiotensin-converting enzyme.

Table 2. Entry values of biochemical parameters and final level of troponin I as a marker of myocardial necrosis

	Group C+ (n = 13)	Group C– (n = 17)	p
Entry level			
troponin I (μg/L)	< 0.5	< 0.5	ns
CRP (mg/L)	5.77 ± 1.81	2.94 ± 0.76	ns
IL-6 (ng/L)	3.27 ± 0.52	2.69 ± 0.19	ns
IL-8 (ng/L)	7.99 ± 2.19	8.2 ± 2.58	ns
Final level			
troponin I (μg/L)	1.29 ± 0.65	1.38 ± 0.38	ns

Data are expressed as mean ± S.E.M. CRP – C-reactive protein; IL-6 – inteleukin-6; IL-8 – interleukin 8.

There were no significant differences in IL-8 between the groups. In group C+ IL-8 level decreased by 0.78 ± 0.79 ng/L, in group C– the level increased by 0.12 ± 1.66 ng/L, p = 0.34 (Fig. 3A). The final level of IL-8 was comparable in both groups (7.21 ± 1.23 vs. 8.32 ± 1.44 ng/L, p = 0.41) (Fig. 3).

Discussion

Our results demonstrate that 0.3 mg cerivastatin, administered at admission, can positively influence the inflammatory parameters CRP and IL-6 in patients with UAP or NQMI 24 h after admission. To our knowledge this is the first clinical report showing the immediate significant effect of just single dose of statin. Our observation is in good agreement with experimental studies. The reduction of IL-6 mRNA and protein was described after 6–24 h of incubation of endothelial

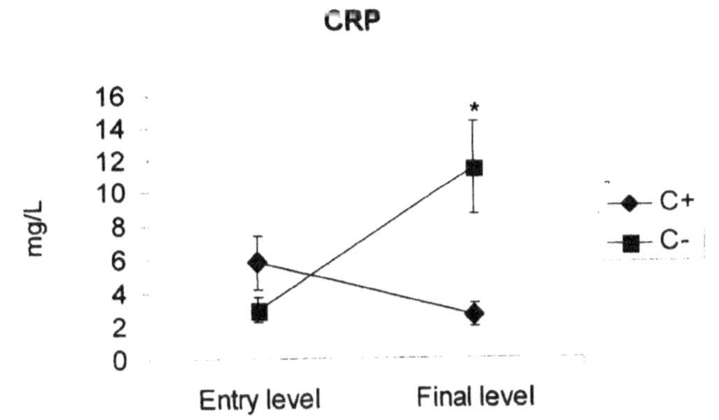

Fig. 1. Effect of cerivastatin on the level of CRP in patients with UAP or NQMI. Group C+ was treated with 0.3 mg cerivastatin at the time of admission in addition to standard therapy; Group C– received just standard therapy. The first sample for CRP determination (entry level) was taken before cerivastatin administration; final level was obtained 24 h later. *p < 0.05.

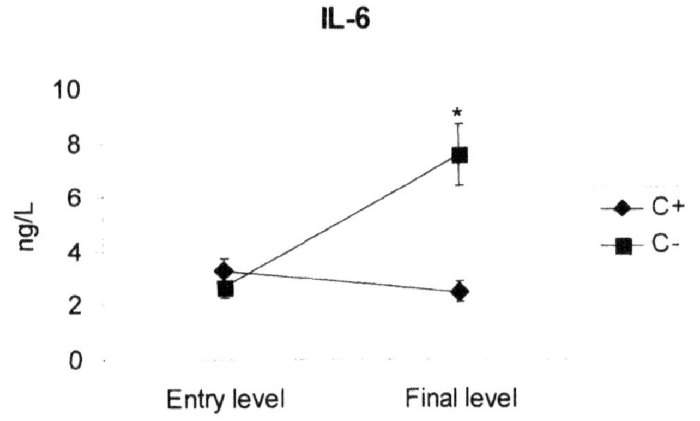

Fig. 2. Effect of cerivastatin on the level of IL-6 in patients with UAP or NQMI. Group C+ was treated with 0.3 mg cerivastatin at the time of admission in addition to standard therapy; Group C– received just standard therapy. The first sample for IL-6 determination (entry level) was taken before cerivastatin administration; final level was obtained 24 h later. $*p < 0.05$.

cells with pravastatin, simvastatin, and fluvastatin *in vitro* [27]. Ando *et al.* [28] observed reduction of TNF-α and IL-1β in lipopolysaccharide (LPS)-induced sepsis in mice after just two doses of cerivastatin 12 and 1 h before LPS injection. Similarly, Romano *et al.* [16] described the effect of 24-h treatment with lovastatin and simvastatin; they observed an inhibition of MCP-1 synthesis in mononuclear and endothelial cells *in vitro*.

In our study cerivastatin was administered at the time of admission. There are no other clinical reports describing such an early statin therapy. Also clinical trials PRINCESS and MIRACL [25], which deal with statin treatment in acute coronary syndromes, were designed for the prevention of recurrent coronary event and patients were randomized to 48–96 h

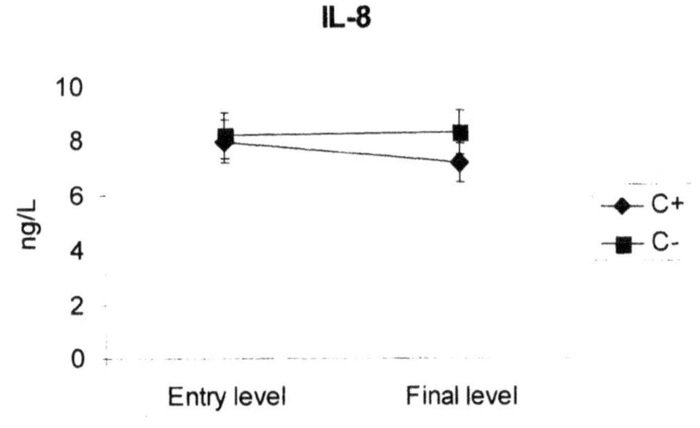

Fig. 3. Effect of cerivastatin on the level of IL-8 in patients with UAP or NQMI. Group C+ was treated with 0.3 mg cerivastatin at the time of admission in addition to standard therapy; Group C– received just standard therapy. The first sample for IL-8 determination (entry level) was taken before cerivastatin administration; final level was obtained 24 h later. $*p < 0.05$.

after admission. Our data draw attention to the possible use of statins not only for primary or secondary prevention but also for treatment of acute coronary syndrome.

We observed a reduction of the CRP level after cerivastatin treatment. Our results are in good agreement with studies describing a decrease of CRP after long-term statin therapy. Albert *et al.* [29] described the reduction in CRP level by pravastatin in both primary and secondary prevention. Similarly, Horne *et al.* [30] observed a beneficial effect of statin therapy in patients with higher CRP. Also Strandberg *et al.* [31] demonstrated that simvastatin and atorvastatin lowered CRP in hyperlipidemic patients. Significant reduction in CRP level after pravastatin treatment not related to the alteration of lipid parameters and improved prognosis after decrease of CRP by pravastatin was described by CARE trial investigators [14]. Pravastatin, simvastatin, and atorvastatin also significantly depressed CRP levels in the report by Jialal *et al.* [32] and cerivastatin therapy-induced CRP reduction was demonstrated by Ridker *et al.* [33].

We observed also a reduction in IL-6 level after cerivastatin treatment. Similarly as with CRP we have no comparison with other clinical studies concerning the immediate effect of a single dose of statin. There are conflicting data about the effect of long-term statin therapy on the IL-6 level. Jialal *et al.* [32] observed a reduction of CRP after statins but without influence on the IL-6 level. Similarly, Deskur-Smielecka *et al.* [34] demonstrated that the IL-6 level remained unchanged after 12 weeks of simvastatin treatment. On the other hand, Brull *et al.* [19] described a lower IL-6 level after coronary artery bypass surgery (CABG) in patients treated with statins. A decrease in the IL-6 level after pravastatin therapy was also observed by Rosenson *et al.* [22]. These conflicting results may be partly explained by the low number of patients in these studies. In addition, we can expect a different influence of IL-6 in stable patients and under actual conditions of activated immune system such as acute coronary syndrome or CABG. Decreased IL-6 production has been reported also in experimental studies. Inoue *et al.* [27] described a reduction in IL-6 mRNA and protein in endothelial cells by pravastatin, simvastatin, fluvastatin, and cerivastatin.

Surprisingly, our observations have shown that single-dose cerivastatin therapy did not influence the IL-8 level. There are insufficient clinical data on the possible effect of statins on the IL-8 level. On the other hand, several experimental studies have been published dealing with statins and IL-8. Takata *et al.* [35] described suppressed IL-8 synthesis by pravastatin in endothelial cells, Kothe *et al.* [36] observed decreased IL-8 production by cerivastatin in infected endothelial cells or macrophages. Also Terkeltaub *et al.* [37] demonstrated a reduction of IL-8 after lovastatin treatment. The reasons for differences between our results and experimental studies are not clear: different inflammatory pathways may be triggered in acute coronary syndromes in comparison with experimental settings; the low number of participants must also be taken into consideration.

Our results for all inflammatory parameters measured should not be influenced by differences in the extent of myocardial necrosis, a potent activator of inflammation, since the final level of troponin I was comparable in both groups of patients.

In conclusion, our data show that a single dose of cerivastatin 0.3 mg at the time of admission of patients with UAP or NQMI decreased CRP and IL-6 levels 24 h after statin administration but did not influence the level of IL-8. Because higher level of both CRP and IL-6 correlates with poorer prognosis for patients with acute forms of ischemic heart disease [38], reduction of these factors by cerivastatin may also decrease the risk of another coronary event and probably also the instability of patients with UAP or NQMI. Although further studies with a higher number of participants are necessary to confirm our results, it seems that statins may be useful not only for prevention, but also for the therapy of acute coronary syndromes.

Acknowledgement

This project was partly supported by the grant NA/5639-3, IGA Ministry of Health of the Czech Republic.

References

1. Whicher J, Biasucci L, Rifai N: Inflammation, the acute phase response and atherosclerosis. Clin Chem Lab Med 37: 495–503, 1999
2. Barath P, Fishbein MC, Cao J, Berenson J, Helfant RH, Forrester JS: Detection and localization of tumor necrosis factor in human atheroma. Am J Cardiol 65: 297–302, 1990
3. Serneri GG, Abbate R, Gori AM, Attanasio M, Martini F, Giusti B *et al.*: Transient intermittent lymphocyte activation is responsible for the instability of angina. Circulation 86: 790–797, 1992
4. Biasucci LM, Vitelli A, Liuzzo G, Altamura S, Caligiuri G, Monaco C *et al.*: Elevated levels of interleukin-6 in unstable angina. Circulation 94: 874–877, 1996
5. Blake GJ, Ridker PM: Novel clinical markers of vascular wall inflammation. Circ Res 89: 763–771, 2001
6. Morrow DA, Rifai N, Antman EM, Weiner DL, McCabe CH, Cannon CP *et al.*: C-reactive protein is a potent predictor of mortality independently of and in combination with troponin T in acute coronary syndromes: A TIMI 11A substudy. Thrombolysis in Myocardial Infarction. J Am Coll Cardiol 31: 1460–1465, 1998
7. Haverkate F, Thompson SG, Pyke SD, Gallimore JR, Pepys MB: Production of C-reactive protein and risk of coronary events in stable and unstable angina. European Concerted Action on Thrombosis and Disabilities Angina Pectoris Study Group. Lancet 349: 462–466, 1997
8. Liuzzo G, Biasucci LM, Gallimore JR, Grillo RL, Rebuzzi AG, Pepys MB *et al.*: The prognostic value of C-reactive protein and serum amyloid a protein in severe unstable angina. N Engl J Med 331: 417–424, 1994
9. Rifai N, Ridker PM. High-sensitivity C-reactive protein: A novel and promising marker of coronary heart disease. Clin Chem 47: 403–411, 2001
10. Goldstein JL, Brown MS: Regulation of the mevalonate pathway. Nature 343: 425–430, 1990
11. The Scandinavian Simvastatin Survival Study (4S): Randomised trial of cholesterol lowering in 4444 patients with coronary heart disease. Lancet 344: 1383–1389, 1994
12. Sacks FM, Pfeffer MA, Moye LA, Rouleau JL, Rutherford JD, Cole TG *et al.*: The effect of pravastatin on coronary events after myocardial infarction in patients with average cholesterol levels. Cholesterol and Recurrent Events Trial investigators. N Engl J Med 335: 1001–1009, 1996
13. Shepherd J, Cobbe SM, Ford I, Isles CG, Lorimer AR, Macfarlane PW *et al.*: Prevention of coronary heart disease with pravastatin in men with hypercholesterolemia. West of Scotland Coronary Prevention Study Group. N Engl J Med 333: 1301–1307, 1995
14. Ridker PM, Rifai N, Pfeffer MA, Sacks FM, Moye LA, Goldman S *et al.*: Inflammation, pravastatin, and the risk of coronary events after myocardial infarction in patients with average cholesterol levels. Cholesterol and Recurrent Events (CARE) Investigators. Circulation 98: 839–844, 1998
15. Ridker PM, Rifai N, Clearfield M, Downs JR, Weis SE, Miles JS *et al.*: Measurement of C-reactive protein for the targeting of statin therapy in the primary prevention of acute coronary events. N Engl J Med 344: 1959–1965, 2001
16. Romano M, Diomede L, Sironi M, Massimiliano L, Sottocorno M, Polentarutti N *et al.*: Inhibition of monocyte chemotactic protein-1 synthesis by statins. Lab Invest 80: 1095–1100, 2000
17. Ikeda U, Shimpo M, Ohki R, Inaba H, Takahashi M, Yamamoto K *et al.*: Fluvastatin inhibits matrix metalloproteinase-1 expression in human vascular endothelial cells. Hypertension 36: 325–329, 2000
18. Fukumoto Y, Libby P, Rabkin E, Hill CC, Enomoto M, Hirouchi Y *et al.*: Statins alter smooth muscle cell accumulation and collagen content in established atheroma of watanabe heritable hyperlipidemic rabbits. Circulation 103: 993–999, 2001
19. Brull DJ, Sanders J, Rumley A, Lowe GD, Humphries SE, Montgomery HE: Statin therapy and the acute inflammatory response after coronary artery bypass grafting. Am J Cardiol 88: 431–433, 2001
20. Niwa S, Totsuka T, Hayashi S: Inhibitory effect of fluvastatin, an HMG-CoA reductase inhibitor, on the expression of adhesion molecules on human monocyte cell line. Int J Immunopharmacol 18: 669–675, 1996
21. Ikeda U, Shimada K: Statins and monocytes. Lancet 353: 2070, 1999
22. Rosenson RS, Tangney CC, Casey LC: Inhibition of proinflammatory cytokine production by pravastatin. Lancet 353: 983–984, 1999
23. Ferro D, Parrotto S, Basili S, Alessandri C, Violi F: Simvastatin inhibits the monocyte expression of proinflammatory cytokines in patients with hypercholesterolemia. J Am Coll Cardiol 36: 427–431, 2000
24. Bustos C, Hernandez-Presa MA, Ortego M, Tunon J, Ortega L, Perez F *et al.*: HMG-CoA reductase inhibition by atorvastatin reduces neointimal inflammation in a rabbit model of atherosclerosis. J Am Coll Cardiol 32: 2057–2064, 1998
25. Schwartz GG, Olsson AG, Ezekowitz MD, Ganz P, Oliver MF, Waters D *et al.*: Effects of atorvastatin on early recurrent ischemic events in acute coronary syndromes: The MIRACL study: A randomized controlled trial. JAMA 285: 1711–1718, 2001
26. Braunwald E, Antman EM, Beasley JW, Califf RM, Cheitlin MD, Hochman JS *et al.*: ACC/AHA guidelines for the management of patients with unstable angina and non-ST-segment elevation myocardial infarction. A report of the American College of Cardiology/American Heart Association Task Force on Practice Guidelines (Committee on the Management of Patients With Unstable Angina). J Am Coll Cardiol 36: 970–1062, 2000

27. Inoue I, Goto S, Mizotani K, Awata T, Mastunaga T, Kawai S *et al.*: Lipophilic HMG-CoA reductase inhibitor has an anti-inflammatory effect: Reduction of MRNA levels for interleukin-1beta, interleukin-6, cyclooxygenase-2, and p22phox by regulation of peroxisome proliferator-activated receptor alpha (PPARalpha) in primary endothelial cells. Life Sci 67: 863–876, 2000
28. Ando H, Takamura T, Ota T, Nagai Y, Kobayashi K: Cerivastatin improves survival of mice with lipopolysaccharide-induced sepsis. J Pharmacol Exp Ther 294: 1043–1046, 2000
29. Albert MA, Danielson E, Rifai N, Ridker PM: Effect of statin therapy on C-reactive protein levels: The pravastatin inflammation/CRP evaluation (PRINCE): A randomized trial and cohort study. JAMA 286: 64–70, 2001
30. Horne BD, Muhlestein JB, Carlquist JF, Bair TL, Madsen TE, Hart NI *et al.*: Statin therapy, lipid levels, C-reactive protein and the survival of patients with angiographically severe coronary artery disease. J Am Coll Cardiol 36: 1774–1780, 2000
31. Strandberg TE, Vanhanen H, Tikkanen MJ. Effect of statins on C-reactive protein in patients with coronary artery disease. Lancet 353: 118–119, 1999
32. Jialal I, Stein D, Balis D, Grundy SM, Adams-Huet B, Devaraj S: Effect of hydroxymethyl glutaryl coenzyme a reductase inhibitor therapy on high sensitive C-reactive protein levels. Circulation 103: 1933–1935, 2001
33. Ridker PM, Rifai N, Lowenthal SP: Rapid reduction in C-reactive protein with cerivastatin among 785 patients with primary hypercholesterolemia. Circulation 103: 1191–1193, 2001
34. Deskur-Smielecka E, Wykr TA, Kempa M, Furmaniuk J, Wysocki H: The influence of short-term treatment with simvastatin on the inflammatory profile of patients with hypercholesterolaemia. Coron Artery Dis 12: 143–148, 2001
35. Takata M, Urakaze M, Temaru R, Yamazaki K, Nakamura N, Nobata Y *et al.*: Pravastatin suppresses the interleukin-8 production induced by thrombin in human aortic endothelial cells cultured with high glucose by inhibiting the p44/42 mitogen activated protein kinase. Br J Pharmacol 134: 753–762, 2001
36. Kothe H, Dalhoff K, Rupp J, Muller A, Kreuzer J, Maass M *et al.*: Hydroxymethylglutaryl coenzyme A reductase inhibitors modify the inflammatory response of human macrophages and endothelial cells infected with Chlamydia pneumoniae. Circulation 101: 1760–1763, 2000
37. Terkeltaub R, Solan J, Barry M Jr, Santoro D, Bokoch GM: Role of the mevalonate pathway of isoprenoid synthesis in IL-8 generation by activated monocytic cells. J Leukoc Biol 55: 749–755, 1994
38. Koukkunen H, Penttila K, Kemppainen A, Halinen M, Penttila I, Rantanen T *et al.*: C-reactive protein, fibrinogen, interleukin-6 and tumour necrosis factor-alpha in the prognostic classification of unstable angina pectoris. Ann Med 33: 37–47, 2001

Molecular and Cellular Biochemistry **246**: 51–56, 2003.

Decreased activity of lecithin:cholesterol acyltransferase and hepatic lipase in chronic hypothyroid rats: Implications for reverse cholesterol transport

Martha Franco,[1] Graciela Castro,[2] Luis Romero,[3] Juan Carlos Regalado,[3] Aida Medina,[4] Claudia Huesca-Gómez,[3] Serafín Ramírez,[3] Luis F. Montaño,[5] Carlos Posadas-Romero[4] and Oscar Pérez-Méndez[3]

Departments of [1]Nephrology; [3]Physiology; [4]Endocrinology; [5]Cellular Biology, Instituto Nacional de Cardiología 'Ignacio Chávez', México; [2]Atherosclerosis and INSERM U325, Institut Pasteur de Lille, Lille, France

Abstract

Chronic hypothyroidism is frequently associated with atherosclerosis due to increased cholesterol plasma levels; nevertheless, the contribution of impaired reverse cholesterol transport (RCT) in this process has not been completely elucidated. The aim of this study was to evaluate the effect of thyroidectomy (Htx) upon the main stages of RCT in rats. Plasma lipid alterations induced by thyroidectomy showed a slight, but significant, reduction of total plasma triglycerides, a 300% increase of LDL-cholesterol and a 25% decrease in HDL-cholesterol compared to control rats. We evaluated the first stage of RCT determining ^{3}H-cholesterol efflux in Fu5AH cells. The capacity of HDL obtained from Htx rats to promote cholesterol efflux was similar to that of controls. Lecithin:cholesterol acyltransferase (LCAT) activity, the second stage and the driving force of RCT was 30% lower in Htx animals compared to controls, as determined by reconstituted HDL used as an external substrate. Lipoproteins are remodeled by hepatic lipase; the mean activity of this enzyme in postheparin plasma of Htx animals was reduced by 30% compared to controls, thus suggesting an impaired HDL remodeling by this enzyme in the hypothyroid status. In contrast, lipoprotein lipase activity in the Htx group was unchanged. In summary, this study demonstrates that chronic hypothyroidism in the rat induced an impaired RCT mainly at the cholesterol esterification, and HDL remodeling mediated by hepatic lipase. The latter probably results in an abnormal HDL structure, i.e. phospholipid enrichment, which contributes to decrease HDL-apo AI fractional catabolic rates. (Mol Cell Biochem **246**: 51–56, 2003)

Key words: reverse cholesterol transport, cholesterol efflux, high-density lipoproteins, lecithin:cholesterol acyltranferase, hepatic lipase, hypothyroidism

Introduction

Low plasma levels of high-density lipoproteins (HDL) correlate with a high risk of atherosclerotic heart disease [1, 2]. Such a diminished concentration of HDL affect reverse cholesterol transport (RCT) [3, 4], which is the metabolic pathway responsible for the movement of cholesterol excess from peripheral tissues to the liver for lipoprotein recycling or excretion. The RCT pathway could be defined as a progression of closely interconnected events [3], among which four are considered as the most relevant: (a) the free cholesterol efflux from extrahepatic cells and its uptake by HDL [4]; (b) the free cholesterol esterification by lecithin:cholesterol acyltranferase (LCAT); (c) the transfer of cholesterol to other apolipoproteins by the cholesterol ester transfer protein (CETP) or directly to the liver by hepatic scavenger re-

Address for offprints: O. Pérez-Méndez, Physiology Department, Instituto Nacional de Cardiología, 'Ignacio Chávez' Juan Badiano 1, Sección XVI, 14080 México (E-mail: opmendez@yahoo.com)

ceptor B type I (SR-BI) [5]; and (d) the remodelling of HDL by the phospholipid transfer protein and hepatic lipase to re-initiate the cholesterol efflux from peripheral tissues [6, 7]. As a consequence, physiopathological conditions that could alter the HDL plasma levels, the RCT pathway, or both, may contribute to atherosclerosis.

Thyroid hormones seem to have a direct impact on HDL synthesis, since these hormones up-regulate the *in vitro* biosynthesis of apo A-I, [8] the major protein of HDL. Thyroid hormone deficiency lowers intracellular apo A-I m-RNA in rats [9]. Despite of the role of thyroid hormones in the up-regulation of the apo A-I gene, HDL-cholesterol (HDL-C) plasma levels are normal or even increased in human hypothyroidism [10–12]. In addition, T4 therapy does not seem to increase serum HDL-C [13]. Low cholesteryl ester transfer protein (CETP) activity present in human hypothyroidism [14, 15] provide an explanation for these paradoxical results. Low CETP activity induces cholesterol accumulation in HDL particles that result in high HDL-C plasma levels [16, 17]. The latter probably counterbalance the decreased synthesis of apo A-I induced by the hypothyroid status, as it has been demonstrated *in vitro* [8, 9]; consequently, HDL-C plasma levels become normal or increased in hypothyroid patients [10–12]. As a consequence, low HDL-C plasma levels could be expected in the absence of CETP activity. Using chronic hypothyroid rats, a species lacking CETP expression, we have recently demonstrated that low HDL-apo AI plasma levels, are the result of a delayed synthesis that outweighs the low catabolism observed in these animals [18]; such a metabolic behavior is in part the result of larger apo E-enriched HDL. Because of the abnormal HDL structure, an impaired RCT pathway contributing to an increased risk of coronary heart disease frequently observed in hypothyroid patients could be expected.

In order to elucidate whether the low plasma levels and structural changes of HDL affect RCT in hypothyroid rats, we evaluated the first two stages of the process, the HDL capacity to promote cholesterol efflux from cells, and LCAT activity. We further determined the activity of hepatic and lipoprotein lipases in order to establish their contribution to HDL structural abnormalities.

Materials and methods

Induction of hypothyroidism

Male Wistar rats weighing 250–300 g underwent surgical thyroidectomy with parathyroid reimplant, as previously described [19]. Briefly, under ether anesthesia, the trachea was exposed, and the parathyroid glands were dissected from the thyroid gland, and reimplanted into the neck muscles. The thyroid gland was then dissected and completely excised. The effectiveness of this procedure was assessed by evaluation of serum calcium, phosphate, and T4 concentrations in 10 sham rats (Ctrl) and 10 thyroidectomized (Htx) rats using standard techniques (Ca 10.1 ± 0.7 Ctrl vs. 9.7 ± 0.7 mg/dL Htx, p = NS; phosphate 6.9 ± 0.4 Ctrl vs. 6.5 ± 1.1 mg/dL Htx, p = NS; T4 5.1 ± 0.3 Ctrl vs. 1.3 ± 0.2 µg/ml Htx, $p < 0.05$, Diagnostic Products Corporation RIA). The animals were used for further studies performed 16 weeks after thyroidectomy. Animals were maintained in a 12:12-h light-dark cycle and fed Purina rat chow throughout the study. All control animals were sham operated. All the animals used in the experiments were treated according to our institutional guidelines for laboratory research.

Blood samples

Arterial blood from hypothyroid and normal sham male Wistar rats was obtained after 12 h fast by abdominal aortic puncture under pentobarbital anaesthesia (30 mg/Kg) and collected in EDTA (1 mg/mL). Plasma was immediately separated by centrifugation at 4°C. Lipoprotein fractionation was started within 2 h of isolation. To obtain samples for post-heparin lipase activity determination, rats were anesthetized with sodium pentobarbital (30 mg/kg, i.p.); polyethylene tubing (PE-50) was used to catheterize the left carotid artery to take samples and to infuse heparin. Blood samples (2 mL) were taken (preheparin sample) and the volume was replaced with blood from a normal or hypothyroid donor rat. Heparin (100 mU/g of body wt) diluted in a saline solution to a final volume of 0.25 mL were infused by the catheter, and washed with 0.5 ml of saline solution. A 5 mL post-heparin sample was taken 10 min later and the rat was killed with an excess of pentobarbital.

Analytical methods

Plasma cholesterol, triglycerides, phospholipids and free cholesterol were determined by commercially available procedures (Menarini, Torino Italy and Boehringer-Mannheim, Mannheim Germany). Total lipoprotein protein was estimated by a modification of the Lowry procedure [20].

Isolation of Htx or normal rat HDL

HDL were separated by ultracentrifugation in a Beckman optima TLX table centrifuge at 110,000 rpm at 10°C in 3.2 mL polycarbonate tubes as previously described [18]. HDL were dialyzed against 10 mM phosphate buffer (PBS), pH 7.5. As we have already demonstrated HDL from the Htx group were larger and had a significant reduction in the amount of apo A–IV, and were enriched in phospholipids and apo E [18].

Cholesterol efflux

Cellular cholesterol efflux was determined using rat Fu5AH hepatoma cells following the procedure described by de la Llera Mora *et al.* [21]. Briefly, the cells were maintained in minimal essential medium (MEM) containing 5% calf serum; 20,000 Fu5AH cells/mL were plated on 2.4-cm multiwell plates using 2 mL/well. Two days after plating, membranes were labelled with ^{3}H-cholesterol (1 μCi/well) for 60-h. To allow equilibration of the radioactive label, cells were rinsed and incubated for 24 h in MEM containing 0.5% bovine serum albumin. To determine cholesterol efflux, the cells were washed with PBS and incubated at 37°C for 3 h with serum or HDL. For this, the serum of Htx or control rats was diluted into de culture medium (5% v/v). HDL were diluted to 50 μg/mL (expressed as total HDL protein concentration) into the culture medium. At the end of the efflux period, the medium was removed and centrifuged to discard cell debris. The cell monolayer was washed three times with iced PBS and harvested with 0.5 mL of 0.1 mol/L NaOH. Finally, radioactivity was measured in both the medium and the cells by liquid scintillation in a Beckman LS 600C counter. Cholesterol efflux was expressed as the percentage of total label released. Each sample was tested by triplicate.

LCAT activity assays

α-LCAT activity was measured by the method of Chen and Albers [22]. Briefly: apoAI/phosphatidilcholine/^{14}C-cholesterol complexes (HDLr) were incubated with either Htx or control rat plasma in a shaking water bath for 1 h at 37°C (esterification was linear during this time). The reaction was stopped, and lipids were extracted with chloroform-methanol 2:1. Esterified and unesterified cholesterol were separated by thin-layer chromatography, and the radioactivity was counted.

Lipases activity

Lipases activity was assessed using a glycerol stabilized phospholipid-triglyceride emulsion as substrate [23]. Briefly, 1.2 g of purified olive oil was mixed with 36 mg of phosphatidylcholine and 12.5 g of glycerol. The mixture was sonicated for 5 min at 20 watts. Total lipase activity was determined using 100 μL of postheparine plasma in 900 μL of 150 mM Tris, 150 mM NaCl, pH 8.5 buffer containing 50 mU Heparin, 3.75% human albumin, 5% of heat-inactivated rat serum as apo CII source and 50 μL of the triglyceride emulsion. The reaction mixture was incubated 1 h at 37°C and the reaction was stopped and the fatty acids produced by the lypolysis were extracted by the method of Belfrage and Vaughan [24]. The fatty acids in the organic phase were estimated by a manual titration method using 1.25 mM NaOH in absolute ethanol and thymol blue as final-point indicator. Quantitative results were obtained using a standard curve of palmitic acid between 2 and 8 mM. Basal free fatty acids were determined in preheparin plasmas and subtracted from lipases activities. Hepatic lipase (HL) activity was determined as described for total lipase activities, but using a reaction buffer that lacked apo CII source, and containing 1 M of NaCl. This buffer inhibits lipoprotein lipase (LPL). LPL activity was calculated by subtracting HL activity from total lipase activity. One unit of lipase activity is defined as 1 mmol of fatty acid produced per hour.

Statistical analysis

Data are expressed as the mean ± S.D. Comparison of experimental vs. control groups was made using the Student's *t*-test, and Pearson's correlation analysis. Differences were considered significant when $p < 0.05$.

Results

Lipid profile

Plasma lipid alterations induced by thyroidectomy were characterized by a slight but significant reduction of total plasma triglycerides (Table 1). We also observed a 300% increase in LDL-cholesterol and a 25% decrease in HDL-cholesterol compared to control group (Table 1). We corroborate our previous observation that plasma cholesterol although modified, did not reach significant differences [18].

Table 1. Lipids and apolipoproteins plasma profile (mg/dL) in hypothyroid and control rats

Group	Hypothyroid (n = 10)	Control (n = 10)
Plasma triglycerides	42.8 ± 7.2*	52.4 ± 12.3
Plasma cholesterol	54.0 ± 13.7	48.1 ± 11.2
LDL-Cholesterol	25.3 ± 8.4*	8.5 ± 3.8
HDL-Cholesterol	23.4 ± 6.1*	32.5 ± 8.2

Values are expressed as mean ± S.D. *Significantly different from control ($p < 0.05$).

Cholesterol efflux and LCAT activity

The first stage of RCT, that is, the removal of cholesterol from peripheral tissues, showed that the proportion of cellular cholesterol released by serum of Htx rats vs. control serum was similar after 3 h of incubation (Fig. 1). Similarly, HDL from Htx rats promoted an average cholesterol efflux of 10%, comparable to the capacity of HDL-Ctrl to remove cholesterol from cell membranes (10.4%) under the same conditions (Fig. 1). Positive controls of human HDL_3 at 100 μg/mL promoted 15.7% of cholesterol efflux, comparable to the efflux promoted using rat HDL (data not shown).

Activities of enzymes involved in RCT

Once cholesterol is integrated to HDL surface, the next stage in RCT is cholesterol esterification by LCAT. Enzyme activity was assessed using phospholipids/apo AI/cholesterol-reconstituted HDL as external substrate, which is indicative of the enzyme concentration [22]. Mean plasma LCAT activities of Htx rats determined by this method, were about 30% lower compared to control group (Table 2). After cholesterol esters are brought to the liver, HDL remodeling take place in part by the hepatic lipase. Using an external source of triglycerides, mean HL activity in Htx group was reduced by 30% compared to control group, meaning that chronic hypothyroid status results in a decreased expression of the enzyme. HL activities did not show any significant correlation

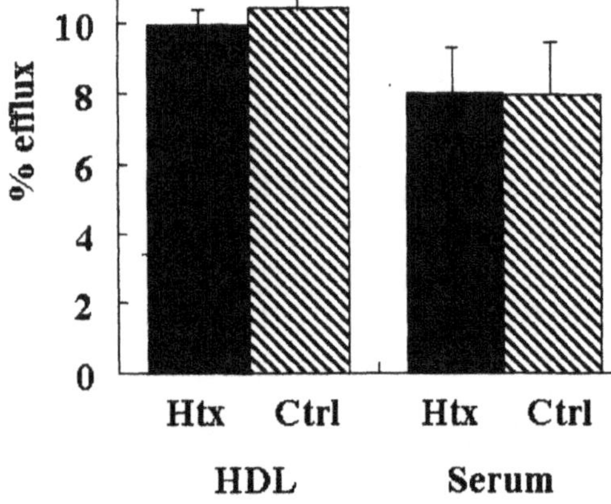

Fig. 1. Bar graph showing the percentage of relative efflux of cholesterol ± S.D. from Fu5AH cells incubated with HDL (50 μg/mL final concentration in culture medium) or serum (5% in culture medium) of Htx (solid bars, n = 8) and control rats (dashed bars, n = 8). Membrane cells were loaded with ^{3}H-cholesterol and incubated for 3 h, as described in the 'Materials and methods' section.

Table 2. Enzyme activity of lecithin cholesterol acyltransferase (LCAT), hepatic lipase (LH), and lipoprotein lipase (LPL) in Htx and Control rats

Enzyme	Group	
	Htx (n = 8)	Control (n = 8)
LCAT (% esterification/h/100μl plasma)	12.2 ± 3.0*	17.3 ± 5.3
LH (mU/mL plasma)	318.57 ± 101.81*	451.16 ± 23.88
LPL (mU/mL plasma)	251.85 ± 155.41	247.50 ± 80.10

*p < 0.05 vs. Control.

with plasma lipids nor with HDL-lipid contents (data not shown). Since triglycerides in HDL have their origin mainly in passive exchange with chylomicrons and VLDL, LPL that catalyze the trigliceride contained in those lipoproteins may have an indirect effect on HDL structure. In this study, Htx rats presented low triglyceride plasma levels, which suggest high LPL activities. However, the LPL activity in Htx animals was similar to those of controls (Table 2). These results indicate that the enzyme has a minimal contribution to altered HDL structure found in chronic hypothyroidism in rats.

Discussion

Atherosclerosis is a common feature in chronic hypothyroidism usually attributed to an increase in cholesterol plasma levels since thyroid hormones up-regulate LDL-receptor expression [25, 26]. High LDL-cholesterol levels were in fact the most striking lipoprotein abnormality in our hypothyroid animal model. Nevertheless, the role of an impaired RCT in this physiopathological condition has not been considered, because HDL plasma levels are frequently normal or even elevated [10–12]. One important difference between HDL metabolism in the rat versus primates is the absence of CETP in the former [27]; as a result, the rat does not transfer the cholesterol ester-laden core of HDL to VLDL, thus providing a suitable model for the study of HDL metabolism without the influence of variations in the CETP activity. Using the rat model, we have recently demonstrated that hypothyroidism results in significantly larger HDL, enriched in apo E and phospholipids, and poor in apo AIV content [18]. Such HDL abnormalities contribute to decrease HDL-apo AI catabolism suggesting an impaired RCT in chronic hypothyroid rats that may contribute to the development of atherosclerosis in the hypothyroid status. In the present work, we analyzed some stages of RCT, in order to establish if there are other abnormalities in this pathway associated with chronic hypothyroidism.

The first RCT stage is the cholesterol efflux from peripheral cells to HDL particles. Since HDL size determines their capacity to accept cholesterol [4], i.e. small apo AI-containing HDL promote more cholesterol efflux and HDL are larger in Htx rats, it could be expected that cholesterol efflux would be impaired in these animals. However, the capacity of serum of Htx rats as well as HDL from Htx rats to promote cholesterol efflux from culture cells is similar to that of control serum or HDL isolated from control rats. These results strongly suggest that structural modifications of HDL from Htx rats [18], do not alter the capacity to accept cholesterol from cells. SR-BI expression is abundant in Fu5AH cells [28] suggesting that HDL from Htx rats interacted normally with the SR-BI efflux pathway. Such SR-BI-mediated cholesterol efflux is strongly dependent on the content and composition of phospholipids in the HDL fraction [29]. Since HDL from Htx animals were 18% enriched in phospholipids [18], the expected decrease in cholesterol efflux, promoted by large HDL from Fu5AH cells, might be counterbalanced by the high HDL-phospholipid content.

Once cholesterol is incorporated into HDL, it is esterified by LCAT. Two different activities have been described for LCAT; α-LCAT activity corresponds to cholesterol esterification in HDL, and β-LCAT indicates cholesterol esterification in VLDL/LDL [30]. Our results showed that α-LCAT activity was reduced in Htx rats using an external HDLr for cholesterol esterification, which is indicative of low LCAT concentration [22]. As LCAT is generally considered the driving force of RCT, this abnormality can be added to the dysfunction of the whole metabolic pathway. It could be argued that LCAT mass does not always reflect *in vivo* esterification rates because they depend upon the nature and concentration of plasma lipoproteins. Nevertheless, high plasma levels of LDL particles and low concentration of HDL from Htx rats are favorable for the β-LCAT over the α-LCAT activity *in vivo* [31]. As a consequence, it can be stated that chronic hypothyroidism results in an impaired RCT at the HDL-cholesterol esterification stage.

Concerning the general lipoprotein metabolism, low LCAT activities has been recently demonstrated in patients with high LDL-cholesterol levels [31]. Our animal model presents a similar feature, low LCAT activity/high LDL-cholesterol, thus supporting the hypothesis that low LCAT activity is a consequence of hypercholesterolemia [31]. Such hypothesis implies that high LDL-cholesterol plasma concentrations contribute at two different levels to the development of atherosclerosis, firstly by cholesterol deposition into the subendothelial space, and secondly, by decelerate RCT rate at the cholesterol esterification stage. Whether or not plasma cholesterol down-regulates LCAT expression remains to be elucidated.

After cholesterol reaches the liver, HDL are remodeled particularly by the phospholipid transport protein (PLTP) and by HL which also possess a phospholipase activity [32]. Since HDL from Htx rats were significantly enriched in phospholipids [18], we expected an impaired HL or phospholipid transfer protein (PLTP) activity in Htx group. PLTP activity was measured, using an external phospholipid donor and no difference was observed between groups (data not shown). On the other hand, HL activity, determined by an external substrate (reconstituted HDL) method that is indicative of HL mass, was decreased in agreement with previous reports in hypothyroidism [15, 33]. In those reports, plasma levels of HDL subfractions correlated with HL activities [15] and T4 treatment resulted in a decrease of HDL_2-phospholipids in hypothyroid patients [33]. In our animal model we found low HL activities and both high phospholipid content in HDL and low plasma levels of those lipoproteins. Together these results suggest that a low HL postheparine plasma level in hypothyroidism increases the phospholipid content of HDL; however, we failed to demonstrate a correlation between postheparin HL activities and phospholipids in HDL, probably because the HDL-phospholipids from Htx rats had only slight variations between individuals [18] and the variation in HL activities is wide in each group. Conversely, HL triglyceride lipase activity seems to have a minor effect in HDL structure since HDL-triglyceride content remains constant, as it has been observed in this work and in previous reports [15, 18, 33].

In summary, this study demonstrates that chronic hypothyroidism in the rat, an animal that lacks CETP activity, induced an impaired RCT, mainly in cholesterol esterification and HDL remodeling by HL. The latter probably results in an abnormal HDL structure, i.e. phospholipid enrichment, which contributes to decreased HDL-apo AI fractional catabolic rates as it has been previously demonstrated [18]. Whether such RCT impairment contributes to atherosclerosis development in hypothyroid status remains to be elucidated.

Acknowledgements

This work was supported by CONACYT grant No. J28942M. We are indebted to Muriel Boully for the determination of PLTP activities.

References

1. Miller NE: High-density lipoprotein: A major risk factor for coronary atherosclerosis. Baillieres Clin Endocrinol Metab 1: 603–622, 1987
2. Parra HJ, Arveiler D, Evans AE, Cambou JP, Amouyel P, Bingham A, McMaster D, Schaffer P, Douste-Blazy P, Luc G: A case-control study of lipoprotein particles in two populations at contrasting risk for coronary heart disease. The ECTIM Study. Arterioscler Thromb 12: 701–707, 1992
3. Fielding CJ, Fielding PE: Molecular physiology of reverse cholesterol transport. J Lipid Res 36: 211–228, 1995

4. Castro GR, Fielding CJ: Early incorporation of cell-derived cholesterol into pre-β-migrating high-density lipoprotein. Biochemistry 27: 25–29, 1988
5. Matveev S, Uittenbogaard A, van Der Westhuyzen D, Smart EJ: Caveolin-1 negatively regulates SR-BI mediated selective uptake of high-density lipoprotein-derived cholesteryl ester. Eur J Biochem 268: 5609–5616, 2001
6. Jiang X, Francone OL, Bruce C, Milne R, Mar J, Walsh A, Breslow JL, Tall AR: Increased prebeta-high density lipoprotein, apolipoprotein AI, and phospholipid in mice expressing the human phospholipid transfer protein and human apolipoprotein AI transgenes. J Clin Invest 98: 2373–2380, 1996
7. Barrans A, Collet X, Barbaras R, Jaspard B, Manent J, Vieu C, Chap H, Perret B: Hepatic lipase induces the formation of pre-beta 1 high density lipoprotein (HDL) from triacylglycerol-rich HDL2. A study comparing liver perfusion to *in vitro* incubation with lipases. J Biol Chem 269: 11572–11577, 1994
8. Taylor AH, Wishart P, Lawless DE, Raymond J, Wong NC: Identification of functional positive and negative thyroid hormone-responsive elements in the rat apolipoprotein AI promoter. Biochemistry 35: 8281–8288, 1996
9. Apostolopoulos JJ, Marshall JF, Howlett GJ: Triiodothyronine increases rat apolipoprotein A-I synthesis and alters high-density lipoprotein composition *in vivo*. Eur J Biochem 194: 147–154, 1990
10. Valdemarsson S, Hedner P, Nilsson-Ehle P: Dyslipoproteinaemia in hypothyroidism of pituitary origin: Effects of L-thyroxine substitution on lipoprotein lipase, hepatic lipase, and on plasma lipoproteins. Acta Endocrinol 103: 192–197, 1983
11. Muls E, Rosseneu M, Blaton V, Lesaffre E, Lamberigts G, De Moor P: Serum lipids and apolipoproteins A-I, A-II and B in primary hypothyroidism before and during treatment. Eur J Clin Invest 14: 12–15, 1984
12. Scottolini AG, Bhagavan NV, Oshiro TH, Abe SY: Serum high-density lipoprotein cholesterol concentrations in hypo- and hyperthyroidism. Clin Chem 26: 584–587, 1980
13. Danese MD, Ladenson PW, Meinert CL, Powe NR: Clinical review 115: Effect of thyroxine therapy on serum lipoproteins in patients with mild thyroid failure: A quantitative review of the literature. J Clin Endocrinol Metab 85: 2993–3001, 2000
14. Dullaart RP, Hoogenberg K, Groener JE, Dikkeschei LD, Erkelens DW, Doorenbos H: The activity of cholesteryl ester transfer protein is decreased in hypothyroidism: A possible contribution to alterations in high-density lipoproteins. Eur J Clin Invest 20: 581–587, 1990
15. Tan KC, Shiu SW, Kung AW: Effect of thyroid dysfunction on high-density lipoprotein subfraction metabolism: Roles of hepatic lipase and cholesteryl ester transfer protein. J Clin Endocrinol Metab 83: 2921–2924, 1998
16. Tall AR: Plasma cholesteryl ester transfer protein and high-density lipoproteins: New insights from molecular genetic studies. J Intern Med 237: 5–12, 1995
17. Yamashita S, Maruyama T, Hirano KI, Sakai N, Nakajima N, Matsuzawa Y: Molecular mechanisms, lipoprotein abnormalities and atherogenicity of hyperalphalipoproteinemia. Atherosclerosis 152: 271–285, 2000
18. Huesca-Gomez C, Franco M, Luc G, Montaño LF, Masso F, Posadasa-Romero C, Pérez-Méndez O: Chronic hypothyroidism induces abnormal structure of high-density lipoproteins and impairs kinetics of apolipoprotein A-I in the rat. Metabolism 51: 443–450, 2002
19. Franco M, Tapia E, Martinez F, Davila ME, Grimaldo JI, Medina K, Herrera-Acosta J: Adenosine regulates renal nitric oxide production in hypothyroid rats. J Am Soc Nephrol 10: 1681–1688, 1999
20. Lowry OH, Rosebrough HJ, Farr AL, Randall RJ: Protein measurement with the Folin phenol reagent. J Biol Chem 193: 265–275, 1951
21. de la Llera Mora M, Atger V, Paul JL Fournier N, Moatti N, Giral P, Friday KE, Rothblat G: A cell culture system for screening human serum for ability to promote cellular cholesterol efflux. Relations between serum complement, esterification and transfer. Arterioscler Thromb 14: 1056–1065, 1994
22. Chen C, Albers JJ: Characterization of proteoliposomes containing apoprotein A-I. A new substrate for the measurement of lecithin cholesterol acyltransferase activity. J Lipid Res 23: 680–691, 1882
23. Nilsson-Ehle, Schotz MC: A stable, radioactive substrate emulsion for assay of lipoprotein lipase. J Lipid Res 17: 536–41, 1976
24. Belfrage P, Vaughan M: Simple liquid-liquid partition system for isolation of labeled oleic acid from mixtures with glycerides. J Lipid Res 10: 341–344, 1969
25. Hudig F, Bakker O, Wiersinga WM: Tri-iodothyronine prevents the amiodarone-induced decrease in the expression of the liver low-density lipoprotein receptor gene. J Endocrinol 152: 413–421, 1997
26. Wiseman SA, Powell JT, Humphries SE, Press M: The magnitude of the hypercholesterolemia of hypothyroidism is associated with variation in the low-density lipoprotein receptor gene. J Clin Endocrinol Metab 77: 108–112, 1993
27. Oschry Y, Eisenberg S: Rat plasma lipoproteins: Re-evaluation of a lipoprotein system in an animal devoid of cholesteryl ester transfer activity. J Lipid Res 23: 1099–1106, 1982
28. Jian B, de la Llera-Moya M, Ji Y, Wang N, Phillips MC, Swaney JB, Tall AR, Rothblat GH: Scavenger receptor class B type I as a mediator of cellular cholesterol efflux to lipoproteins and phospholipid acceptors. J Biol Chem 273: 5599–5606, 1998
29. Fournier N, Paul JL, Atger V, Cogny A, Soni T, de la Llera-Moya M, Rothblat G, Moatti N: HDL phospholipid content and composition as a major factor determining cholesterol efflux capacity from Fu5AH cells to human serum. Arterioscler Thromb Vasc Biol 17: 2685–2691, 1997
30. Carlson LA, Holmquist L: Evidence for the presence in human plasma of lecithin:cholesterol acyltransferase activity (β-LCAT) specifically esterifying free cholesterol of combined preβ- and β-lipoproteins: Studies of fish eye patients and control subjects. Acta Med Scand 218: 197–205, 1985
31. Lee M, Kim JQ, Oh H, Park M: Studies on plasma lipid profiles, and LCAT and CETP activities according to hyperlipoproteinemia phenotypes (HLP). Atherosclerosis 159: 381–389, 2001
32. Hime NJ, Barter PJ, Rye KA: Evidence that apolipoprotein A-I facilitates hepatic lipase mediating phospholipid hydrolysis in reconstituted HDL containing apolipoprotein AII. Biochemisty 40: 5496–5505, 2001
33. Pazos F, Alvarez JJ, Rubies-Prat J, Varela C, Lasuncion MA: Long-term thyroid replacement therapy and levels of lipoprotein(a) and other lipoproteins. J Clin Endocrinol Metab 80: 652–656, 1995

Molecular and Cellular Biochemistry **246**: 57–67, 2003.

Angiogenesis and myogenesis as two facets of inflammatory post-ischemic tissue regeneration

Dimitri Scholz,[1,2] Sylvia Thomas,[1] Sigrun Sass[1] and Thomas Podzuweit[1]
[1]*Department of Experimental Cardiology, Max-Planck-Institute;* [2]*Department of Thoracic and Cardiovascular Surgery, Kerckhoff-Clinic, Bad Nauheim, Germany*

Abstract

Ischemia is known to promote angiogenesis, and the molecular mechanisms and growth factors involved have been thoroughly investigated. Less attention has been paid to the cellular interactions between proliferating capillaries, inflammatory cells and muscle tissue. In the present investigation, we examined ischemia-induced cell injury and subsequent spontaneous regeneration in relation to angiogenesis and myogenesis. Balb/C mice were anesthetized, and the right femoral artery was occluded proximal to the origin of the arteria poplitea; the left femoral artery was sham-operated. One hour to 3 weeks after occlusion, the ischemic burden was assessed by quantifying the lower hindlimb muscle content of high energy phosphates, nucleosides, NAD+, glycogen and lactate. *In vivo* uptake of propidium iodide was used as a marker of cell injury. Proliferation of different cell types was evaluated immunomorphologically. As soon as 4 h after occlusion, morphological alterations in lower leg muscle occurred. Rupture of sarcolemma and loss of the contractile filaments were followed by plasma exudation, edema and infiltration of leukocytes, which target myofibers. During the first 12 h after occlusion, high energy phosphate and glycogen content decreased gradually. Levels remained low until day 3 and recovered almost completely until day 21. Nucleoside and lactate content peaked between 6 and 12 h following occlusion. Three days after occlusion, mitotic activity began in endothelial and muscle satellite cells, resulting in new formation of capillaries and muscle fibers. Angiogenesis and myogenesis occur concomitantly in regenerating skeletal muscle because of ischemia-induced cell death and inflammation. (Mol Cell Biochem **246**: 57–68, 2003)

Key words: angiogenesis, myogenesis, ischemia, cell injury, inflammation, regeneration

Abbreviations: CP – creatinphosphate; ATP – adenosine triphosphate; ADP – adenosine diphosphate; AMP – adenosine monophosphate; ADO – adenosine; INO – inosine; HX – hypoxanthine; XAN – xanthine; NAD+ – nicotineamide-adenine-dinucleotide (oxidized form); NADH – nicotineamide-adenine-dinucleotide (reduced form); PTCA – percutaneous transluminal coronary agioplasty; TMLR – transmyocardial laser revascularization; LDI – laser Doppler imaging

Introduction

Chronic occlusion of the femoral artery is a clinical entity, that may cause intermittent claudication and, at worst, necrosis of the lower limb. Following occlusion of a major artery, two different types of vascular growth participate in the salvage of ischemic tissue, first, sprouting of capillaries (angiogenesis) and secondly, growth of collateral arteries from pre-existing arterioles (arteriogenesis) [1–6].

We have shown previously that following femoral artery occlusion in rabbits or mice, collateral arteries developed only in the upper leg, whereas angiogenesis took place only in lower leg where reduction of blood flow was most pronounced [7–9].

The molecular mechanisms of angiogenesis have been thoroughly investigated. In this complex setting, the transcription factor Hypoxia-Inducible Factor-1 (HIF-1) seems of particular importance, which activates hypoxia-responsive genes that encode for Vascular Endothelial Growth Factor (VEGF) and inducible nitric oxide synthase (iNOS) and others, promoting angiogenesis [3, 10–23].

In response to skeletal muscle injury, the quiescent myogenic precursor cells (satellite cells) become activated, pro-

Address for offprints: D. Scholz, Max-Planck-Institute, Department of Experimental Cardiology, Benekestrasse 2, D-61231 Bad Nauheim, Germany (E-mail: d.scholz@kerckhoff.mpg.de)

liferate, fuse with each other to form myotubes, and mature into myofibers [24, 25]. Several Myogenic Regulatory Factors (MRF) are described to be responsible for the activation of satellite cells and myogenesis [26].

However, the relationship and cellular interaction between angio- and myogenesis are only poorly understood. In particular, the role of inflammation is grossly underestimated. Moreover, published findings do not yield information as to how quick ischemic injury in skeletal muscle develops, nor about for how long muscle fibers and/or vessels survive an ischemic insult caused by obstruction of the femoral artery.

In the present study, we focused on ischemia-induced injury and subsequent spontaneous recovery of ischemic lower limb muscle in mice, using physiological, morphological, and biochemical methods. Our aim was to reveal the time course and relationship of inflammation, angiogenesis and myogenesis following femoral artery occlusion.

Materials and methods

Surgery

The present study was performed with permission of the State of Hessen, Regierungspräsidium Darmstadt, according to section 8 of the German 'Law for the Protection of Animals'. The experiments were conducted using 11 weeks old male Balb/C mice from Charles River Laboratories (Sulzfeld, Germany). Mice were anesthetized by intraperitoneal injection of a mixture of ketamin (50 mg/kg body wt) and xylasin (5 mg/kg body wt). The right femoral artery was occluded proximal to the orifice of the arteria poplitea. The left femoral artery was sham-operated. Blood flow and tissue oxygen saturation of hindlimb feet were measured before, immediately after, and (again under anesthesia) in intervals until 3 weeks after occlusion (for methods see below). Ratios of blood flow and tissue oxygen saturation between ischemic and non-ischemic feet were calculated for each mouse and timepoint and statistically evaluated.

In a subset of experiments, the descending thoracic aorta was cannulated for the infusion of dye or contrast medium and subsequent perfusion fixation (see below).

Analysis of purines, creatine phosphate, lactate and glycogen

In another series of experiments, skeletal muscles were excised and quick-frozen in liquid nitrogen for subsequent analysis of high energy phosphates and their degradation products, NAD+, lactate and glycogen. Following freeze-drying, samples were stored in evacuated vials at –20°C until use. Aliquots of dried skeletal muscles were extracted into 5% perchloric acid with the aid of an ultrasonic disintegrator (Branson model B-15 with standard 1/2 inch microtip). Following 2 min of centrifugation in a microcentrifuge (Eppendorf model 5415 C), the acid supernatants were neutralized with 5 N KOH. The $KClO_4$ precipitate was removed by centrifugation.

Extracts were analyzed for purine nucleotides (ATP, ADP, AMP, NAD+), nucleosides (adenosine, inosine) and nucleobases (hypoxanthine, xanthine, urate) by reversed-phase high-performance liquid chromatography (HPLC) as described previously [27].

The creatine phosphate, lactate and glycogen content was quantitated by enzymatic analysis [28].

Laser-Doppler Imaging (LDI)

The animals were placed for 5 min at 37°C in a climatized chamber before the onset of measurements to avoid vasoconstriction by anesthetic heat loss. A red Laser Doppler imager (MLDI 5063, Moor Instruments Ltd., Devon, UK) was used to estimate relative blood flow. Measurements were performed at a pixel resolution of 256 × 256 and at a scan rate of 4 msec/pixel. Ratios of occluded over non-occluded values were compared after subtraction of the background (value for dead mice).

Oxygen saturation of hemoglobin

Oxygen saturation of hemoglobin was determined in the same climatized chamber at 37°C, by laser backscatter between the wavelength range of 500–620 nm (AbTisSpec™ spectrometer). A specially designed circular probe was constructed by LEA Medizintechnik GmbH for placement on the mouse foot.

Perfusion fixation

Mice were anesthetized, the thoracic aorta cannulated anterogradely and the right atrium severed. The circulatory system was rinsed free of blood with isotonic buffer containing 140 mM NaCl, 2 mM $CaCl_2$, 2 mM $MgCl_2$, 0.1% adenosine and 10 mM Tris-HCl, pH 7.4 (Fluka, Steinheim, Germany) under a perfusion pressure of 100 mmHg. To detect cells with damaged plasma membranes, a bolus of 40 µl buffer containing 0.1% propidium iodide was injected into the aortic cannula. This was followed by 10 min infusion at 100 mmHg of freshly prepared 1% formaldehyde in buffer. Finally, contrast medium (colloidal bismuth chloride in gelatine) was infused.

Staining and light microscopy

After perfusion fixation and cryoprotection with 20% sucrose overnight, M. gastrocnemius was dissected and snap-frozen in methylbutane (chilled to –130°C by immersion in liquid nitrogen). Cryosections of 5–10 µm thickness were cut with the Leica CM 3000 Cryocut (Bensheim, Germany). Endothelial cells were stained with TRITC-labeled lectin (BS-I). Ten µm cryosections were incubated with an antibody against the cellular proliferation marker Ki-67 to assess the mitotic activity. For this purpose, the polyclonal rabbit antibody raised against the mouse Ki-67 antigen was followed by biotinylated secondary goat anti-rabbit antiserum (Dianova, Hamburg, Germany) and Cy2-Streptavidin (Rockland, USA). Finally, nuclei were stained with 0.01% 7-actinomycin D (7AAD, Molecular Probes, Leiden, The Netherlands). Micrographs were obtained with the Leica DM microscope using an oil immersion 20/0.70 Planapo objective and Leica CM200 digital camera and analyzed with the NIH Image software. The total number of capillaries was counted on 10 viewfields for each muscle preparation. Results were expressed as number of capillaries per square millimeter.

To prepare high quality histological sections, tissue was postfixed in 3% glutaraldehyde for 4 h and in 1% OsO_4 for 2 h, dehydrated in graded alcohols and embedded in Epon using the Lynx Microscopy Tissue Processor (Reichert-Jung, Bensheim, Germany). Semi thin sections (0.5 µm thickness) were cut with the Ultracut microtome (Reichert-Jung, Bensheim, Germany), stained with toluidine blue, and analyzed with a Leica DAM (Leica, Bensheim, Germany) microscope using an oil immersion 63/1.4 Planapo objective.

Statistics

Results were expressed as mean values (± S.E.M.). Data were analyzed statistically with the Anova and Bonferoni tests. Results were considered to be statistically significant at $p < 0.05$.

Results

Femoral artery occlusion causes breakdown of aerobic metabolism in lower leg muscle

In order to demonstrate ischemic conditions of skeletal muscle in lower hindlimb of Balb/C mice; gastrocnemius muscle was analyzed for its content of high energy phosphates and degradation products, as well as NAD+, glycogen and lactate.

As of 60 min following femoral artery occlusion, there was significant decrease of creatine phosphate and modest accumulation of 5′-AMP, whereas tissue glycogen and lactate content remained constant (Fig. 1). ATP content was normal and there was no accumulation of nucleosides (adenosine, inosine) or nucleobases (hypoxanthine, xanthine, urate). The sum of skeletal muscle purines (nucleotides, nucleosides, nucleobases) remained constant during the first hour of ischemia, indicating fairly intact metabolism.

Six hours following femoral artery occlusion, the ATP content of gastrocnemius muscle was decreased approx. 35% compared to the sham-operated left hindlimb (Fig. 1B). Creatine phosphate was decreased by approximately 75% (Fig. 1A), and there was significant accumulation of adenosine, inosine, hypoxanthine, xanthine and urate (Figs 1E and 1F). Skeletal muscle glycogen content was depleted by approximately 70% (Fig. 1I). Tissue lactate content was temporarily increased (Fig. 1J), indicating that the compromised skeletal muscle had switched from aerobic to anaerobic metabolism.

The energy status of the muscle continued to decrease during the following hours and deprivation of high energy phosphates (ATP, creatine phosphate) was complete by day one. Accumulation of nucleosides and bases (distal catabolites of ATP) was maximal between 6 and 12 h after femoral artery occlusion. The combined gastrocnemius nucleotide, nucleoside and nucleobase content was lowest between 24 h and day 3 after the onset of ischemia, as was tissue glycogen content.

There was no recovery of high energy phosphates until at least day 3 post-occlusion. As of day 7, skeletal muscle ATP and creatine phosphate content increased significantly and 3 weeks after the onset of ischemia, gastrocnemius content of both high energy phosphates was barely different from the sham-operated control limb. Glycogen content had recovered fully by day 21.

Despite chronic femoral artery occlusion, accumulation of nucleosides and bases was no longer apparent after 3 weeks and the relation of tissue content of nucleotides, nucleosides and bases was normal, indicating that the adenylate kinase (myokinase) reaction was at equilibrium.

The skeletal muscle content of the nicotinamide-adenine-dinucleotide NAD+ (Fig. 1H) followed a similar pattern as the ATP (Fig. 1B) content and the sum of ATP-derived purines (Fig. 1G).

After the first week of femoral artery occlusion, the skeletal muscle content of the compounds tested recovered substantially, and after 3 weeks, all examined markers of ischemia approached baseline levels.

Blood flow and oxygen saturation of hemoglobin

Immediately after occlusion of the femoral artery, blood flow in the ischemic foot measured by LDI decreased to about 2% and oxygen saturation to about 9% of control values (Fig. 2). In the following, blood flow and oxygen saturation recovered gradually to approximately 6 and 27%, respectively at day 3, to 17 and 54% at day 7, to 41 and 72% at day 14 and to 50 and 83% at day 21.

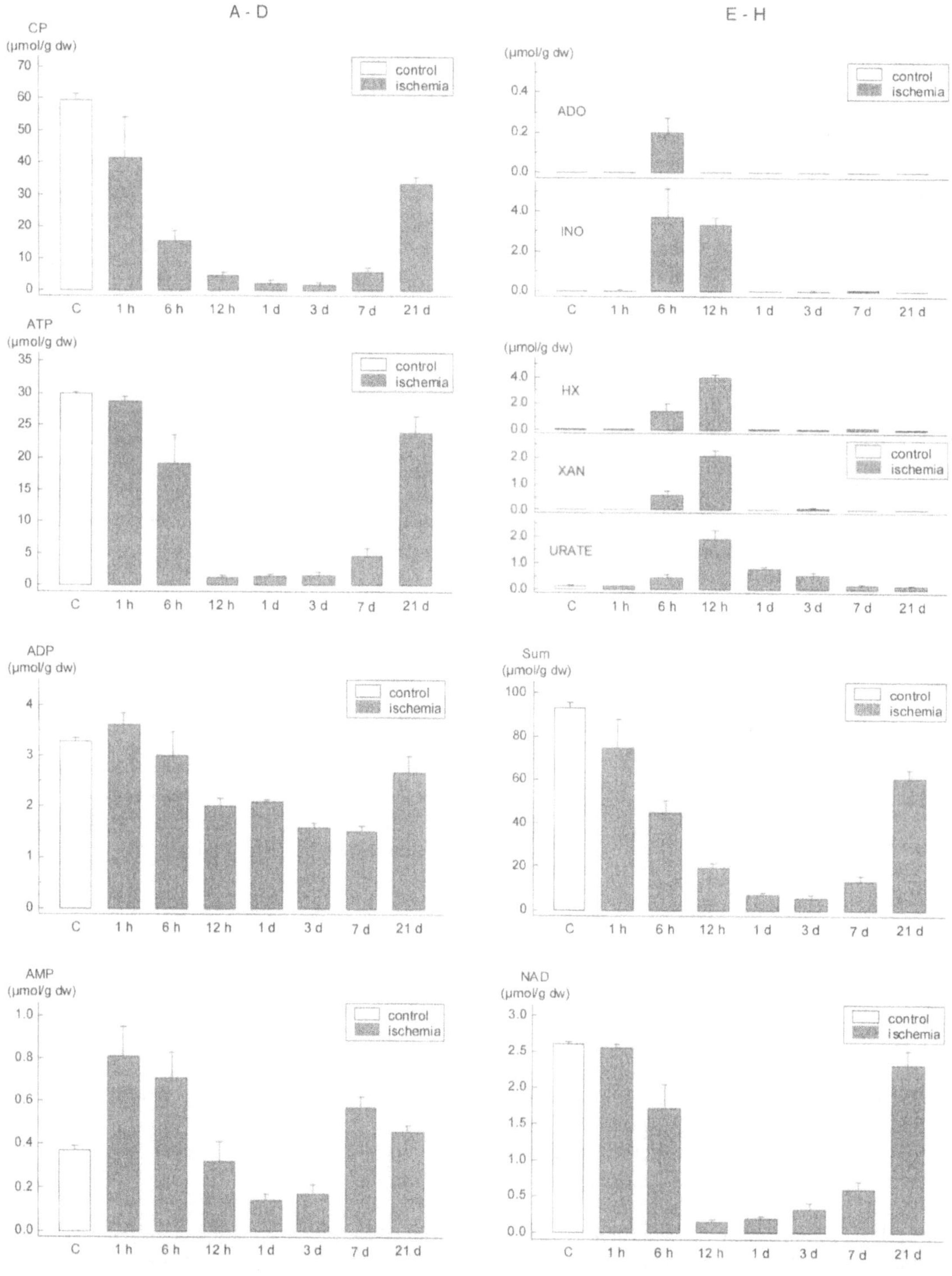

Fig. 1. Content of different markers of ischemia in m. gastrocnemius before and after chronic occlusion of the femoral artery in mice. Panels A–D: High energy phosphates (CP, ATP, ADP, AMP). Panels E and F: Nucleosides (adenosine, inosine) and nucleobases (hypoxanthine, xanthine and urate). Panel G: Sum of creatine phosphate, nucleotides, nucleosides and bases. Panel H: Nicotineamide-adenine-dinucleotide (NAD+) content. Panels I and J: Glycogen and lactate content. Statistical significance of biochemical data are shown in Table 1.

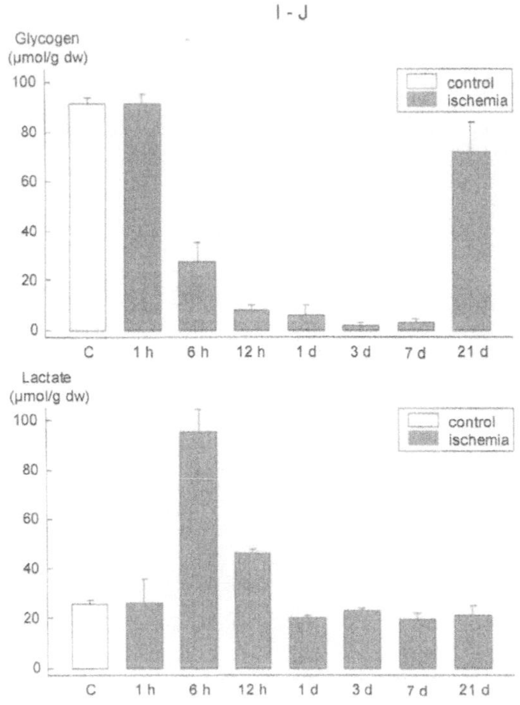

Fig. 1. Continued.

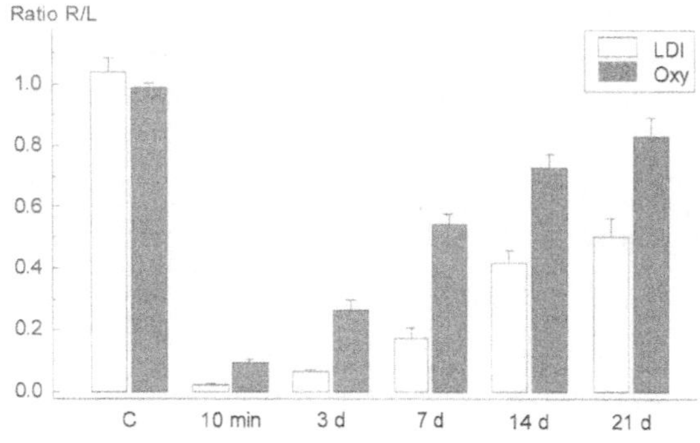

Fig. 2. Time course of blood flow measured by laser Doppler imaging (LDI) and oxygen saturation of hemoglobin (Oxy) in mouse feet before and after occlusion of the femoral artery. N.B.: these results correlated well with the biochemical data for the gastrocnemius muscle (Fig. 1).

Cellular injury and inflammation

Twelve to 72 h following femoral artery occlusion, ischemia-induced injury occurred in M. gastrocnemius (Fig. 3). *In vivo* infusion of propidium iodide revealed cells with damaged cell membranes (Figs 3F and 3G); their number was maximal at day 3. Disarray and loss of the contractile filaments were accompanied by plasma exudation and leukocyte infiltration (Figs 3B–3D). Severely damaged myofibers were targeted and subsequently removed by monocytes. At later timepoints, we did not find inflammatory cells (Fig. 5), indicating that they survive only a few days in tissue and presumably die by apoptosis.

Cellular proliferation

Three days after occlusion, mitotic activity was found in endothelial cells and muscular satellite cells (Fig. 4). The number of capillaries per square mm increased within 3 weeks after occlusion from 827 ± 76 to 1374 ± 62 (1.7-fold, $p < 0.01$), whereas there was only slight change from 800 ± 125 to 932 ± 182 (1.16-fold, non-significant) in the sham operated leg. Newly developed muscle fibers made about 50% of the whole cell population. They were smaller than normal muscle fibers and their nuclei were situated centrally (Figs 3H, 5B and 5D).

Discussion

In the present study, we examined the time course of ischemic injury, structural degeneration and regeneration and physiological recovery of the compromised hindlimb after chronic femoral artery occlusion in mice.

Up to 3 weeks following femoral artery occlusion, the extent of the ischemic insult inflicted to the lower hindlimb was examined by measuring the muscle content of high energy phosphates (creatine phosphate, ATP, ADP, AMP) and their degradation products – nucleosides (adenosine, inosine) and nucleobases (hypoxanthine, xanthine, urate) – as well as tissue glycogen and lactate content.

The ischemia-induced damage developed gradually, as evident from the decay of high energy phosphates and tissue glycogen, as well as morphological observations. However, there was a discrepancy between the two. Whereas immediately after occlusion of the femoral artery, blood flow to the foot was almost completely interrupted and tissue oxygen saturation decreased to below 10% of control, the levels of high-energy phosphates, glycogen and lactate in the M. gastrocnemius were almost unchanged during the first hour after occlusion (only AMP content increased 2-fold and CP content decreased by approximately 30%), indicating that the energy status was not yet severely compromised. However, 6 h after occlusion, the situation had changed. Although ATP content was decreased by a mere 30% compared to control, cellular injury had already begun as was shown histologically. The combined nucleoside and nucleobase content peaked 30–50 fold compared to control between 6 and 12 h after occlusion, which indicates marked breakdown of high energy phosphates and predicts exacerbated ischemic cell death, which corresponds well to the histological data. The low levels of high energy phosphates and glycogen content of the ischemic hindlimb would predict that the majority of cells had

Table 1. Statistical significance of biochemical data

	CP	ATP	ADP	AMP	ADO	INO	HX	XAN	URATE	Sum	NAD	Glycogen	Lactate
C vs. 1 h	*			**						*			
C vs. 6 h	***	***		**	***	***	***	***	*	***	***	***	***
C vs. 12 h	***	***	***			***	***	***	***	***	***	***	***
C vs. 1 day	***	***	***	**			***		*	***	***	***	
C vs. 3 days	***	***	***	**				***	***	***	***	***	
C vs. 7 days	***	***	***	***		***	***		*	***	***	***	*
C vs. 21 days	***	***	***							***	**		
1 h vs. 6 h					*	*		*	*			***	**
1 h vs. 12 h	*		**	*		***	***	***	**	**	***	***	
1 h vs. 1 day		***	**	***					**	**	***	***	
1 h vs. 3 days	*	***	***	**				*	*	***	***	***	
1 h vs. 7 days	**	***	***							***	***	***	
1 h vs. 21 days		***		*									
6 h vs. 12 h	*			*	*		*	***	**	**	**		**
6 h vs. 1 day	*	**		*		*				**	*		***
6 h vs. 3 days	***	*	*	**	**	**	*	*		***	**	**	***
6 h vs. 7 days	**	**	**		***	**	*	**	*	***	**	***	***
6 h vs. 21 days	**	**			*	*		*				*	***
12 h vs. 1 day						***	***	***	*	**			***
12 h vs. 3 days	*					*	***	***	**	***	*	**	***
12 h vs. 7 days			*	*		**	***	**	**		*		***
12 h vs. 21 days	***	***				***	***	***	**	***	***	**	***
1 day vs. 3 days			*				***						
1 day vs. 7 days			*	**		*			***				
1 day vs. 21 days	***	***		***					**	***	***	**	
3 days vs. 7 days	*			***			*	**	**		*		
3 days vs. 21 days	*	***	**	**					*	***	*	***	
7 days vs. 21 days	***	***	**							***	***	***	

*< 0.05–0.01; **< 0.01–0.001; ***< 0.001; all others n.s.

died 24 h following occlusion, as adequate functioning of ATP dependent ion channels and ion homeostasis seems unlikely with maximally depleted ATP. However, although tissue ATP content was decreased to less than 5% of control by day 1 after femoral artery occlusion, obviously far more than 5% of the total cell population survived the first 3 weeks.

It cannot be answered, at present, how cells managed to survive under such conditions. The discrepancy between energy status of skeletal muscle and histological data may be explained by assuming that cells switched from a 'functional' to a dormant 'survival' modus, comparable to 'hibernating' myocardium [29–31]. This prediction corresponds well with the animals behavior, as most mice did not make use of the operated leg during the first days after femoral artery occlusion (mice developed approximately 30% of power on the treadmill exercise test at day 7, unpublished observations). In addition, temperature of the operated hindlimb was markedly reduced after femoral artery occlusion [32], presumably contributing to cell survival.

Three days after femoral artery occlusion, the ischemic burden was still very high as indicated by minimal levels of high energy phosphates and glycogen. At this timepoint, the tissue glycogen content was still maximally depleted (only 1% of control value and approximately 10% from the 1 or 7 day values). However, the histological data showed that the proliferation of endothelial and muscular satellite cells peaked around day 3. We have shown mitotic activity both by histological detection of chromosomes and immunomorphologically with the proliferation marker Ki-67.

Twenty-one days after occlusion, no dead cells were found histologically, indicating that cell death had stopped and cell debris had already been removed by phagocytosis.

As shown previously, angiogenesis and myogenesis can both be induced by non-ischemic injury. For instance, transmyocardial laser revascularization (TMLR) is known to stimulate angiogenesis in heart muscle. TMLR is used in patients with end-stage ischemic heart disease not amenable to coronary bypass grafting or PTCA. The principal reason for using TMLR was to mimic the coronary circulation of certain amphibians which lack coronary arteries and to produce artificial non-thrombogenic channels in ischemic tissue which could be covered by intact endothelium in order to create new functional blood vessels.

Experimental studies and clinical post-mortem histological examinations revealed that acute blood flow was not improved by TMLR [33]. Moreover, laser channels did not convert into patent blood vessels and the arterial vessels originating from the channels during the weeks after TMLR were small, and plenty capillaries developed. The mechanism of new vessel formation following TMLR is considered to be the result of a non-specific response to injury [34–39].

In addition, chronic electrical stimulation of skeletal muscles of the lower hind limb of rats induced angiogenesis [40, 41]. These investigators proposed a fairly complex mechanism of angiogenesis, involving an increase of shear stress and secretion of nitric oxide (NO). However, their data did not exclude a more simple mechanism of capillary growth, such as micro-injuries caused by electrical stimulation that are followed by inflammation and subsequent cellular proliferation.

Mechanical stretch was also found to induce angiogenesis in rat hindlimbs [42]. Proliferation of satellite cells and formation of muscle fibers was induced by myotrauma caused by resistance exercise [43] or by limb immobilisation followed by re-activation [44].

Thus, focal cell damage, caused by different mechanisms including or excluding ischemia could induce similar cellular reactions, which involved inflammation and formation of new capillaries and skeletal muscle fibers.

The following sequence of events is proposed (Fig. 5). Cell injury causes inflammation. Infiltrating blood leukocytes target muscle fibers damaged by the ischemic insult. This inflammatory cellular reaction causes edema and stimulates tissue regeneration. Inflammatory leukocytes are known to secrete a cocktail of growth factors and cytokines [45–48]. Therefore, tissue repair is thought to be due to the concerted action of such peptide cocktail from infiltrating leukocytes rather than to one or two major factors produced by ischemic cells themselves. Stimulated neutrophiles were shown to secret proinflammatory mediators such as VEGF and induce angiogenesis *in vitro* [49]. Thus, no singular growth factor, but the powerful inflammatory reaction is likely responsible for angiogenesis as well as myogenesis.

If minor cell damage would stimulate angiogenesis in general, this may have implications to tumor growth. Proliferation of many tumors depends on angiogenesis and interventions that promote tumor cell damage and enhance angiogenesis should be avoided in tumor therapy.

Ischemia-induced lactate accumulation from endogenous glycogen stores

Lactate accumulation during ischemia indicates anaerobic energy production through glycogenolysis and glycolysis, whereas glucose oxidation via Krebs cycle does not produce lactate.

We did not distinguish between glycolysis and glucose oxidation. However, comparison of the amounts of skeletal muscle glycogen and lactate found in skeletal muscle at different time points following femoral occlusion allows semi-quantitative analysis of glycogen derived lactate accumulation. Twelve hours post-occlusion, tissue lactate content above control equalled the amount of glycogen lost during ischemia. As two lactate molecules are formed from one glucose molecule, it may be concluded that anaerobic glycolysis was responsible for at least 50% of glycogen consumption during the first 12 h following femoral artery occlusion, which indicates severe ischemic conditions.

It is feasible that the other half of glycogen was consumed by glucose oxidation, however, the oxygen content of the ischemic leg was extremely low during the first 12 h of ischemia when glycogen stores were depleted. Thus, it is likely that most of the glycogen was metabolized to lactate and that part of the lactate formed was subsequently washed out from the ischemic skeletal muscle into the circulation.

It is clear that glycogen would not have been wasted in the presence of sufficient oxygen and a functioning collateral circulation. Thus, the loss of glycogen and the temporary accumulation of lactate following femoral artery occlusion indicate that the perfusion of the lower limb of Balb/C mice was severely compromised and that metabolism was largely anaerobic.

Circulating glucose was lacking in ischemic muscle. To this end, glycogenolysis was the prime source of energy that helped to preserve ATP levels in the early phase of ischemia. This is why the decrease of ATP content paralleled the loss of glycogen stocks.

Glycogen resynthesis occurred only after day 7, a little later than *de novo* synthesis of ATP, that is only after oxidative phosphorylation had resumed (Fig. 1I).

The pattern of consumption and recovery of NAD+ was comparable to that of the high energy phosphates ATP and creatine phosphate, although NAD+ itself is not a high energy phosphate. The prime function of the nicotinamide-adenine-dinucleotide (NAD+) is to act as hydrogen transferring coenzyme (NAD+ + 2H = NADH + H+) and thus, is indirectly linked to energy production. NAD+ is formed from NADH as pyruvate is reduced to lactate. Thus, one might expect NAD+ to increase during ischemia. However, the opposite was the case. Despite the strong accumulation of lactate at 6 h after femoral artery occlusion, NAD+ levels fell, presumably because oxidation of NADH to NAD+ in the respiratory chain was limited because of the shortage of oxygen, causing NADH to accumulate at the expense of NAD+. Conversely, the gradual increase of NAD+ as of day 3 following femoral artery occlusion may indicate restart of oxidative metabolism.

In the present investigation, purine catabolism was examined stepwise from ATP to urate. We observed increased levels of xanthine and urate in ischemic hindlimb as of 6 h after femoral artery occlusion, indicating that xanthine: oxygen oxidoreductase activity (EC 1.1.3.22) was present in mouse skeletal muscle.

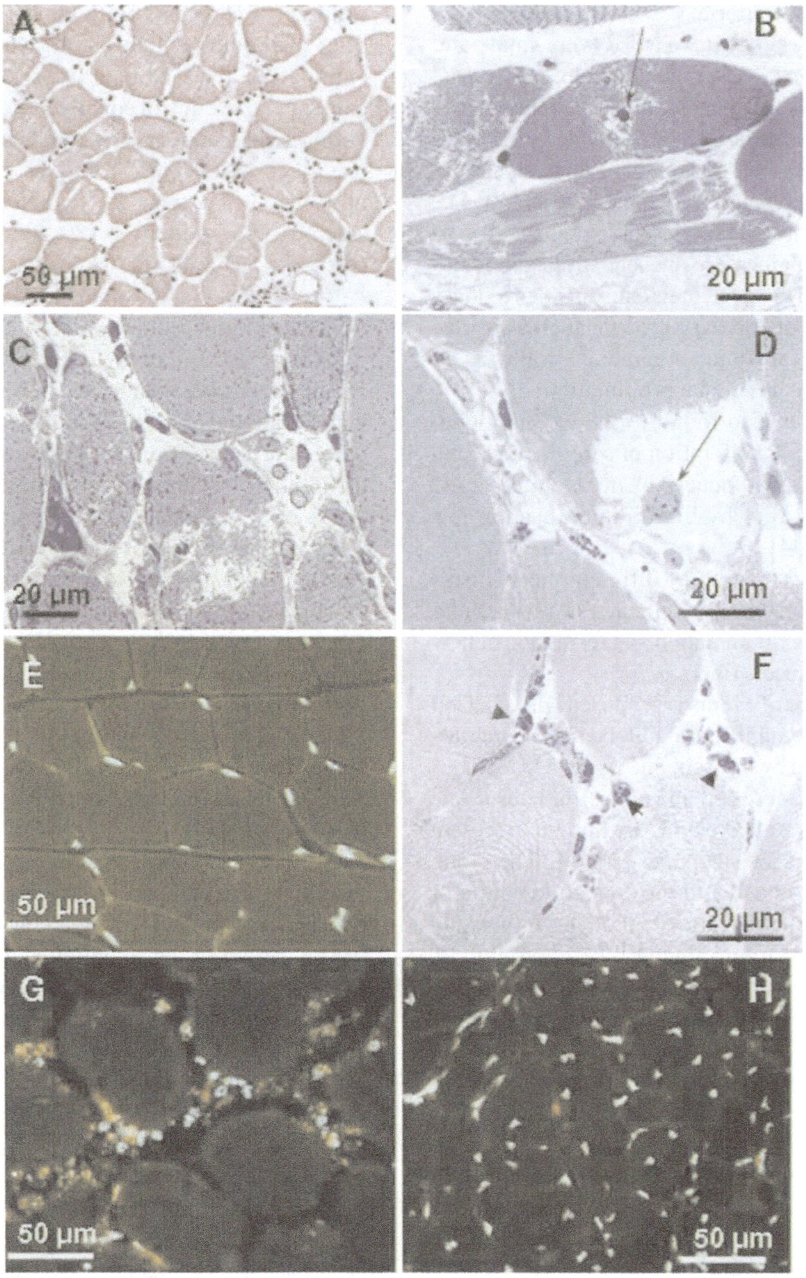

Fig. 3. Morphological alterations in the murine M. gastrocnemius 12–72 h after occlusion of the femoral artery. (A) Infiltration of leukocytes 24 h after occlusion. Cryosection 5 µm thick, stained with hematoxylin-eosin. Magn. ×200. (B) Edema, rupture of sarcolemm 12 h after occlusion. Lymphocyte targeting myofiber is labeled with arrow. Semithin section 0.5 µm thick, stained with toluidine blue. Magn. ×630. (C, D) More leukocytes target myofibers 48 h after occlusion. Mononuclear cell inside a myofiber is labeled with arrow. Semithin section 0.5 µm thick, stained with toluidine blue. Magn. C: ×630, D: ×1000. (E) Control tissue from the non-occluded side after *in vivo* propidium iodide infusion. No nucleus is stained red. Cryosection 5 µm thick. Vessels are labeled green with FITC-lectin, nuclei counterstained blue with DAPI. Magn. ×400. (F) Three days after occlusion of the femoral artery. Destroyed nuclei probably underlying apoptosis (arrowheads) belonged to muscle fibers, capillaries and fibroblasts. (Corresponds to the fluorescent picture shown at G). Semithin section, stained with toluidine blue. Magn. ×1000. (G) Three days after occlusion of the femoral artery. Propidium iodide infusion. Red labeled nuclei belonging to dead cells of different types. Cryosection 5 µm thick. Nuclei are counterstained blue with DAPI. Magn. ×400. (H) Three weeks after occlusion of the femoral artery. Propidium iodide infusion. Only some muscle fiber nuclei are still stained red. Some nuclei are situated centrally, marking newly developed myofibers. Cryosection 5 µm thick. Nuclei are counterstained blue with DAPI. Magn. ×400.

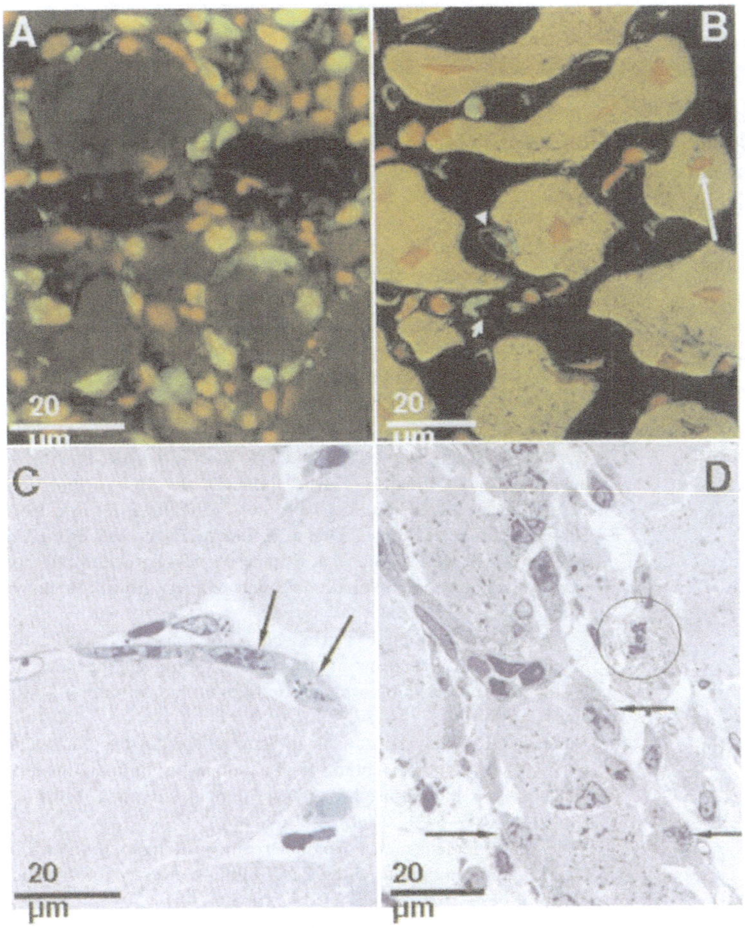

Fig. 4. Proliferative activity in the M. gastrocnemius three days after occlusion of the femoral artery. (A) Nuclei of muscular satellite cells are labelled green, nuclei counterstained red with 7-AAD. Cryosection 5 µm thick, Magn. ×700. (B) Nuclei of capillary endothelial cells (arrowheads) and one middle-standing nucleus (arrow) of muscle fibre are labelled green, nuclei counterstained red with 7-AAD. Magn. ×600. (C) Satellite cell with activated nucleus (arrow). Dark structures labeled with arrowheads represent probably apoptotic leukocytes. Semithin section 0.5 µm thick stained with toluidine blue. Magn. ×1000. (D) Mitotic chromosomes inside a myofiber (circle), representing a proliferating satellite cell. Other satellite cells with activated nuclei are labeled with arrows. Semithin section 0.5 µm thick, stained with toluidine blue. Magn. ×630.

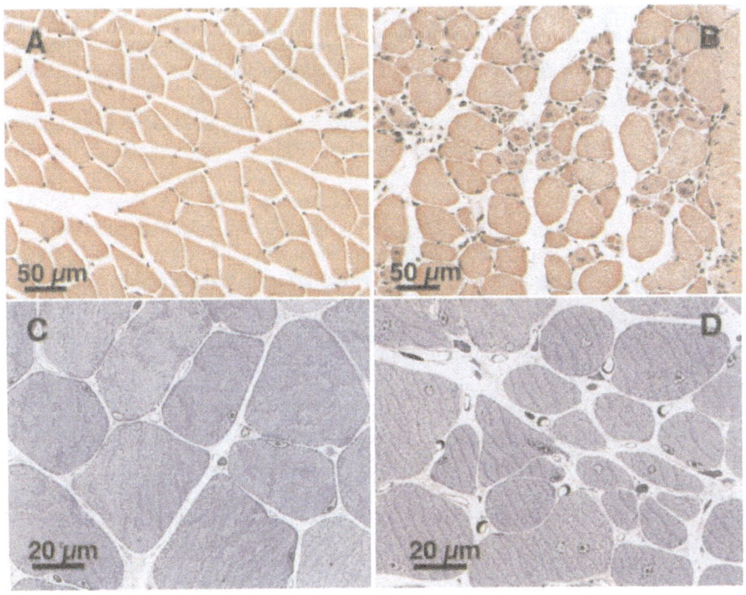

Fig. 5. Newly developed muscle fibers 21 days after occlusion of the femoral artery. (A) Cross-section of M. gastrocnemius from the sham-operated side demonstrating normal structure. Cryosection 5 µm thick, stained with hematoxylin-eosin. Magn. ×200. (B) Cross-section of M. gastrocnemius from the occluded side 3 weeks after occlusion. Newly developed muscle fibers can be recognized on their smaller size and medial nuclei. Cryosection 5 µm thick, stained with hematoxylin-eosin. Magn. ×200. (C) Cross-section of M. gastrocnemius from the sham-operated side demonstrating normal structure. Semithin section 0.5 µm thick, stained with toluidine blue. Magn. ×1000. (D) Cross-section of M. gastrocnemius from the occluded side 3 weeks after occlusion. Semithin section 0.5 µm thick, stained with toluidine blue. Magn. ×630

Increased levels of urate were found in ischemic M. gastrocnemius as well as in the sham operated control (not shown) at a time (days 1 and 3) when hypoxanthine and xanthine levels had returned to basal levels, suggesting that fairly high concentrations of urate were present in peripheral blood.

Xanthine oxidase derived oxygen radicals may be involved in ischemia-induced angiogenesis in skeletal muscle. In accordance with this concept, there was a correlation between the occurrence of xanthine and urate and the onset of mitotic activity in our study. However, the presence of these catabolites *per se* does not prove that oxygen free radicals were formed, for two reasons. First, oxygen (the electron acceptor required by xanthine oxidase) is lacking in ischemic skeletal muscle of Balb/C mice. Secondly, xanthine and urate may have been formed in a reaction catalyzed by the enzyme xanthine dehydrogenase, which makes use of NAD+ instead of oxygen as electron acceptor and does not yield oxygen free radicals [27].

Conclusion

Ischemic cell death in lower leg muscle triggers a cascade of events leading to both, angiogenesis and myogenesis, which reflects two facets of the same regenerative process. We propose that tissue regeneration is due to inflammation after ischemia-induced cell damage.

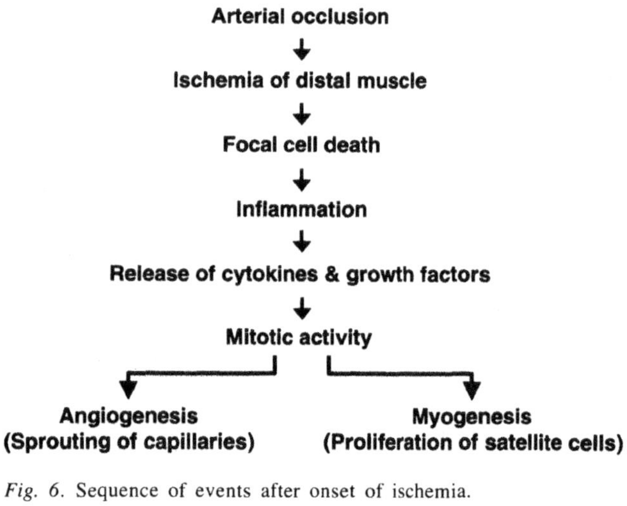

Fig. 6. Sequence of events after onset of ischemia.

References

1. Folkman J, D'Amore PA: Blood vessel formation: What is its molecular basis? Cell 87: 1153–1155, 1996
2. Risau W: Mechanisms of angiogenesis. Nature 386: 671–674, 1997
3. Carmeliet P: Mechanisms of angiogenesis and arteriogenesis. Nature Med 6: 389–395, 2000
4. Schaper W: Quo vadis collateral blood flow? A commentary on a highly cited paper. Cardiovasc Res 45: 220–223, 2000
5. Scholz D, Ziegelhoeffer T, Friedrich C: Collateral circulation – theory and experimental observations. In: P. Lanzer, E.J. Topol (eds). Theory and Practice of Vascular Medicine. Springer, New York/Heidelberg, 2001
6. Scholz D, Cai WJ, Schaper W: Arteriogenesis, a new concept of vascular adaptation in occlusive disease. Angiogenesis: 2002 (in press)
7. Ito WD, Arras M, Scholz D, Winkler B, Htun P, Schaper W: Angiogenesis but not collateral growth is associated with ischemia after femoral artery occlusion. Am J Physiol 273: H1255–H1265, 1997
8. Scholz D, Ito W, Fleming I, Deindl E, Sauer A, Babiak A, Bühler A, Wiesnet M, Busse R, Schaper J, Schaper W: Ultrastructure and molecular histology of rabbit hindlimb collateral artery growth. Virchows Arch 436: 257–270, 2000
9. Scholz D, Ziegelhoeffer T, Helisch A, Wagner S, Friedrich C, Schaper W: Contribution of arteriogenesis and angiogenesis to postocclusive hindlimb perfusion in different mouse strains. J Mol Cell Cardiol 34: 775–787, 2002
10. Semenza GL, Wang GL: A nuclear factor induced by hypoxia via *de novo* protein synthesis binds to the human erythropoietin gene enhancer at a site required for transcriptional activation. Mol Cell Biol 12: 5447–5454, 1992
11. Wang GL, Semenza GL: Characterization of hypoxia-inducible factor 1 and regulation of the DNA binding activity by hypoxia. J Biol Chem 268: 21513–21518, 1993
12. Semenza GL, Roth PH, Fang HM, Wang GL: Transcriptional regulation of genes encoding glycolytic enzymes by hypoxia-inducible factor 1. J Biol Chem 269: 23757–23763, 1994
13. Levy AP, Levy NS, Wegner S, Goldberg MA: Transcriptional regulation of the rat vascular endothelial growth factor by hypoxia. J Biol Chem 270: 13333–13340, 1995
14. Forsythe JA, Jiang BH, Iyer NV, Agani F, Leung SW, Koos RD, Semenza GL: Activation of vascular endothelial growth factor gene transcription by hypoxia-inducible factor 1. Mol Cell Biol 16: 4604–4613, 1996
15. Richard DE, Berra E, Pouyssegur J: Angiogenesis: How a tumor adapts to hypoxia. Biochem Biophys Res Commun 266: 718–722, 1999
16. Vincent KA, Shyu KG, Luo Y, Magner M, Tio RA, Jiang C, Goldberg MA, Akita GY, Gregory RJ, Isner JM: Angiogenesis is induced in a rabbit model of hindlimb ischemia by naked DNA encoding an HIF-1alpha/VP16 hybrid transcription factor. Circulation 102: 2255–2261, 2000
17. Jin KL, Mao XO, Nagayama T, Goldsmith PC, Greenberg DA: Induction of vascular endothelial growth factor and hypoxia-inducible factor-1alpha by global ischemia in rat brain. Neuroscience 99: 577–585, 2000
18. Lee SH, Wolf PL, Escudero R, Deutsch R, Jamieson SW, Thistlethwaite PA: Early expression of angiogenesis factors in acute myocardial ischemia and infarction. New Engl J Med 342: 626–633, 2000
19. Wurzel J, Goldman BI: Angiogenesis factors in acute myocardial ischemia and infarction. New Engl J Med 343: 148–149, 2000

20. Marti HJ, Bernaudin M, Bellail A, Schoch H, Euler M, Petit E, Risau W: Hypoxia-induced vascular endothelial growth factor expression precedes neovascularization after cerebral ischemia. Am J Pathol 156: 965–976, 2000
21. Li J, Post M, Volk R, Gao Y, Li M, Metais C, Sato K, Tsai J, Aird W, Rosenberg RD, Hampton TG, Sellke F, Carmeliet P, Simons M: PR39, a peptide regulator of angiogenesis. Nature Med 6: 49–55, 2000
22. Gorlach A, Camenisch G, Kvietikova I, Vogt L, Wenger RH, Gassmann M: Efficient translation of mouse hypoxia-inducible factor-1alpha under normoxic and hypoxic conditions. Biochim Biophys Acta 1493: 125–134, 2000
23. Kawata H, Yoshida K, Kawamoto A, Kurioka H, Takase E, Sasaki Y, Hatanaka K, Kobayashi M, Ueyama T, Hashimoto T, Dohi K: Ischemic preconditioning upregulates vascular endothelial growth factor mRNA expression and neovascularization via nuclear translocation of protein kinase C epsilon in the rat ischemic myocardium. Circ Res 88: 696–704, 2001
24. Reimann J, Irintchev A, Wernig A: Regenerative capacity and the number of satellite cells in soleus muscles of normal and mdx mice. Neuromusc Dis 10: 276–282, 2000
25. Barbero A, Benelli R, Minghelli S, Tosetti F, Dorcaratto A, Ponzetto C, Wernig A, Cullen MJ, Albini A, Noonan DM: Growth factor supplemented matrigel improves ectopic skeletal muscle formation – a cell therapy approach. J Cell Physiol 186: 183–192, 2001
26. Sabourin LA, Rudnicki MA: The molecular regulation of myogenesis. Clin Genet 57: 16–25, 2000
27. Podzuweit T, Beck H, Müller A, Bader R, Görlach G, Scheld HH: Absence of xanthine oxidoreductase activity in human myocardium. Cardiovasc Res 25: 820–830, 1991
28. Bergmeyer HU: Methods of Enzymatic Analysis. Weinheim, 1985, VI and VIII
29. Rahimtoola SH: The hibernating myocardium. Am Heart J 117: 211–221, 1989
30. Schaper W: 'Hibernating Myocardium'. Zeit für einen Paradigmenwechsel? Z Kardiol 80: 712–715, 1991
31. Elsässer A, Schaper J: Hibernating myocardium: adaptation or degeneration? Basic Res Cardiol 90: 47–48, 1995
32. Sauer A, Scholz D, Babiak A, Schaper W: Infra-red thermography is a very sensitive method to study arteriogenesis. J Mol Cell Cardiol 31: A103, 1999
33. Götzfried S, Thomas S, Reutebuch OT, Bauer EP, Klövekorn W-P, Podzuweit T: Circulation 100: A1980, 1999
34. Fleischer KJ, Goldschmidt Clermont PJ, Fonger JD, Hutchins GM, Hruban RH, Baumgartner WA: One-month histologic response of transmyocardial laser channels with molecular intervention. Ann Thorac Surg 62: 1051–1058, 1996
35. Malekan R, Reynolds C, Narula N, Kelley ST, Suzuki Y, Bridges CR: Angiogenesis in transmyocardial laser revascularization. A nonspecific response to injury. Circulation 98(suppl 19): II62–II65; discussion II66, 1998
36. Gassler N, Rastar F, Hentz MW: Angiogenesis and expression of tenascin after transmural laser revascularization. Histol Histopathol 14: 81–87, 1999
37. Bridges CR: Angiogenesis in myocardial laser 'revascularization'. Herz 25: 579–588, 2000
38. Simons M: Therapeutic coronary angiogenesis: a fronte praecipitium a tergo lupi? Am J Physiol – Heart Circ Physiol 280: H1923–H1927, 2001
39. Domkowski PW, Biswas SS, Steenbergen C, Lowe JE: Histological evidence of angiogenesis 9 months after transmyocardial laser revascularization. Circulation 103: 469–471, 2001
40. Hudlicka O, Brown MD, Silgram H: Inhibition of capillary growth in chronically stimulated rat muscles by N(G)-nitro-l-arginine, nitric oxide synthase inhibitor. Microvasc Res 59: 45–51, 2000
41. Linderman JR, Kloehn MR, Greene AS: Development of an implantable muscle stimulator: Measurement of stimulated angiogenesis and poststimulus vessel regression. Microcirculation 7: 119–128, 2000
42. Egginton S, Zhou A-L, Brown MD, Hudlicka O: Unortodox angiogenesis in skeletal muscle. Cardiovasc Res 49: 634–646, 2001
43. Vierck J, O'Reilly B, Hossner K, Antonio J, Byrne K, Bucci L, Dodson M: Satellite cell regulation following myotrauma caused by resistance exercise. Cell Biol Int 24: 263–272, 2000
44. Wanek LJ, Snow MH: Activity-induced fiber regeneration in rat soleus muscle. Anat Rec 258: 176–185, 2000
45. Wahl SM, Wong H, McCartney-Francis N: Role of growth factors in inflammation and repair. J Cell Biochem 40: 193–199, 1989
46. Hubner G, Brauchle M, Smola H, Madlener M, Fassler R, Werner S: Differential regulation of pro-inflammatory cytokines during wound healing in normal and glucocorticoid-treated mice. Cytokine 8: 548–556, 1996
47. Moore K: Cell biology of chronic wounds: The role of inflammation. J Wound Care 8: 345–348, 1999
48. Gillitzer R, Goebeler M: Chemokines in cutaneous wound healing. J Leuk Biol 69: 513–521, 2001
49. McCourt M, Wang JH, Sookhai S, Redmond HP: Proinflammatory mediators stimulate neutrophil-directed angiogenesis. Arch Surg 134: 1325–1331; discussion 1331–1332, 1999

Molecular and Cellular Biochemistry **246**: 69–74, 2003.

Impact and mechanism for oxidized and glycated lipoproteins on generation of fibrinolytic regulators from vascular endothelial cells

Garry X. Shen
Diabetes Research Group, University of Manitoba, Winnipeg, Manitoba, Canada

Abstract

Thrombogenesis depends on the balance between coagulation and fibrinolysis in vasculature. Vascular endothelial cells (EC) synthesize activators and inhibitors for fibrinolysis, tissue and urokinase plasminogen activators (tPA and uPA) and plasminogen activator inhibitor-1 (PAI-1). Increased levels of PAI-1 with various levels of tPA have been frequently found in plasma of patients with coronary heart disease (CHD) or diabetes mellitus (DM). Dyslipidemia is common feature in patients with CHD or DM, which is characterized by elevated levels of total cholesterol, triglycerides, low or very low density lipoproteins (LDL or VLDL) and decreased levels of high density lipoprotein (HDL). LDL and VLDL stimulated the generation of PAI-1 from cultured EC. LDL and lipoprotein(a) [Lp(a)], another lipoprotein risk factor for CHD, reduced the generation of tPA from EC. HDL did not greatly alter the release of PAI-1 from EC. Oxidative modification by copper, ultraviolet or long exposure to EC enhanced the effect of LDL on the generation of PAI-1 and tPA from EC. Glycation amplified the effect of LDL and Lp(a) on the changes in the generation of the fibrinolytic regulators from EC. Treatment with antioxidants or HDL normalized glycated LDL-induced changes in the generation of fibrinolytic regulators from EC. Activation of protein kinase C is required for oxidized LDL or Lp(a)-induced PAI-1 production in EC. VLDL, but not LDL or its oxidized form, stimulated PAI-1 production through the activation of the VLDL-responsive element in the PAI-1 promoter. Plasma levels of fibrinolytic regulators in CHD or DM patients may be normalized by HMG-CoA reductase inhibitors and angiotensin II converting enzyme inhibitors. This review summarizes the up-to-date information on effects, mechanism and management for disorders in EC-derived fibrinolytic regulators induced by modified lipoproteins. (Mol Cell Biochem **246**: 69–74, 2003)

Key words: fibrinolytic regulators, endothelial cells, lipoproteins, oxidation, glycation

Introduction

Intravascular thrombosis anticipates the development of atherosclerosis, vascular remodeling, occlusive vascular disease and tissue ischemia [1, 2]. These processes play critical roles in the pathophysiology of coronary heart disease (CHD) and diabetic vascular complications. Intravascular thrombogenesis depends on the balance between coagulation and fibrinolysis. Vascular endothelial cells (EC) are one of the major sources of fibrinolytic regulators in the body particularly in blood circulation. Dyslipidemia and endothelial dysfunction are common biomedical features in these disorders. Oxidation may increase the thrombogenicity as well as atherogenicity of the lipoproteins [3]. Glycation enhances the oxidative stress of the lipoproteins [4]. This review focuses on the impact of oxidized and glycated lipoproteins on the generation of fibrinolytic regulators from vascular EC.

Fibrinolytic system

The biologically active product of the fibrinolytic system is plasmin. It functions in dissolving fibrin clot in blood circulation and maintains the fluency of blood flow. The precur-

Address for offprints: G.X. Shen, Diabetes Research Group, University of Medidine, 835-715 McDermot Avenue, Winnipeg, Manitoba, Canada, R3E 3P4 (E-mail: gshen@ms.umanitoba.ca)

sor of plasmin is plasminogen, which is abundant in most of body fluid and usually serves as a limitless supply. The generation of plasmin is mainly regulated by tissue and urokinase plasminogen activators (tPA and uPA). The activities of tPA and uPA are modulated by their major physiological inhibitor, plasminogen activator inhibitor-1 (PAI-1) (Fig. 1). Vascular EC and smooth muscle cells synthesize tPA, uPA and PAI-1 [5]. The generation of these fibrinolytic regulators from vascular EC and smooth muscle cells may be modulated by a group of biological activators, including thrombin, interleukin-1, angiotensin II (AII), endotoxin, transformation growth factor (TGF)-β, glucocorticoid, insulin and multiple types of plasma lipoproteins [6].

Changes of fibrinolytic activity in patients with cardiovascular disease and diabetes

Impaired fibrinolytic activity characterized by elevated levels of PAI-1 with or without a reduction in tPA has been frequently found in CHD patients [7, 8]. Increases in PAI-1 protein and mRNA were detected in atherosclerotic and thrombotic lesions in arteries or veins [9, 10]. Tests of fibrinolytic activity after vascular stimulation indicate that the release of fibrinolytic activity under stress was impaired in diabetic subjects [11, 12]. Increased plasma levels of PAI-1 were found in diabetic patients [13–17]. Strong correlations were found between PAI-1 and plasma insulin, body mass index, triglycerides or apoB in type 2 diabetes [18, 19]. Immunohistochemical analysis revealed the presence of PAI-1 antigen in human coronary atherosclerotic lesions [20]. Increased plasma PAI activity, which represents the imbalance between PAI-1 and tPA in blood circulation, has been considered as a new risk factor for CHD.

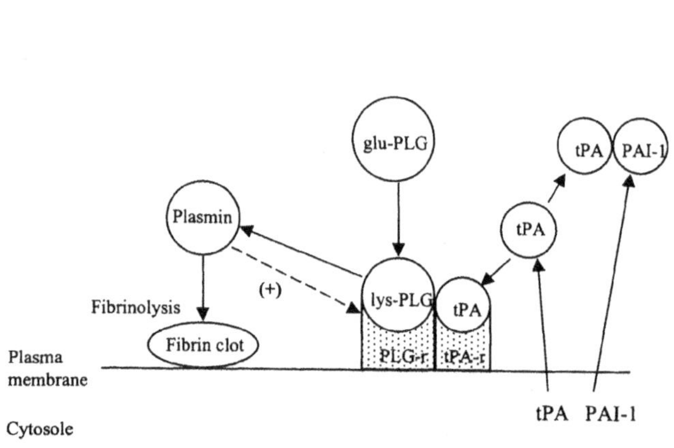

Fig. 1. Scheme for fibrolytic system. PLG – plasminogen; tPA – tissue plasminogen activator; PAI-1 – plasminogen activator inhibitor-1; r – receptor.

Impact of plasma lipoproteins on vascular EC-derived fibrinolytic regulators

The first report on the effect of lipoproteins on the generation of fibrinolytic regulators from EC appeared in 1990. Very low density lipoprotein (VLDL) from hypertriglyceridemic subjects increased PAI-1 antigen from human umbilical vein EC (HUVEC) [21]. Low density lipoprotein (LDL) and its structural homologues, lipoprotein(a) [Lp(a)], stimulated the transcription and secretion of PAI-1 from EC [22–26]. The effect of Lp(a) on EC-derived PAI-1 was > 10-times stronger than LDL [26, 27]. The sizes of LDL particles were inversely correlated to PAI-1 levels in a large tri-ethnic population [28]. Addition to LDL and Lp(a) reduced the release of tPA from HUVEC [29]. High density lipoprotein (HDL) did not evidently increase the generation of PAI-1 from EC [30]. The findings on the effects of HDL on the generation of tPA from EC were not homogenized [29, 30].

Effect of oxidized lipoproteins on EC-derived fibrinolytic regulators

LDL oxidatively modified by ultraviolet or copper induced significantly greater production of PAI-1 from EC compared to native LDL [22–26]. Oxidation also enhanced the effect of Lp(a) on the production in EC [26]. Recent studies suggest that O2*-HO* free radical- oxidized LDL and Lp(a) decreased the release of PAI-1 from EC compared to copper oxidized LDL an Lp(a). Further observation demonstrated that O2*-HO* free radical-oxidized LDL contained lower levels of oxidized lipid components (phosphatidylcholine hydroxyperoxides, lysophosphatidylcholine, 7-ketocholesterol, 7β-hydroxycholesterol, 5,6β-epoxycholesterol compared to copper oxidized LDL [31]. Lipoproteins may be modified by prolonged incubation with EC, smooth muscle cells or monocytes. PAI-1 and tPA release in EC induced by LDL was related to EC modification of the lipoproteins. Increased formation of lipid peroxidation products was found in EC-modified LDL. Co-treatment of LDL with antioxidants, butylated hydroxytoluene or vitamin E, attenuated the changes in the generation of the fibrinolytic regulators induced by LDL in EC as well as lipid peroxidation products in LDL [30]. This suggests that oxidation of lipid components in LDL by chemical or cells may contribute to the changes in the generation of fibrinolytic regulators induced by oxidatively modified LDL.

Impact of glycation on lipoproteins-induced generation of fibrinolytic regulators from EC

The levels of glycated LDL increased in patients with diabetes and correlated to glucose control [32, 33]. Our group is the first one described the effects of glycated lipoproteins on the generation of fibrinolytic regulators from EC. Glycation of LDL with ≥ 25 mM of glucose *in vitro* for ≥ 2 weeks resulted in significant changes in generation of fibrinolytic regulators in EC compared to native LDL. Glycated LDL stimulated the production of PAI-1 in EC at mRNA level. Glycated LDL decreased the de novo synthesis and release of tPA but without parallel reduction in tPA mRNA, in EC [34]. Glycation also enhanced the effect of Lp(a) on the generation of PAI-1 and tPA from EC [35]. Glycated HDL moderately increased the production of PAI-1 in EC but did not significantly affect tPA release compared to native HDL. Cotreatment of native or glycated HDL prevented glycated LDL-induced changes in the generation of PAI-1 or tPA from EC [30]. Our recent studies demonstrated that LDL from type 1 and type 2 diabetic patients and VLDL from type 2 diabetic patients increased the generation of PAI-1, and decreased the release of tPA from cultured EC compared to corresponding lipoproteins from healthy subjects [36]. The findings suggest that atherogenic lipoproteins glycated *in vitro* or *in vivo* may affect the generation of fibrinolytic regulators from EC, which contributes, at least in part, to attenuated fibrinolytic activity in diabetic patients.

Transcription and signal transduction mechanism for lipoproteins-induced PAI-1 production in EC

Results from previous studies suggest that atherogenic lipoproteins modulate the production of PAI-1 in EC at transcription level. The sequence of human PAI-1 promoter was documented in 1988 [37, 38]. Proximal region of the PAI-1 promoter contains a responsive element for phorbol myristate acetate (PMA), a potent protein kinase C (PKC) agonist, and a homologue for cAMP-responsive element [39]. Two distinct regions in the PAI-1 promoter located at –791/–546 and –328/–186 bp are involved in TGF-β-induced PAI-1 transcription in HepG2 cells (40). In Hep3B cells, two regions in the PAI-1 promoter (–740/–636 and –87/–487 bp) were activated by TGF-β [41]. One of them (–87/–48 bp) was distinct from those elicited by TGF-β in HepG2 cells. A 5′-flank region between –1212 and –1187 bp of PAI-1 promoter is essential for glucocorticoid-induced activation of the PAI-1 promoter in rat HTC hepatoma cells [42]. The VLDL-RE was detected between –672 and –657 bp of the –657 bp of the PAI-1 promoter [43]. The VLDL-RE is adjacent to the region of 4G/5G polymorphism (44]. This polymorphism is characterized by missing or inserting a G at –675 bp of the PAI-1 promoter to form a sequence of GGGG (4G) or GGGGG (5G). Homozygous 4G allele polymorphism of the PAI-1 promoter is associated with higher plasma levels of PAI-1 compared to 5G/5G or 4G/5G allele. Correlation between 4G allele and CAD was found in some studies, but not in the others [45–47]. The binding site (–682/–675 bp) of a transcription activator (factor A) is adjacent to 4G/5G region and is present in both 4G or 5G alleles. The activation of the factor A is involved in the upregulation of PAI-1 transcription. A transcription repressor (factor B, –676/–670 bp) in the PAI-1 promoter is 5G allele-specific and its sequence is incomplete in 4G-allele. VLDL activated the PAI-1 promoter with 4G allele in an extent greater than that in 5G allele. This may be due to that the PAI-1 promoter with 4G-allele lacks a functional transcription repressor. Unsaturated fatty acids, especially 18:3 and 20:5, activated the VLDL-RE in EC [48]. VLDL, Lp(a) and insulin, but not interleukin-1 activated –1106/+1 bp region of the PAI-1 promoter transfected in EC in a genotype-specific pattern [49]. Our group has investigated the effects of oxidized LDL on the activation of the PAI-1 promoter. The preliminary results suggested that oxidized LDL and VLDL activates the PAI-1 promoter through a responsive element distinct from the VLDL-RE [50]. Signal transduction mechanism involved in lipoprotein-induced PAI-1 production has little been known. PKC activation is required for LDL, Lp(a) and their oxidized forms induced PAI-1 production in venous or arterial EC. The activation of PKC-β is crucial for oxidized LDL and Lp(a) induced PAI-1 overproduction in EC [51].

Therapeutic modulation of fibrinolytic activity in blood circulation of humans

Previous studies indicated that fibrinolytic activity in the blood circulation of patients may be improved by various types of medications. Captopril, the first clinically available angiotensin converting enzyme (ACE) inhibitor, did not significantly affect the levels of PAI-1 or tPA in healthy subjects [52], but reduced PAI-1 and tPA levels in uncompleted myocardial infarction patients [53]. Acute administration of losartan, a AT1 antagonist, reduced PAI-1 and increased tPA in patients with chronic heart failure [54]. However, enalapril and losartan had no significant effect on PAI-1 patients with essential hypertension [55]. Pravastatin, a HMG-

CoA reductase inhibitor, reduced PAI-1 levels in hypercholesterolemic patients [56]. Our recent studies demonstrated that simvastatin reduced PAI-1 levels in type 2 diabetic patients. The levels of PAI-1 antigen in plasma of the diabetic patients correlated with total cholesterol, LDL cholesterol and hemoglobulin Alc. This suggests the effect of simvastatin on PAI-1 is secondary to its effects of reduction of LDL-cholesterol in type 2 diabetic patients [57]. Estrogen lowered plasma PAI-1 in hypercholesterolemic post-menopausal women [58]. PAI-1 in plasma may come from other sources, such as adipose tissue and hepatocytes. Significant increases in PAI-1 antigen and activity were found in obese individuals. Weight loss corrected the increase of level of PAI-1 and the attenuation of fibrinolytic activity in obese subjects [59].

Pharmacological modulation of fibrinolytic regulators in cultured EC

The production of PAI-1 on cultured EC at basal or stimulated conditions may be reduced by the treatment with several types of drug possibly improving endothelial function including AII receptor (AT) antagonists, statins or antioxidants. AT1 antagonist (DuP 735), but not AT2 antagonist (PD123319), inhibited AII induced overexpression of PAI-1 vascular EC and smooth muscle cells [60]. Constitutive or tumor necrosis factor-α-induced production of PAI-1, but not tPA, in EC may be reduced by atorvastatin or fluvastatin [61]. Simvastatin reduced platelet-derived growth factor-induced PAI-1 secretion from EC [62]. Antioxidants, BHT and vitamin E, normalized native and glycated LDL-increased increases in PAI-1 production and the decrease in tPA generation from EC [30]. Pre-enrichment of oxidized or glycated LDL enriched with vitamin E PAI-1 production in EC [63]. The results suggest that EC-derived fibrinolytic regulators may be pharmacologically modulated.

Summary

Oxidized and glycated atherogenic lipoproteins alter the generation of fibrinolytic regulators in vascular EC, which may lead to attenuation of fibrinolytic activity in blood circulation and thrombotic tendency. The overproduction of PAI-1 in EC induced by oxidized LDL is mediated by the activation of PKC-β and an undefined responsive element in the PAI-1 promoter distinct from the VLDL-responsive element in the PAI-1 promoter distinct from the VLDL-responsive element. The generation of fibrinolytic regulators in EC induced by modified atherogenic lipoproteins may be modulated through treatment with cholesterol lowering drugs and antioxidants.

Acknowledgements

The author is thankful for the contributions from trainees in his laboratory, including Dr. Kevin Cockell, Song Ren, Jianying Zhang, Lin Lu, Limei Hu, Shalini Shatada, Dr. Donfeng Sun, Dr. Sudharshan Dharmalingham, and his collaborators, including Dr. Liam J. Murphy, Dr. Harvey Lee and Sora Ludwig to relevant studies. The studies have been financially supported by Canadian Institute of health Research, Canadian Diabetes Association, Heart and Stroke Foundation of Canada, Manitoba Medical Service Foundation, Manitoba Health Research Council, Health Sciences Centre Foundation, University of Manitoba, Dr. Paul HT Thorlakson Foundation and St. James Kwanis Club.

References

1. Worthley SG, Osende JI, Helft G, Badimon JJ, Fuster V: Coronary artery disease: Pathogenesis and acute coronary syndrome. Mt Sinai J Med 68: 167–181, 2001
2. Badimon L: Atherosclerosis and thrombosis: Lessons from animal models. Thromb Haemost 86: 356–365, 2001
3. Naruszewicz M, Giroux LM, Davignon J: Oxidative modification of Lp(a) causes changes in the structure and biological properties of Lp(a). Chem Phys Lipids 67: 67–68, 1994
4. Kobayashi K, Watanabe J, Umeda F, Nawata H: Glycation accelerate the oxidation of low density lipoprotein by copper ions. Endocrine J 42: 461–465, 1995
5. Shen GX: Vascular cell-derived fibrinolytic regulators and atherothrombotic vascular disorders. Int J Mol Med 1: 399–408, 1998
6. Saksela O, Rifkin DB: Cell-associated plasminogen activation: Regulation and physiological functions. Ann Rev Cell Biol 4: 93–126, 1988
7. Hamsten A, D Faire U, Walldius G, Dahlen G, Szamosi A, Blomback M, Landou C, Wiman B: Plasminogen activator inhibitor in plasma: Risk factor for recurrent myocardial infarction. Lancet 2: 3–9, 1987
8. Cortellaro M, Cafrancesco E, Boshetti C, Mussoni L, Ponati MB, Cardillo M, Garbrielli L, Lombard B, Secchia G: Increased fibrin turnover and high PAI-1 antigen as predicator of ischemic event in atherosclerotic patients. The PLAT group. Arterioscler Thromb Vasc Biol 13: 1412–1417, 1993
9. Yorimitsu K, Saito T, Toyozaki T, Ishide T, Ohnuma N, Inagaki Y: Immunohistochemical localization of plasminogen activator inhibitor-1 in human coronary atherosclerosis lesions involved in acute myocardial infarction. Heart Vessels 8: 160–162, 1993
10. Arnman V, Nilson A, Stemme S, Risberg B, Rymo L: Expression of plasminogen activator inhibitor-1 mRNA in healthy, atherosclerotic and thrombotic arteries and veins. Thromb Res 74: 487–499, 1994
11. Geiger M, Binder BR: Plasminogen. Enzyme 40: 149–157, 1988
12. Auwerx J, Bouillon R, Collen D, Geboers J: Tissue-type plasminogen activator antigen and plasminogen activator inhibitor in diabetes mellitus. Arteriosclerosis 8: 68–72, 1988
13. Matsuo T, Kadowaki S, Okada K, Matsuo O: Activity of tissue plasminogen activator and plasminogen activator inhibitor in noninsulin-dependent diabetes mellitus. J Diabet Complications 4: 119–121, 1990
14. Gough SC, Rice PJ, McCormack L, Chapman C, Grant PJ: The relationship between plasminogen activator inhibitor-1 and insulin resistance in newly diagnosed type 2 diabetes mellitus. Diabet Med 10: 638–642, 1993

15. Vicari AM, Vigano-D'Angelo S, Testa S, Comi G, Galardi G, Orsi E, D'Angelo A: Normal tissue plasminogen activator and plasminogen activator inhibitor activity in plasma from patients with type 1 diabetes mellitus. Horm Metab Res 24: 516–519, 1992
16. Ceriello A, Quatraro A, Marchi E, Barbanti M, Giugliano D: Impaired fibrinolytic response to increased thrombin activation in type activation in type 1 diabetes mellitus: Effects of the glycosaminoglycan sulodexide. Diabetes Metab 19: 225–259, 1993
17. Carmassi F, Morale M, Puccett R, De-Negre F, Monzani F, Navalesi R, Mariani G: Coagulation and fibrinolytic system impairment in insulin dependent diabetes mellitus. Thromb Res 67: 643–654, 1992
18. Davi G, Violi F, Catalano I, Giammarresi C, Putignano E, Nicolosi G, Barbagallo M, Notarbartolo A: Increased plasminogen activator inhibitor antigen levels in diabetic patients with stable angina. Blood Coagul Fibrinolysis 2: 41–45, 1991
19. Nagi K, Mohammed AV, Jain SK, Walji S, Yudkin JS: Plasminogen activator inhibitor (PAI-1) activity is elevated in Asian and Caucasian subjects with non-insulin-dependent (type-2) diabetes but not in those with impaired glucose tolerance (IGT) or non-diabetic Asians. Diabet Med 13: 59–64, 1996
20. Jokl R, Klein RL, Lopes-Virella MF, Colwell JA: Release of platelet plasminogen activator inhibitor 1 in whole blood is increased in patients with type II diabetes. Diabetes Care 18: 1150–115455, 1995
21. Striko-Rahm A, Wiman B, Hamsten A, Nilsson J: P Secretion of plasminogen activator inhibitor-1 from cultured human umbilical vein endothelial cells is induced by very low density lipoprotein. Arteriosclerosis 10: 1067–1073, 1990
22. Tremoli E, Camera M, Maderna P, Sironi L, Prati L, Colli S, Piovella F, Bernini F, Corsini A, Mussoni: Increased synthesis of plasminogen activator inhibitor-1 by cultured human endothelial cells exposed to native and modified LDLs: An LDL receptor-independent phenomenon. Arterioscler Thromb 13: 338–346, 1993
23. Latron Y, Chautan M, Anfosso F, Salessi MC, Lafont H, Juhan-Vague I: Stimulating effect of oxidized low density lipoprotein on plasminogen activator-1 synthesis by endothelial cells. Arteroscler Thromb 11: 182–189, 1991
24. Chauton M, Latron Y, Anfosso F, Alessi MC, Lafont H, Juhan-Vague I, Nalbone G: Phosphatidylinositol turnover during stimulation of plasminogen activator inhibitory secretion induced by oxidized low density lipoproteins in human endothelial cells. J Lipid Res 34: 101–110, 1993
25. Kugiyama K, Sakamoto T, Ohgushi M, Ogawa H, Horiguchi M, Yasue H: Transferable lipids in oxidized low density lipoprotein stimulate plasminogen activator inhibitory and inhibit tissue plasminogen activator release from endothelial cells. Circ Res 73: 335–343, 1993
26. Ren S, Man RYK, Angel A, Shen GX: Oxidative modification enhances lipoprotein(a)-induced overproduction of plasminogen activator inhibitory in cultured vascular endothelial cells. Atherosclerosis 128: 1–10, 1997
27. Etingin OR, Hajjar KA, Harpel PC, Nachman RL: Lipoprotein(a) regulates plasminogen activator inhibitor- I expression in endothelial cells. J Biol Chem 266: 2456–2465, 1991
28. Festa A, D'Agostino R Jr, Mykkanen L, Tracy R, Howard BV, Haffner SM: Low-density lipoprotein particle size is inversely related to plasminogen activator inhibitor- 1 levels. The Insulin Resistance Atherosclersis Study. Arterioscler Thromb Vasc Biol 19: 605–610, 1999
29. Levin EG, Miles LA, Fless GM, Scanu AM, Baynhrna P, Curtiss LK, Plow EF: Lipoproteins inhibits the secretion of tissue plasminogen activator from human endothelial cells. Arterioscler Thromb 14: 438–42,1994
30. Ren S, Shen GX: Impact of antioxidant and HDL on glycated LDL-induced generation of fibrinolytic regulators from vascular endothelial cells. Arterioscler Thromb Vasc Biol 20: 1688–1693, 2000
31. Beaudeux JL, Therond P, Bonnefont-Rousselot D, Gardes-Albert M, Legrand A, Delattre J, Peynet J: Comparison of the effects of O2*-/HO* free radical- and copper-oxidized LDL or lipoprotein(a) on the endothelial cell release of tissue plasminogen activator and plasminogen activator inhibitory. Life Sci 69: 2371–2382, 2001
32. Bunn HF, Gabbay KH, Gallo PM: The glycosylation of haemoglobin: Relevance to diabetes mellitus. Science 200: 21–27, 1978
33. Lyons TJ, Baynes JW, Patrick JS, Colwell JA, Lopes-Virella MF: Glycosylation of low density lipoprotein in patients with type I (insulin-dependent) diabetes: Correlation with other parameters of glycaemic control. Diabetologia 29: 685–689, 1986
34. Zhang JY, Ren S, Sun DF, Shen, GX: Influence of glycation on LDL-induced generation of fibrinolytic regulators in vascular endothelial cells. Arterioscler Thromb Vasc Biol 18: 1140–1148, 1998
35. Zhang JY, Ren S, Shen, GX: Glycation amplifies lipoprotein(a)-induced alterations in the generation of fibrinolytic regulators from human vascular endothelial cells. Atherosclerosis 150: 299–308, 2000
36. Ren S, Lee H, Hu L, Lu L, Shen GX: Impact of diabetes-associated lipoproteins on the generation of fibrinolytic regulators from vascular endothelial cells. J Clin Endocrin Metab 87: 286–291, 2002
37. Bosma PJ, van der Berg E, Kooistra T, Siemieniak DR, Slightom JL: Human plasminogen activator inhibitory. Promoter and structural gene nucleotide sequences. J Biol Chem 263: 9129–9141, 1988
38. Strandberg L, Awrence D, Ny T: The organization of the human-plasminogen-activator-inhibitor 1 gene. Implication on the evolution of the serine-protease inhibitor family. Eur J Biochem 176: 609–616, 1988
39. Deschemaeker KA, Wyns S, Nelles L, Auwerx J, Ny T, Collen D: Interaction of AP-1, AP-2, and Spl-like proteins with two distinct sites in the upstream regulatory regions of the plasminogen activator inhibitory gene mediates the phorbol 12-myristate 13-acetate response. J Biol Chem 267: 15086–15091, 1992
40. Westerhausen DR Jr, Hopkins WE, Billadello JJ: Multiple transformation growth factor-β-inducible elements regulate expression of the plasminogen activator inhibitor type-1 gene in Hep G2 cells. J Biol Chem 266: 1092–1100, 1991
41. Keeton MR, Curriden SA, van Zonneveld AJ, Loskutoff DJ: Identification of regulatory of sequences in the type-1 plasminogen a gene responsive to transformation growth factor-β. J Biol Chem 266: 23048–23052, 1991
42. Bruzdzinski CJ, Johnson MR, Goble CA, Winograd SS, Gelenrter TD: Mechanism of glucocorticoid induction of the rat plasminogen activator inhibitory gene in HTC Rat hepatoma cells: Identification of cis-acting regulatory elements. Mol Endocrinol 7: 1169–1177, 1993
43. Eriksson P, Nilsson L, Karpe F, Hamsten A: Very-low-density lipoprotein response element in the promoter region of the human plasminogen activator inhibitor- 1 gene implicated in the impaired fibrinolysis of hypertriglyceridemia. Arterioscler Thromb Vasc Biol 18: 20–26, 1998
44. Dawson S, Wiman B, Hamsten A, Green F, Humphries SE, Hanney AM: The two allele sequences of a common polymorphism in the promoter of the plasminogen activator inhibitory (gene) respond differently to interleukin-1 in HepG2 cells. J Biol Chem 268: 10739–10745, 1993
45. Erikson P, Kallin B, van't Hooft FM, Bavenholm P, Hamsten A: Allele-specific increase in basal transcription of the plasminogen-activator inhibitor- I gene is associated with myocardial infarction. Proc Natl Acad Sci USA 92: 1851–1855, 1995
46. Margaglione M, Cappucci G, Colaizzo D, Giuliani N, Vecchione G, Grandone E, Pennelli O, Minno GD: The PAI-1 gene locus 4G/5G polymorphism is associated with a family history of coronary artery disease. Arterioscler Thromb Vasc Biol 8: 152–156,1998.

47. Ye S, Green FR, Scarabin PY, Nicaud V, Bara L, Dawson SJ, Humphries SE, Evans A, Luc G, Cambou JP, Arveiler D, Henney AM, Cambien F: The 4G/5G genetic polymorphism in the promoter of the plasminogen activator inhibitor- I (PAI-1) gene is associated with differences in plasma PAI-I activity but not with risk of myocardial infarction in the ECTIM study. Thromb Haemost 74: 837–841, 1995
48. Nilsson L, Banfi C, Diczfalusy U, Tremoli E, Eriksson P: Unsaturated fatty acids increase plasminogen activator inhibitory expression in endothelial cells. Arterioscler Thromb Vasc Biol 18: 1679–1685,1998.
49. Allison BA, Nilsson L, Karpe F, Hamsten A, Eriksson P: Effects of native, triglyceride-enriched, and oxidatively modified LDL on plasminogen activator inhibitor-1 expression in human endothelial cells. Arterioscler Thromb Vasc Biol 19: 1354–1360, 1999
50. Lu L, Hu L, Ren S, Shen GX: Oxidized LDL and VLDL stimulates plasminogen activator inhibitory promoter through a mechanism independent from the VLDL-response element Arterloscler Thromb Vasc Biol 21: (abstr) 641, 2001
51. Ren S, Shatadal S, Shen GX: Protein kinase C-β mediates lipoprotein-induced generation of PAI-I from vascular endothelial cells. Am J Physiol: Endocrinol Metab 278: E656–E662, 2000
52. Lottenuiser K, Wostmann B, Weisser B, Hertfelder HJ, Schmitz U, Vetter H, Dusing R: Effects of catopril on fibrinolytic function in healthy humans. Eur J Med Res 4: 31–34, 1999
53. Wright RA, Flapan AD, Alberti KG, Ludlam CA, Fox KA: Effects of captopril therapy on endogenous fibrinolysis in men with recent, uncomplicated myocardial infarction. J Am Coll Cardiol 24: 67–73, 1994
54. Goodfield N-E, Newby DE, Ludlam CA, Flapan AD: Effects of acute angiotensin II type 1 receptor antagonism and angiotensin converting enzyme inhibition on plasma fibrinolytic parameters in patients with heart failure. Circulation 99: 2983–2985, 1999
55. Li-Saw-Hee FL, Beever DG, Lip GY: Effect of antihypertensive therapy using enalapril or losartan on haemostatic markers in essential hypertension: A pilot prospective randomized double blind parallel group trial. Int J Cardiol 78: 241–246, 2001
56. Wada H, Mori Y, Kaneko T, Wakita Y, Nakase T, Minamikawa K, Ohiwa M, Tamaki, Tanigawa M, Kageyma S: Elevated plasma levels of vascular endothelial cell markers in patients with hypercholesterolemia. Am J Hematol 44: 112–116, 1993
57. Dhannalingam S, Ludwig S, Ren S, Fenton JW II, Velthius HT, Ofosu FA, Shen G: Simvastatin reduces plasma levels of plasminogen activator inhibitor-1 in patients with type 2 diabetic patients. Can J Diabetes Care 25: (abstr) 194, 2001
58. Koh KK, Cardillo C, Bui NIN, Hathaway L, Csalo G, Waclawiw MA, Panza JA, Cannon RO III: Vascular effects of estrogen and cholesterol lowering therapies in hypercholesterolemic postmenopausal women. Circulation 99: 3 54–3 60, 1999
59. Estelles A, Dalmau J, Falco C, Berbel O, Castello R, Espnan F, Aznar J: Plasma PAI-I levels in obese children-effect of weight loss and influence of PAI-1 promoter 4G/5G gene type. Thromb Haemost 86: 647–652, 2001
60. Feener EP, Northrup JM, Aiello LP, King GL: Angiotensin 11 induces plasminogen activator inhibitor-1 and -2 expression in vascular EC and smooth muscle cells. J Clin Invest 95: 125 3 –1262, 1995
61. Lopez S, Peiretti F, Bonardo B, Juhan-Vague I, Nalbone G: Effect of atorvastatin and fluvastatin on the expression of plasminogen activator inhibitor type-1 in cultured human endothelial cells. Atherosclerosis 152: 359–366, 2000
62. Boucier T, Libby P: HMG-COA reductase inhibitors reduce plasminogen activator inhibitors expression by human vascular smooth muscle and endothelial cells. Arterioscler Thromb Vasc Biol 20: 556–562, 2000
63. Jenkins AJ, Li W, Moller K, Klein RL, Fu MX, Baynes JW, Thorpe SR, Lyons TJ: Preenrichment of modified low density lipoproteins with alpha tocopherol mitigates adverse effect on cultured retinal capillary cells. Cur Eye Res 19: 137–145, 1999

Molecular and Cellular Biochemistry **246:** 69–74, 2003.

Impact and mechanism for oxidized and glycated lipoproteins on generation of fibrinolytic regulators from vascular endothelial cells

Garry X. Shen
Diabetes Research Group, University of Manitoba, Winnipeg, Manitoba, Canada

Abstract

Thrombogenesis depends on the balance between coagulation and fibrinolysis in vasculature. Vascular endothelial cells (EC) synthesize activators and inhibitors for fibrinolysis, tissue and urokinase plasminogen activators (tPA and uPA) and plasminogen activator inhibitor-1 (PAI-1). Increased levels of PAI-1 with various levels of tPA have been frequently found in plasma of patients with coronary heart disease (CHD) or diabetes mellitus (DM). Dyslipidemia is common feature in patients with CHD or DM, which is characterized by elevated levels of total cholesterol, triglycerides, low or very low density lipoproteins (LDL or VLDL) and decreased levels of high density lipoprotein (HDL). LDL and VLDL stimulated the generation of PAI-1 from cultured EC. LDL and lipoprotein(a) [Lp(a)], another lipoprotein risk factor for CHD, reduced the generation of tPA from EC. HDL did not greatly alter the release of PAI-1 from EC. Oxidative modification by copper, ultraviolet or long exposure to EC enhanced the effect of LDL on the generation of PAI-1 and tPA from EC. Glycation amplified the effect of LDL and Lp(a) on the changes in the generation of the fibrinolytic regulators from EC. Treatment with antioxidants or HDL normalized glycated LDL-induced changes in the generation of fibrinolytic regulators from EC. Activation of protein kinase C is required for oxidized LDL or Lp(a)-induced PAI-1 production in EC. VLDL, but not LDL or its oxidized form, stimulated PAI-1 production through the activation of the VLDL-responsive element in the PAI-1 promoter. Plasma levels of fibrinolytic regulators in CHD or DM patients may be normalized by HMG-CoA reductase inhibitors and angiotensin II converting enzyme inhibitors. This review summarizes the up-to-date information on effects, mechanism and management for disorders in EC-derived fibrinolytic regulators induced by modified lipoproteins. (Mol Cell Biochem **246**: 69–74, 2003)

Key words: fibrinolytic regulators, endothelial cells, lipoproteins, oxidation, glycation

Introduction

Intravascular thrombosis anticipates the development of atherosclerosis, vascular remodeling, occlusive vascular disease and tissue ischemia [1, 2]. These processes play critical roles in the pathophysiology of coronary heart disease (CHD) and diabetic vascular complications. Intravascular thrombogenesis depends on the balance between coagulation and fibrinolysis. Vascular endothelial cells (EC) are one of the major sources of fibrinolytic regulators in the body particularly in blood circulation. Dyslipidemia and endothelial dysfunction are common biomedical features in these disorders. Oxidation may increase the thrombogenicity as well as atherogenicity of the lipoproteins [3]. Glycation enhances the oxidative stress of the lipoproteins [4]. This review focuses on the impact of oxidized and glycated lipoproteins on the generation of fibrinolytic regulators from vascular EC.

Fibrinolytic system

The biologically active product of the fibrinolytic system is plasmin. It functions in dissolving fibrin clot in blood circulation and maintains the fluency of blood flow. The precur-

Address for offprints: G.X. Shen, Diabetes Research Group, University of Medidine, 835-715 McDermot Avenue, Winnipeg, Manitoba, Canada, R3E 3P4 (E-mail: gshen@ms.umanitoba.ca)

The degree of coenzyme Q10 deficiency has been found to correlate directly with the degree of impairment in left ventricular function [15] and degree of oxidation of LDL [11, 12]. Coenzyme Q10 is also regenerable, bioenergetic, powerful antioxidant and ATP sparing agent capable of protecting mitochondria, SMC and platelets and lipids from oxidative damage during plaque disruption as well as during ischemia and reperfusion injury [15–18]. We have reported that coenzyme Q10 administration early after AMI and thrombolysis was safe, and reduced cardiac events within 4 weeks of follow-up [14]. In this study, we have examined the possible role of coenzyme Q10 on risk factors of atherosclerosis in patients with AMI during a follow-up of 1 year.

Materials and methods

The subjects (n = 144) for the present study were from our previous study conducted in 1996–1997 [14]. It was a double-blind, randomized, parallel group, placebo-controlled trial. The subjects were stratified for location of infarction (anterior or posterior) and complications (hypotension, arrhythmia and heart failure) and randomized by selecting cards marked group A and B kept in sealed envelopes. Subjects admitted to the cardiac care unit with classical manifestations qualified for the diagnosis of AMI such as chest pain (> 30 min duration) electrocardiographic changes and serum concentration of creatine kinase and MB isoform (double the rise of normal) consistent with the diagnosis (n = 113). The electrocardiographic changes were symmetric ST segment elevation of 1 mm from baseline in limb leads or of 2 mm in chest leads or T wave inversions with or without Q waves. Possible myocardial infarction was diagnosed by the presence of a convincing history of cardiac chest pain accompanied by an increase of less than twice the upper limit of normal values of cardiac enzymes and electrocardiographic abnormalities that were not diagnostic but suggestive of AMI (n = 18). Unstable angina was diagnosed by cardiac chest pain lasting for > 30 min without a significant increase in enzymes, absence of diagnostic electrocardiographic changes or a combination (n = 13). The study was approved by the ethics committee of our centre and all the subjects gave informed consent in writing. Exclusion criteria were non-cardiac chest pain (n = 8), angina pectoris with cardiac chest pain < 30 min (n = 5), death before randomization (n = 7), diarrhea, vomiting (n = 3), cardiogenic shock (n = 4), blood urea nitrogen > 20 mg/dl (n = 2), malignancy (n = 1), as described elsewhere [14].

All the patients presented within 72 h of the onset of symptoms of AMI. All the patients were advised aspirin and the decision for heparin, thrombolysis, beta blockers, ACE inhibitors was left entirely to the judgement of the admitting physician. Subsequently after randomization, subjects were administered either two capsules of coenzyme Q10 twice daily (120 mg of coenzyme Q10/day) to group A or two capsules twice daily of B vitamins to control group B (each capsule containing thiamine mononitrate 1P 3 mg, riboflavine 1P 3 mg, pyridoxine hydrochloride 1 mg and nicotinamide 1P 25 mg). The active and control capsules were not truly identical, therefore both groups of patients were instructed not to show and compare their capsules with those of other patients and not to show them to their doctor and both groups met separately. Both groups were told that they are given vitamin capsules for the treatment of AMI to maintain the blind nature of the trial. The test capsules were provided to patients by the pharmacist in sealed containers that were identical in all respects. Compliance was monitored by counting of number of capsules returned by the patients on follow-up visits during the period of 12 months.

End points were defined as cardiac events such as cardiac death, reinfarction, unstable angina, stroke, coronary angioplasty and bypass surgery. However, these end points were noted in both groups without discontinuation of treatment. Information on 1-year survival was also obtained from the hospital, the patient's home, and health authority records of the corporation. Total cardiac events included non-fatal myocardial infarction as diagnosed earlier and cardiac deaths where a history or record indicating AMI/angina was available.

Laboratory data

A blood sample after at least 10 h of fasting was drawn at entry and then at 4, 24 and 48 weeks of follow-up in the morning and analysed for blood counts, hemoglobin, urea, glucose, vitamin E and C, thiobarbituric acid reactive substances (TBARS), malondialdehyde and diene conjugates [19–23]. Total cholesterol and high density lipoprotein cholesterol (HDL) were estimated enzymatically. HDL was assayed from the supernatant after selective precipitation of VLDL and LDL with magnesium chloride/phosphotungstic acid. Triglycerides were also estimated enzymatically (GPO-PAP reagents). LDL was evaluated by the Friedwald's formula (LDL-C = total – HDL-C – TG/6, valid for TG < 400 mg/dl) [24–27]. Technician was blind to groups.

Statistical analysis

All discrete variables are expressed as numbers and percentages and continuous variables as mean ± standard deviation (S.D.). Only two-tailed *t*-test and a p-value < 0.05 were considered significant. Analysis of variance followed by Tukey's *post hoc* test were performed to find out differences between the two groups and the level of significance.

Results

Of the 174 patients screened, 144 were randomized to intervention group A (n = 73) or placebo group B (n = 71). Mean age (mean + S.D.) (48.0 + 86 vs. 47.6 + 8.2 years), body weight (65.0 + 6 vs. 64.7 + 6 kg), body mass index (23.8 + 1.6 vs. 23.6 + 1.5 kg/m^2), showed no significant differences. Elapsed time from symptom onset to intervention (42 + 4.6 vs. 46 + 4.8 h) and previous drug therapy showed no significant differences. Heart rate (115 + 18.6 vs. 118 + 21.6 per min) and systolic (138 + 20 vs. 140 + 22 mmHg) and diastolic (88.6 + 10.6 vs. 90.6 + 12.5 mmHg) blood pressures were comparable between the two groups at entry to the study. Peak creatine phosphokinase (696 + 108 vs. 668 + 95 IU) showed no significant difference. Atherosclerotic risk factors such as male sex, overweight, known hypertension, diabetes, ex-smokers, previous myocardial infarction and angina pectoris and left ventricular enlargement were comparable (Table 1). Anterior or universal infarction, posterior infarction as well as AMI, possible MI and unstable angina pectoris showed no significant differences between the two groups. Current smokers were greater in the intervention group compared to control group. Drug therapy administered during the follow-up of 1 year showed no significant differences (Table 1). However, the nitrate consumption was significantly lower in the intervention group compared to control group (30.5 + 5.8 vs. 62.6 + 9.5 mg/day, $p < 0.01$) and the incidence of lovastatin intakes were similar in the two groups.

Plasma levels of vitamin E and high density lipoprotein cholesterol showed significant increase in the CoQ group compared to control group (Table 2). However TBARS, malondialdehyde, diene conjugates and indicators of oxidative image showed marked reduction in the CoQ group compared to B vitamin group. Total and LDL cholesterol were lower in both the groups and T-C/HDL-C ratio was lower in group A, after 1 year of follow-up. Total cholesterol/TBARS ratio, an indicator of unoxidised cholesterol, showed significant increase in the CoQ group than B vitamin group.

Total cardiac events, including cardiac deaths and non-fatal infarction (24.6 vs. 45.0, $p < 0.02$) were significantly lower in the CoQ group compared to control group. Non-fatal myocardial infarction (13.7 vs. 25.3, $p < 0.05$) also were significantly fewer in the CoQ group than B vitamin group. Total

Table 1. Atherosclerotic characteristics at entry to the study and drug therapy given during 1 year

Data	Coenzyme Q10 Group A (n = 73) n (%)	B vitamin Group B (n = 71) n (%)
Male sex	58 (79.4)	57 (80.3)
Overweight (BMI > 25 kg/m^2)	30 (41.0)	29 (40.8)
Known hypertension (> 140/90 mmHg)	30 (41.0)	29 (40.8)
Known diabetes (by record)	22 (30.1)	21 (29.6)
Current smokers (> 1 cigarette/day)	28 (38.3)*	18 (25.3)
Ex-smoker	8 (10.9)	12 (16.6)
Previous myocardial infarction	5 (6.8)	8 (11.2)
Previous angina pectoris	9 (12.3)	10 (14.1)
Left ventricular enlargement	14 (19.2)	7 (9.8)
Anterior or universal infarction	47 (64.4)	47 (66.2)
Posterior/inferior wall infarction	20 (27.4)	17 (23.9)
Final diagnosis:		
Acute myocardial infarction	57 (78.0)	56 (78.8)
Possible myocardial infarction	10 (13.7)	8 (11.2)
Unstable angina	6 (8.2)	7 (9.8)
Drug therapy administered:		
Streptokinase (075–15 mill IU)	6 (8.2)	6 (8.4)
Nitrates (10–60 mg/day)	70 (95.8)	68 (95.7)
Aspirin (75–150 mg/day)	71 (97.2)	70 (98.6)
Diltiazem (40–120 mg/day)	10 (13.7)	14 (19.7)
Metaprolol (50–200 mg/day)	10 (13.7)	14 (19.7)
ACE-inhibitors (2.5–10 mg/day)	10 (13.7)	12 (16.6)
Lovastatin (10–20 mg/day)	36 (49.3)	31 (43.6)

Values are number (%); *$p < 0.05$. P-value was obtained by comparison of group A and B by Z score test for proportions.

Table 2. Atherosclerotic risk factors at entry and changes after 1 year

	Coenzyme Q10		B vitamins	
	At entry	After 1 year	At entry	After 1 year
Vitamin E (umol/L)	16.3 ± 2.7	32.4 ± 4.3*	17.1 ± 2.8	22.1 ± 3.6
Vitamin C (umol/L)	4.9 ± 1.2	35.1 ± 5.3*	5.1 ± 1.3	20.2 ± 3.8
TBARS (pmol/L)	1.86 ± 0.42	1.0 ± 0.31*	1.78 ± 0.38	2.5 ± 0.34
Malondialdehyde (pmol/L)	2.75 ± 0.36	1.9 ± 0.31*	2.68 ± 0.35	3.1 ± 0.32
Diene conjugates (OD units)	28.8 ± 4.3	28.1 ± 3.8*	29.6 ± 4.4	31.4 ± 4.2
Total cholesterol (mmol/L)	5.31 ± 1.0	4.81 ± 0.81*	5.37 ± 1.1	5.01 ± 0.82*
Low density lipoprotein cholesterol (mmol/L)	3.36 ± 0.71	3.01 ± 0.52*	3.38 ± 0.72	3.12 ± 0.61*
High density lipoprotein cholesterol (HDL-C)mmol/L	1.10 ± 0.32	1.26 ± 0.43*	1.11 ± 0.33	1.12 ± 0.32
T-C/HDL-C ratio	4.82 ± 0.31	3.81 ± 0.20*	4.88 ± 0.81	4.47 ± 0.55
Triglycerides (mmol/L)	1.88 ± 0.41	1.80 ± 0.36	1.89 ± 0.42	1.83 ± 0.41
Total cholesterol/TBARS ratio	2.85 ± 2.4	4.81 ± 2.6*	3.01 ± 2.8	2.0 ± 2.4

*$p < 0.05$; p-value was obtained by comparison of values in the two groups after 1 year by analysis of variance. T-C – total cholesterol.

Table 3. Cardiac events in the intervention and control groups

Cardiac events	Coenzyme Q10 (n = 73)	B vitamins (n = 71)	Relative risk (95%) confidence interval
Sudden cardiac death (< 1 h)	3 (4.1)	5 (7.0)	0.58
Fatal myocardial infarctions	5 (6.8)	9 (12.7)	0.53
Non-fatal myocardial infarction	10 (13.7)*	18 (25.3)	0.54
Total cardiac deaths	8 (10.9)	14 (19.7)	0.55
Stroke	–	2 (2.8)	0.46
Angioplasty or CABG	1 (1.3)	3 (4.2)	0.31
Total cardiac events	18 (24.6)**	32 (45.0)	0.54
Total cardiovascular events	18 (24.6)**	34 (47.8)	0.51
Total cardiac end points	19 (25.9)**	37 (52.0)	0.49

CABG – coronary artery bypass grafting. * $p < 0.05$; ** $p < 0.02$. P-value was obtained by comparison of intervention and control group by Z score test of proportions.

Table 4. Circadian pattern of cardiac events during the follow-up of 1 year

Hours	Coenzyme Q (n = 73)	B vitamins (n = 71)
1 – 6 h	4 (5.5)	4 (5.6)
6–12 h	6 (8.2)*	16 (25.5)**
12–18 h	3 (4.1)	5 (7.0)
18–24 h	5 (6.8)	7 (9.8)
Total	18 (24.6)	32 (54.0)

*$p < 0.02$; p-value was obtained by comparison of CoQ group with B vitamins group by Z score test for proportions. **$p < 0.05$; p-value was obtained by intragroup comparison.

cardiovascular events (24.6 vs. 47.8%, $p < 0.02$) which also included stroke, and total cardiac end points (25.9 vs. 52.0, $p < 0.02$) which included, coronary angioplasty and coronary artery bypass grafting (CABG) were significantly lower in the CoQ group compared to control group (Table 3). There was a significant increase in the circadian rhythm of cardiac events in the control group in the second quarter of the day without such circadian occurrence in CoQ group (Table 4). Adverse effects such as nausea (30.1 vs. 9.8%), vomiting (13.7 vs. 11.2%) were more common in the CoQ group whereas fatigue (6.8 vs. 40.8%, $p < 0.01$) was more common in the control group.

Discussion

This study showed for the first time that treatment with coenzyme Q10 was associated with significant decline in total cardiac events (24.6 vs. 45.0, $P < 0.02$) including non-fatal myocardial infarction (13.7 vs. 25.3%, $p < 0.05$) and cardiac deaths, during a follow-up of 1 year (Table 3). Total cardiac events at 28 days of follow-up were also significantly fewer in the CoQ group than B vitamin group (15.0 vs. 30.9%, $p < 0.02$). These findings indicate that in patients at relatively high risk for recurrent coronary events, treatment with coenzyme Q10 reduces the risk possibly due to its rapid protective effects on thrombosis, endothelial function, oxidative damage as well as vascular and myocardial remodeling [28–35). Treatment with oral CoQ was beneficial despite half of the patients in both groups were receiving lovastatin (Table 1) indicating that this treatment may provide additional beneficial effects which may be unrelated to reduction in serum cholesterol [6–8].

No other study has examined the role of CoQ on cardiac events and risk of atherosclerosis in patients with AMI. In one study by Kuklinsky *et al.* [13] in 61 patients of AMI, treatment with CoQ plus selenium was associated with non-significant benefit on QT intervals and cardiac deaths (0 vs. 6 deaths) during a follow-up of 1 year. In our study, CoQ was administered within a mean of 42 h after the onset of symptoms of AMI, which indicates that apart from its beneficial effect on myocardium, [30, 31, 34, 35], it may have provided beneficial effects on platelet size and diminished vitronectin-receptor expression which are important precursors of thrombosis and instability of atheroma [28, 29, 32, 33].

This study also showed a significant increase in plasma level of vitamin E and C and a significant reduction in TBARS, malondialdehyde and diene conjugates in the CoQ group compared to control group (Table 2). It is possible that antioxidant and vitamin E sparing, activities of CoQ10 [4, 40] may be responsible for decrease in all the parameters of oxidative damage and increase in plasma level of vitamin E which are known to inhibit thrombosis and atherosclerosis and improve the quality, composition and stability of atheroma [9–10]. These effects appear to be unrelated to cholesterol lowering.

There was a significant reduction in total and LDL cholesterol in both the groups with a significant increase in beneficial [42], HDL cholesterol and decrease in T-C/HDL-C ratio in the CoQ group than control group. Total cholesterol/TBARS ratio showed significant increase in the CoQ10 group compared to control group. Reduction in cholesterol and oxidised cholesterol are known to decrease lesion cholesterol and make the unoxidised LDL less attractive to macrophages decreasing their activity which may stabilize the atheroma [41], Stocker *et al.* [11] have demonstrated that ubiquinol-10 is the only natural co-antioxidant associated within human LDL, indicating that it may play a protective role against *in vivo* oxidative modification of this lipoprotein and hence its adverse effects in the development of atherosclerosis. Dietary supplementation with coenzyme Q10 increased both the number of ubiquinol-10 molecules per LDL to more than one and the resistance of lipoprotein towards oxidative modification induced by one electron oxidants. These investigators also studied, using apolipoprotein E gene knockout mice as a model of atherosclerosis, the time dependent changes in plasma and aortic lipid parameters including ubiquinol-10 levels upon feeding a high fat diet [41] while lipid parameters showed a significant increase, circulating aortic concentrations of CoQ remain largely unaltered. Supplementation of a high fat diet with CoQ resulted in a concentration dependent increase in the plasma levels of ubiquinol-10 in apo-E gene knockout mice in association with increased protection of mouse plasma lipoprotein lipids against *ex vivo* induced oxidation. One important weakness in our study was that we could not measure plasma levels of CoQ, additional data which is available in the above study.

A lower total cholesterol/TBARS ratio in the CoQ group at entry to the study indicated a more adverse status in this group than control group which responded to treatment with CoQ (Table 2). Increased concentration of TBARS, indicating greater free radical stress and lipid peroxidation and CoQ deficiency have been observed in patients with hypertension, hyperlipidemia, coronary artery disease and diabetes by several workers [22, 32, 43, 45]. Treatment with CoQ has been demonstrated to be useful in angina pectoris, CAD, AMI, heart failure, hypertension and diabetes which are risk factors of thrombosis and instability of atheroma [9, 10, 13, 15, 45].

Increased availability of vitamin E in the plasma in our study indicate that recycling of vitamin E by ubiquinol resulting in a greater availability is an important mechanism for its antioxidant activity [40]. Apart from vitamin E, the antioxidant activity of coenzyme Q10 appear to be mainly due to its fully reduced form QH2 [45]. This is not only because it may react with different types of oxygen radicals but also because QH2 is present in relatively high amounts in the mitochondrial Q-pool and can be readily recycled by electron transport. The capacity of QH2 to recycle vitamin E radicals thereby increases the chain breaking potency of antioxidant. In one *in vitro* experiment [46] vitamin E supplementation alone resulted in an LDL which was more prone to oxidation as compared to the combination of CoQ and vitamin E which increased the resistance to oxidation. In 1997, another study documented a doubling of CoQ10 content in LDL particles after supplementing CoQ10 at 100 mg/day [47] which decreases the peroxidation of LDL cholesterol [48].

HMG-Co A reductase has been reported to be a major regulator of biosynthesis of CoQ10, therefore statins which lower cholesterol by inhibiting this coenzyme, also lower the synthesis of CoQ10 [49, 50]. In our study, one half of the patients in each group were receiving lovastatin indicating that both groups may have developed CoQ deficiency [43, 44] which may have been repaired by CoQ therapy in the CoQ group resulting into better outcome. In the control group, possible CoQ deficiency might have worsen the outcome resulting into higher oxidative damage (Table 2) and oxidised LDL and greater thrombosis and vulnerability of plaque leading to higher total cardiac events (Table 3). In 1992, in a double-blind, controlled trial in 40 hypercholesterolemic patients, Ghirlanda reported 40% drop in blood CoQ level after treatment with either pravastatin or with simvastatin [50]. In a randomized trial with either 20 mg simvastatin or 20 mg simvastatin plus 100 mg CoQ10 over 90 days, there was a significant decrease in CoQ level in plasma (0.8 vs. 1.1 ug/ml) and platelet (90 vs. 104 ng/mg) respectively [51] without such changes in the later group. In 1997. Mortensen observed similar reductions in serum CoQ10 levels in a placebo controlled double-blind trial [52]. There was a significant decrease in serum CoQ level from 1.3 to 1.0 mmol/L with pravastatin and from 1.2 to 0.8 mmol/L with lovastatin. It is possible that enhanced oxidisability of LDL cholesterol due to a decrease in the number of CoQ10 molecules per LDL cholesterol particle and decreased arterial smooth muscle and myocardial CoQ10 level in patients receiving lovastatin may have decreased the benefit of cholesterol reduction in the control group due to development of a vulnerable plaque and left ventricular dysfunction respectively. CoQ deficiency in patients receiving statins has been associated with greater incidence of fatigue, however, in our study, fatigue was more common in the control group indicating that CoQ supplementation may have modulated this adverse effect of statins.

A vulnerable plaque is characterised by a lipid core occupying at least 50% of the overall plaque volume, a high density of macrophages, a low density of smooth muscle cells in the cap, a high tissue factor content and a thin plaque cap in which the collagen structure is disorganised [1–4]. These risk markers of a vulnerable plaque may be the result of macrophage activity and lack of smooth muscle cells in the plaque which are normally rich in CoQ10 [43]. Low overall concentration of CoQ10 in a plaque enhances free radical generation and peroxidation in the plaque which may directly enhance inflammation and macrophage activity and further enhance its vulnerability to rapture [4, 46, 50]. Thus, CoQ10 deficiency [43, 49] enhances, possibly vulnerability of plaque by oxidation of LDL, enhancing immune inflammatory response, activation of macrophages, decreasing the density of smooth muscle cell (due to poor CoQ10 status) which appear to be important in the stabilization of plaque [4]. The cap of the plaque is a dynamic structure within which the connective tissue matrix, upon which its tensile strength depends, is constantly being replaced and maintained by smooth muscle cells [2–4]. However smooth muscle cells possibly with poor CoQ10 status and deficient in ATP and energy generation are not able to fight the inflammatory processes resulting into poor collagen synthesis and death by apoptosis [53]. Macrophages also produce a wide range of metalloproteases and metalloproteinases capable of degrading all the components of connective tissue matrix including collagen which becomes worst in presence of CoQ deficiency [4]. The metalloproteinases are secreted into the tissues in an inactive form that may be activated by plasmin and proinflammatory cytokines [4]. Plaque disruption may be the predominant cause of major coronary thrombi in males [54, 57] with higher oxidised LDL and low high density lipoprotein cholesterol such as our control group (Tables 2 and 3). In contrast, in female patients, endothelial erosion may be responsible for around 50% of major thrombi which are more accessible to fibrinolysis than disruption induced thrombi [54–56].

The risk of any subject, after AMI, developing a future acute event will depend on the number and severity of vulnerable plaques that are present, rather than on the total number of plaques [4, 56]. Patients, however might vary in the number of vulnerable plaques present in the coronary arteries depending upon CoQ status in the blood and arterial smooth muscle cells. It is possible that this variation may explain why control group patients had more total cardiac events compared to CoQ group. This also explains why Indians have greater risk of CAD and cardiac deaths than others [57]. The beneficial effects of CoQ in our study appears to be due to its antioxidant and bioenergetic activity on coronary endothelial and smooth muscle cells and myocardial cells during AMI [58–66]. However, some studies showed no effect of CoQ in heart function [67–69]. In the first study, enough myocardial dysfunction was absent and in the other two studies enough plasma level of CoQ10 was not achieved [67–69].

ACE inhibitors and beta-blockers were not commonly used, as the study was conducted in 1996–1997, which have been shown to be beneficial following AMI. The incremental benefit of ubiquinone over and above these agents would have produced most interesting data in our study.

In brief, the results of our study indicate that treatment with CoQ10 in patients with AMI administered within 72 h of infarction may be associated with significant decline in total cardiac events in presence of optimal, lipid lowering therapy during a follow-up of 1 year. This treatment may also decrease the risk of atherosclerosis causing significant decrease in circadian rhythms of cardiac events, increase in plasma level of vitamin E and C and significant reduction in all the parameters of oxidative damage as well as an increase in high density lipoprotein cholesterol. Larger and long-term follow-up studies would be necessary to confirm our findings.

Acknowledgements

Acknowledgements are due to Tishcon Corporation, Westbury, New York and Centre of Nutrition and Heart, Moradabad, India.

References

1. Davies M: Stability and unstability: Two faces of coronary atherosclerosis. Circulation 94: 2013–2020, 1996
2. Davies MJ: Pathophysiology of acute coronary syndromes. Heart 83: 361–366, 2000
3. Libby P, Aikawa M: New insights in plaque stabilization by lipid lowering. Drugs (suppl 1): 9–13, 1998
4. Weissberg PL: Atherogenesis: Current understanding of the causes of atheroma. Heart 83: 247–252, 2000
5. Toussaint JF, La Muraglia GM, Southern JF *et al.*: Magnetic resonance images lipid, fibrous, calcified, hemorrhagic and thrombotic components of human atherosclerosis *in vivo*. Circulation 94: 932–938, 1996
6. MAAS Investigators: Effect of simvastatin on coronary atheroma: The Multicentre Anti-Atheroma Study (MAAS). Lancet 334: 633–638, 1994
7. Shepherd J, Cobbe S, Ford I *et al.*: Prevention of coronary heart disease with pravastatin in men with hypercholesterolemia. N Engl J Med 333: 1301–1307, 1995
8. LIPID Study Group: Prevention of cardiovascular events and death with pravastatin in patients with coronary heart disease and a broad range of initial cholesterol levels. The Long-term Intervention with Pravastatin in Ischaemic Disease (LIPID) study. N Engl J Med 339: 1349–1357, 1998
9. Singh RB, Singh NK, Rastogi SS, Wander GS, Aslam, Onouchi Z, Kummerow FA, Nangia S: Antioxidant effects of lovastatin and vitamin E on experimental atherosclerosis in rabbits. Cardiovasc Drug Ther 11: 575–580, 1997
10. Singh RB, Shinde SN, Chopra RK, Niaz MA, Thakur AS, Onouchi Z: Effect of coenzyme Q10 on experimental atherosclerosis and chemical composition and quality of atheroma in rabbits. Atherosclerosis 148: 275–282, 2000
11. Thomas SR, Witting PK, Stocker R: A role of reduced coenzyme Q in atherosclerosis. BioFactors 9: 207–224, 1999
12. Tomasetti M, Alleva R, Solenghi MD, Littarru GP: Distribution of antioxidants among blood components and lipoproteins: Significance of lipids/CoQ10 ratio as a possible marker of increased risk of atherosclerosis. BioFactors 9: 225–230, 1999
13. Kuklinski B, Weissenbacher E, Fahnrich A: Coenzyme Q10 and antioxidants in acute myocardial infarction. Mol Aspects Med 15 (suppl): 143–147, 1994
14. Singh RB, Wander GS, Rastogi A, Shukla PK *et al.*: Randomized, double blind, placebo controlled trial of coenzyme Q10 in patients with acute myocardial infarction. Cardiovasc Drug Ther 12: 347–353, 1998
15. Langsjoen PH, Langsjoen AM: Overview of the use of coenzyme Q10 in cardiovascular disease. BioFactors 9: 273–284, 1999
16. Hill MF, Singal PK: Antioxidant and oxidative stress changes during heart failure subsequent to myocardial infarction in rats. Am J Pathol 148: 291–300, 1996
17. Mckay RG, Pfeffer MA, Pasternak RC: Left ventricular remodeling following myocardial infarction: A corollary to infarct expansion. Circulation 74: 693–702, 1986
18. Niibori K, Wroblewski KP, Yokoyama H, Juan A, Crestanello JA, Whitman GJR: Bioenergetic effect of liposomal coenzyme Q10 on myocardial ischaemia reperfusion injury. BioFactors 9: 307–313, 1999
19. Brubacher G, Vuilleumier JP, Inairtis HC, Roth Meds: Vitamin C: Clinical Biochemistry. Principles and Methods, Vol. 2. De Gruyter, Berlin, 1974, pp 989–997
20. Vuilleumier JP, Keller HE, Gysel D, Hunziker F: Clinical chemical methods for the routine assessment of the vitamin status in human population, part 1. The fat soluble vitamin A, E and beta-carotene. Int J Vit Nutr Res 58: 265–272, 1983
21. Esterbaur H, Cheeseman K: Determination of aldehyde peroxidation products MDA and 4 hydroxynonenal. Meth Enzymol 186: 407–421, 1990
22. Niato C, Kawamura M, Yamamoto Y: Lipid peroxides as the initiating factor of atherosclerosis. NY Acad Sci 676: 27–45, 1993

23. Lunec J, Hallovan SP, White AG, Dormandy TL: Free radical oxidation (peroxidation) products in serum and synovial fluids in rheumatoid arthritis. J Rheumatol 8: 233–245, 1981
24. Allain CC, Poon LS, Chan, Richmond W, Fu PC: Enzymatic determination of total serum cholesterol. Clin Chem 20: 470–475, 1974
25. Finley PR, Schifmjan RB, William RJ, Lichti DA: Cholesterol in high density lipoprotein: Use of magnesium/dextran sulphate in its enzymatic measurement. Clin Chem 24: 931–933, 1978
26. Friedwald WT, Levy RI, Fredrickson DS: Estimation of low density lipoprotein cholesterol concentration in plasma without use of the preparative ultracentrifuge. Clin Chem 18: 499–502, 1972
27. Bucolo G, Devi H: Quantitative determination of serum triglycerides by the use of enzymes. Clin Chem 79: 476–482, 1973
28. Herzog WR, Schlossberg ML, Mortensen SA, Serebruany VL: Dietary supplementation with coenzyme Q10 reduces platelet aggregability in swine. Coenzyme Q. Res Biol Med 3: 5–8, 1995
29. Serebruany V, Ordonez JV, Herzog WR, Morten R, Mortensen SA, Folkers K, Gurbel PA: Dietary coenzyme Q10 supplementation alters platelet size and inhibits human Vitronectin (CDSI/CD61) receptor expression. J Cardiovasc Pharmacol 29: 16–22, 1997
30. Kamikawa T, Kobayashi A, Yamashita T, Yahashi H, Yamasaki N: Effects of coenzyme Q10 on exercise tolerance in chronic stable angina pectoris. Am J Cardiol 56: 247–257, 1985
31. Hofman-Bang C, Rehnqvist N, Swedberg K, Astrom H: Coenzyme Q10 as an adjunctive in treatment of congestive heart failure. J Am Coll Cardiol 19 (suppl): 774–776, 1992
32. Kontush A, Reich A, Baum K, Spranger T, Finckh B, Kohlschutter A, Beisiegel U: Plasma ubiquinol is decreased in patients with hyperlipidemia. Atherosclerosis 129: 119–126, 1997
33. Yokoyama H, Lingle DM, Crestanello JA *et al.*: Coenzyme Q10 protects coronary endothelial function from ischaemia reperfusion injury via an antioxidant effect. Surgery 120: 189–196, 1996
34. Baggio G, Gandini R, Plancher AC, Passeri M, Carmosino G: Italian multicenter study on the safety and efficacy of coenzyme Q10 as adjunctive therapy in heart failure. Mol Aspects Med 15 (suppl): 287–294, 1994
35. Soja AM, Mortensen SA: Treatment of congestive heart failure with coenzyme Q10 illuminated by meta-analyses of clinical trials. Mol Aspects Med 18 (suppl): 1591–1568, 1997
36. Beere PA, Glagov S, Zarins CK: Retarding effect of lowered heart rate on coronary atherosclerosis. Science 226: 180–182, 1984
37. Kaplan JR, Manuck SB, Adams MR *et al.*: Inhibition of coronary atherosclerosis by propranolol on behaviorally predisposed monkeys fed an atherogenic diet. Circulation 86: 1364–1372, 1987
38. Kaplan JR, Manuck SB, Clarkson TB: The influence of heart rate on coronary atherosclerosis. J Cardiovasc Pharm 10 (suppl 2): S100–102, 1987
39. Goldberger AL: Non-linear dynamics for clinicians: Chaos theory, fractals and complexity at the bedside. Lancet 347: 1312–1314, 1996
40. Quinn PJ, Fabisiak JP, Kagan VE: Expansion of antioxidant function of vitamin E by coenzyme Q. BioFactors 4: 149–154, 1999
41. Stocker R, Bowry VW, Frei B: Ubiquinol-10 protects human low density lipoproteins more efficiently against lipid peroxidation than does alpha-tocopherol. Proc Natl Acad Sci USA 88: 1646–1650, 1991
42. Jappensen J, Hein HO, Suadicani P, Gyntelberg F: Relation of high triglyceride-low HDL cholesterol and LDL cholesterol to the incidence of ischaemic heart disease: An eight year follow up in the Copenhagen male study. Art Thromb Vasc Biol 17: 1114–1120, 1997
43. Littarru GP, Ho L, Folkers K: Deficiency of coenzyme Q10 in human heart disease II. Int J Vit Nutr Res 42: 291–295, 1972
44. Littarru GP, Ho L, Folkers K: Deficiency of coenzyme Q10 in human heart disease. Int J Vit Nutr Res 42: 291–295, 1972
45. Langsjoen PH, Langsjoen AM: Coenzyme Q10 in cardiovascular disease with emphasis on heart failure and myocardial ischaemia. Asia Pacif Heart J 7: 160–168, 1998
46. Thomas SR, Neuzil J, Stocker R: Co-supplementation with coenzyme Q prevents the prooxidant effect of alphatocopherol and increases the resistance of LDL to transition metal-dependent oxidation initiation. Arterio Thromb Vas Biol 16: 687–696, 1996
47. Aejmelaeus R, Metsa-Ketela T, Laippala P, Alho H, Solakvi T: Ubiquinol-10 and total peroxyl radical trapping capacity of LDL lipoproteins during aging: The effects of Q10 supplementation. Mol Aspects Med 18 (suppl): 113–120, 1997
48. Alleva R, Tomasetti M, Battino M, Curatola G, Littarru GP, Folkers K: The role of coenzyme Q10 and vitamin E on peroxidation of human low density subfractions. Proc Natl Acad Sci USA 92: 9388–9393, 1995
49. Rudney H, Nambudiri AMD, Ranganathan S: The regulation of the synthesis of coenzyme Q in fibroblasts and in heart muscle. In: K. Folkers, Y. Yamamura (eds). Biomedical and Clinical Aspects of Coenzyme Q. Elsevier North Holland Press, Amsterdam, 1981, pp 279–290
50. Ghirlanda G, Oradei A, Manto A *et al.*: Evidence of plasma coenzyme Q10 lowering effect by HMG-CoA reductase inhibitors: A double blind placebo controlled study. J Clin Pharmacol 33: 226–229, 1993
51. Bargossi AM, Battino M, Gaddi A *et al.*: Exogenous coenzyme Q10 preserves plasma ubiquinone levels in patients treated with 3-hydroxy-3-methymglutaryl coenzyme A reductase inhibitors. Int J Clin Lab Res 24: 171–176, 1994
52. Mortensen SA, Leth A, Agner E, Rohde M: Dose related decrease of serum coenzyme Q10 during treatment with HMG-CoA reductase inhibitors. Mol Aspects Med 18 (suppl): 137–144, 1997
53. Lopez-Lluch G, Barroso MP, Martin SF *et al.*: Role of plasma membrane coenzyme Q on the regulation of apoptosis. BioFactors 9: 171–178, 1999
54. Burke A, Farb A, Malcom G *et al.*: Coronary risk factors and plaque morphology in men with coronary disease who died suddenly. N Engl J Med 336: 1276–1282, 1997
55. Ross R: Atherosclerosis, an inflammatory disease. N Engl J Med 340: 115–126, 1999
56. Arbustini E, Dal Bello P, Morbini P *et al.*: Plaque erosion is a major substrate for coronary thrombosis in acute myocardial infarction. Heart 82: 269–272, 1999
57. Anand SS, Yusuf S, Vuksan V *et al.*: Differences in risk factors, atherosclerosis and cardiovascular disease between ethnic groups in Canada: The Study of Health Assessment and Risk in Ethnic Groups (SHARE). Lancet 356: 279–284, 2000
58. Singh RB, Niaz MA, Rastogi SS, Sharma JP, Kumar R, Bishnoi I, Beegom R: Plasma levels of antioxidant vitamins and oxidative stress in patients with suspected acute myocardial infarction. Acta Cardiol 49: 411–452, 1994
59. Grech ED, Jackson M, Ramsdale DR: Reperfusion injury after acute myocardial infarction. Br Med J 310: 477–478, 1995
60. Khaper N. Singal PK: Effects of after load reducing drugs on the pathogenesis of antioxidant changes and congestive heart failure in rats. J Am Coll Cardiol 219: 856–861, 1997
61. Morisco C, Trimarco B, Condorelh M: In: K. Folkers, S.A. Mortension, G.P. Littarru, T. Yamagam, G. Lenaz (eds): Effect of coenzyme therapy in patients with congestive heart failure: A long term multicentre randomized study. Clin Invest 71 (suppl): 134–136, 1993
62. Soja AM, Mortensen SA: Treatment of congestive heart failure with coenzyme Q10 illuminated by metaanalysis of clinical trials. Mol Asp Med 18 (suppl): 159–168, 1997
63. Swedberg K, Hoffman-Bang C, Rehnqvist N, Astrom H: Coenzyme Q10 as adjunctive in treatment of congestive heart failure. J Cardiac Fail 1: 101–107, 1995

64. Shi H, Noguchi N, Niki E: Dynamics of antioxidant action of ubiquinol: A reappraisal. Biofactors 9: 141–148, 1999
65. Mortensen SA, Leth A, Agner E, Rohde M: Coenzyme Q10: Clinical benefits with biochemical correlates suggesting a scientific breakthrough in the management of chronic heart failure. Int J Tissue React 12: 155–162, 1990
66. Beyer R: An analysis of coenzyme Q in free radical generation and as an antioxidant. Biochem Cell Biol 70: 390–403, 1992
67. Permanetter B, Roessy W, Klein G, Weingartner F, Seidl KF, Bloemer H: Ubiquinone (coenzyme Q10) in the long term treatment of cardiomyopathy. Eur Heart J 13: 1528–1533, 1992
68. Watson PS, Scalia GM, Galbraith A, Burstow DJ, Bett N, Aroney CN: Lack of effect of coenzyme Q on left ventricular function in patients with congestive heart failure. J Am Coll Cardiol 33: 1549–1552, 1999
69. Khatta M, Alexander BS, Krichten CM, Freudenberger R, Robinson SW, Gottlieb SS: The effect of coenzyme Q10 in patients with congestive heart failure. Ann Intern Med 132: 636–640, 2000

Molecular and Cellular Biochemistry **246:** 83–90, 2003.

Methodologic challenges in designing clinical studies to measure differences in the bioequivalence of n-3 fatty acids

Diane H. Morris
Mainstream Nutrition, Toronto, Ontario, Canada

Abstract

Although epidemiologic studies suggest a role for alpha-linolenic acid (ALA) in the prevention of coronary heart disease and certain types of cancer, the findings of clinical studies suggest that ALA is inferior biologically to the n-3 long-chain fatty acids because its bioconversion to eicosapentaenoic acid (EPA) and docosahexaenoic acid (DHA) is limited in humans and because the magnitude of its biologic effects is smaller than that of EPA and DHA. This paper reviews several methodologic issues that may confound the findings of clinical studies and complicate our interpretations of them: the ALA and EPA + DHA dietary enrichment levels; the choice of tissue; the choice of lipid species; and the method of reporting fatty acid composition. Although the ALA enrichment levels used in most clinical studies can be achieved by consuming ground flaxseed, flaxseed oil, canola oil and other ALA-rich plants as part of a typical dietary pattern, the EPA + DHA enrichment levels are not practical and can only be obtained from fish oil supplements. The lack of consistency in the choice of lipids species and the reporting of data makes it difficult to compare outcomes across studies. The choice of tissue (blood) for analysis is a limitation that probably cannot be overcome. The use of practical ALA and EPA+ DHA dietary enrichment levels and some standardization of clinical study design would allow for greater comparisons of outcomes across studies and ensure a more realistic analysis of how individual n-3 fatty acids differ in their biologic effects in humans. (Mol Cell Biochem **246**: 83–90, 2003)

Key words: alpha-linolenic acid, bioequivalence, clinical studies, docosahexaenoic acid, eicosapentaenoic acid, n-3 fatty acids, methodology

Introduction

Alpha-linolenic acid (ALA; 18:3n-3) is the parent compound of the n-3 fatty acid family and the precursor of the long-chain n-3 fatty acids, eicosapentaenoic acid (EPA; 20:5n-3) and docosahexaenoic acid (DHA; 22:6n-3). ALA is considered inferior biologically to the long-chain n-3 fatty acids because its bioconversion to EPA and DHA is limited in humans and because the magnitude of its biologic effects is smaller than that of EPA and DHA [1–4].

ALA is converted to EPA and DHA by a series of desaturations and elongations. The extent to which bioconversion occurs is influenced by metabolic factors such as the balance among oxidation, desaturation-elongation and carbon recycling of ALA [5–7], and by dietary factors such as the ratio of dietary linoleic acid (LA; 18:2n-6) to saturated fatty acids [8], the dietary ratio of polyunsaturated to saturated fat [9], the dietary cholesterol level [10, 11], and the dietary level of *trans* fatty acids, LA, EPA and DHA [6, 12–15]. ALA can interfere with its own metabolism when the diet contains more than 12 g ALA/day [16]. Between 0.2% [17] and 6.0% [13] of ALA is converted to EPA in healthy adults. Emken *et al.* [13] reported that about 15% of ALA is converted to the long-chain n-3 fatty acids in seven young adult men. At this conversion rate, a diet containing 2.22 g ALA/day, the current proposed adequate intake level of ALA [18], would supply about 330 mg of long-chain n-3 fatty acids/day. This ALA intake level would meet the daily n-3 fatty acid requirements of some, but not all, adults [13].

Address for offprints: D.H. Morris, Mainstream Nutrition, 904-130 Carlton Street, Toronto, Ontario M5A 4K3, Canada (E-mail: mainnut@primus.ca)

Despite the limited bioconversion of ALA to long-chain n-3 fatty acids in humans, epidemiologic studies suggest a role for ALA in reducing the risk of certain types of cancer [19] and coronary heart disease (CHD). For example, populations with ALA intakes of about 0.9–1.8 g/day had an inverse risk of myocardial infarction [20], reduced mortality from all cardiovascular diseases [21], reduced risk of fatal ischemic heart disease (among women) [22], and a reduced rate of overall mortality and recurrence of myocardial infarction [23, 24]. In the Multiple Risk Factor Intervention Trial (MRFIT), each S.D. increase (0.06%) in the ALA content of serum phospholipids was associated with a 28% decrease in risk of stroke [25].

Whereas the epidemiologic data suggest a role for ALA in reducing chronic disease risk, the clinical data show that the magnitude of ALA's biologic effects is less than that of EPA and DHA. However, certain methodologic issues may confound the findings of clinical studies and limit our understanding of how and to what extent individual n-3 fatty acids differ in their biologic effects. The purpose of this paper is to examine these methodologic issues, which include: (1) the dietary enrichment level of ALA and EPA + DHA; (2) the choice of tissue for analysis; (3) the choice of lipid species for fatty acid composition analysis; and (4) the method of reporting fatty acid data. It is hoped that this paper will launch some discussions of n-3 fatty acid metabolism and nutriture among researchers, policy makers and health professionals.

Methodologic challenges in designing n-3 fatty acid clinical studies

Choice of dietary n-3 fatty acid enrichment level

Our understanding of n-3 fatty acid bioequivalence is based, in part, on clinical studies designed to assess the effects of dietary fatty acids on various biologic outcomes such as the fatty acid composition of tissue lipids [3, 9, 26–31] and the concentrations of lipid mediators like eicosanoids and cytokines [30, 32, 33]; on clinical outcomes such as serum lipids, blood pressure, bleeding time, coagulation factors and platelet aggregation in adults [9, 26–31, 34–42, 45]; and on functional measures of visual, cognitive and neural development in infants [43, 44]. Studies typically compare a vegetable oil rich in ALA, usually obtained in flaxseed oil or canola oil, with a saturated fat diet [41] or with vegetable oils rich in n-6 fatty acids like corn, safflower and sunflower oils [29, 45–49], and/or with monounsaturated vegetable oils like olive oil [26, 49–51]. Effects of ground flaxseed [16, 35, 52], flaxseed oil [16, 38] and canola oil [53, 54] have also been assessed separately. Some studies compared the effects of vegetable oils like flaxseed oil, canola oil and olive oil with EPA + DHA-rich fish oil capsules [3, 9, 27, 28, 30–32, 34, 37, 39, 40] or canned mackerel [42]. One study [55] evaluated the effects of a dietary regimen that included a combination of flaxseed oil, fatty fish and foods enriched with flaxseed oil and fish oil. Many clinical studies involving fish oil have been conducted over the past several decades (see reviews [56–58]).

The level of ALA dietary enrichment used in clinical studies typically ranged from 2.6–13.7 g ALA/day [3, 9, 16, 27, 28, 30–32, 34, 35, 37, 39, 40, 47, 48, 51, 55]. A few studies used dietary ALA enrichment levels that exceeded 20 g/daily [38, 41, 42]. The EPA + DHA dietary enrichment level typically ranged from 0.87–5.96 g EPA + DHA/day [3, 9, 27, 28, 30–32, 34, 37, 39, 40, 42, 45, 59, 60]. In individual studies, the level of dietary ALA enrichment was about 2–8-fold greater than the level of EPA + DHA enrichment. The rationale for the higher ALA enrichment level is that, because its bioconversion to EPA and DHA is limited in humans, a high intake of ALA is required to produce a meaningful result within the relatively short time frame of a clinical study.

A key issue in study design is the level of dietary n-3 enrichment. Table 1 shows the amount of canola oil, flaxseed oil, ground flaxseed and fish that must be consumed to achieve the ALA and EPA + DHA enrichment levels used in some clinical studies. Values representing the low and high ends of dietary n-3 fatty acid enrichment levels used in several clinical studies are given. ALA enrichment levels can be achieved by consuming between 4.6 and 35 g flaxseed oil daily, which represents about 1 tsp to 2½ tbsp of flaxseed oil added to salad dressings, fruit smoothies and other foods. Canola oil and ground flaxseed incorporated into the diet at levels up to about 10–12 g ALA/day are also reasonable. For example, McDonald *et al.* [33] fed healthy students about 90 g (6 tbsp) of canola oil, which provided ~10 g ALA/day, for 18 days, and Cunnane *et al.* [16, 52] fed healthy adults 50 g ground flaxseed (~12 g ALA/day) mixed into muffins for 4 weeks. An ALA intake of 20 g/day can be achieved by consuming flaxseed oil or a combination of flaxseed oil and ground flaxseed, but not by consuming canola oil.

By comparison, the dietary EPA + DHA enrichment levels used in many clinical studies are not realistic. To achieve the lowest enrichment level of 0.87 g EPA + DHA/day [59], adults must eat daily about 70 g cooked salmon, 322 g of tuna packed in water or 552 g of cod, based on the n-3 fatty acid content of these fish reported in USDA's nutrient database [61]. It is virtually impossible for humans to eat enough fish to achieve dietary EPA + DHA enrichment levels greater than 2.7 g/day.

Thus, the dietary EPA + DHA enrichment levels used in clinical studies are not realistic within the context of North American eating patterns and cannot be achieved by eating foods. In most studies, they exceed both the proposed ad-

Table 1. Amount of canola oil, flaxseed oil, ground flaxseed and fish that must be eaten daily to achieve the dietary ALA and EPA + DHA enrichment levels reported in selected clinical studies

Dietary n-3 fatty acid enrichment levels in clinical studies		Amount of canola oil, flaxseed oil, ground flaxseed and fish that must be eaten daily to achieve the dietary n-3 enrichment levels											
ALA[1]	EPA + DHA[2]	Canola oil[3]		Flaxseed oil[4]		Ground flaxseed[4]		Salmon[5]		Tuna canned in water[5]		Cod[5]	
g/d	g/d	g/d	number of servings/day	g/d	number of servings/day	g/d	number of servings/day	g/d	number of 3-oz servings/day	g/d	number of 3-oz servings/day	g/d	number of 3-oz servings/day
2.6[6]	–	24	1¾ tbsp	4.6	~1 tsp	12	1½ tbsp	–	–	–	–	–	–
13.7[7]	–	124	~½ cup	24	1¾ tbsp	61	½ cup	–	–	–	–	–	–
20.0[8]	–	182	~¾ cup	35	2½ tbsp	89	~2/3 cup	–	–	–	–	–	–
–	~0.87[9]	–	–	–	–	–	–	70	~1	322	~4	552	6½
–	2.7[7]	–	–	–	–	–	–	210	2½	998	12	1713	20
–	5.96[10]	–	–	–	–	–	–	465	5½	2203	26	3780	44

[1]The sources of ALA in the clinical studies were flaxseed oil [32, 41] and canola oil [54]. [2]The sources of EPA + DHA in the clinical studies were fish oil capsules. [3]Source of ALA content data: Canola Council of Canada. Proximate analysis of canola oil conducted by POS Pilot Plant Corp., Winnipeg; Canola Council of Canada, 1994. [4]Source of ALA content data: Flax Council of Canada. Proximate analysis of flaxseed conducted by the Canadian Grain Commission; Winnipeg, Flax Council of Canada, 2001. [5]The following EPA + DHA values were obtained from USDA's nutrient database, release 14 [61]: salmon, pink, cooked, dry heat, 1.09 g EPA + DHA/85 g serving; tuna, light, canned in water, drained solids, 0.23 g EPA + DHA/85 g serving; cod, Atlantic, cooked, dry heat, 0.134 g EPA + DHA/85 g serving. [6]Bierenbaum *et al.* [54]. [7]Caughey *et al.* [32]. [8]Nestel *et al.* [41]. [9]GISSI-Preventione Investigators [59]. [10]Sanders and Roshanai [31].

equate intake of 0.87 g EPA + DHA/day for adults [18] and the current mean EPA + DHA intake of 0.1 g/day among US adults [62], shown in Table 2. Adding two servings of fish per week to the diet, as recommended by the American Heart Association [63], increases the EPA + DHA intake by 0.17 g/day, an amount that when added to the current US adult intake, gives an intake that is about one-third of the proposed adequate intake for adults. Adding one omega-3-enriched egg to the daily diet would increase the EPA + DHA intake by 0.2–0.5 g/day, bringing the total intake closer to but still lower than the proposed adequate intake.

Table 2. Proposed recommended vs. actual n-3 fatty acids intakes of US adults

Fatty acid	Proposed recommended adequate intake[1]	Current mean intake[2]
	g/day	
ALA[3]	2.22	1.3
EPA	0.65	<0.05
DHA	0.22	<0.05

Source: [1]Simopoulos *et al.* [18]. [2]USDA [62]. [3]Abbreviations: ALA – alpha-linolenic acid; DHA – docosahexaenoic acid; EPA – eicosapentaenoic acid

Table 3. Comparison of the effects of flaxseed oil and fish oil on the concentration of cytokines and eicosanoids in mononuclear cells[1,2]

Variable	week 0	week 4	week 8	Decrease from baseline	
				Flax oil period (wk 0–4)	Flax oil + fish oil period (wk 4–8)
		μg/L		%	%
TNFα	30.1 ± 6.2[a,3]	22.1 ± 5.4[b]	7.1 ± 3.2[c]	(26)	(76)
IL-1β	31.5 ± 6.2[a]	21.8 ± 8.6[b]	5.9 ± 3.1[c]	(31)	(81)
TXB_2	12.8 ± 2.6[a]	8.9 ± 3.1[b]	5.4 ± 1.6[c]	(30)	(58)
PGE_2	13.1 ± 5.2[a]	8.8 ± 4.1[b]	6.3 ± 2.4[b]	(33)	(52)

[1]Source: Caughey *et al.* [32]. Twenty-eight healthy men were randomly assigned to consume a flaxseed oil-rich diet or a sunflower oil-based diet for 8 weeks. At week 4, subjects added 9 fish oil capsules to their assigned diet. Only data from the flaxseed oil arm are shown. Flaxseed oil provided 13.7 g ALA/day; fish oil provided 2.7 g EPA + DHA/day. [2]Abbreviations: IL-1β – interleukin-1β; PGE_2 – prostaglandin E_2; TNFα – tumor necrosis factor α; TXB_2 – thromboxane B_2. [3]Values with different superscripts are significantly different, $p < 0.05$. Adapted with permission by the *American Journal of Clinical Nutrition.*

One consequence of this design issue is that clinical studies using large doses of EPA and DHA obtained from fish oil capsules may produce biologic or clinical outcomes that are only achievable through the regular use of fish oil supplements. Table 3 shows the decrease from baseline that occurred in the concentrations of two cytokines and two eicosanoids measured in mononuclear cells of 28 healthy men [32]. The subjects were randomly assigned to consume either a flaxseed oil-based diet or a sunflower oil-based diet for 8 weeks; at week 4, subjects added to their assigned diets 9 fish oil capsules daily. Only data from the flaxseed oil arm are provided in the table. The flaxseed oil diet provided 13.7 g ALA/day, while the fish oil capsules provided 2.7 g EPA + DHA/day, giving a dietary ALA to EPA + DHA ratio of 5:1. The concentrations of TNFα, IL-1β, TXB2, and PGE2 decreased significantly (26–33%) from baseline during the flaxseed oil period. The decreases from baseline in cytokine and eicosanoids concentrations were greater during the fish oil period (weeks 4–8) than during the flaxseed oil period (weeks 0–4).

One interpretation of these findings is that, even though the diet contained 5 times more ALA than EPA + DHA, the biologic effects of EPA and DHA were superior to those of ALA. However, the biologic effects of ALA in this study were achieved by consuming a reasonable level (24 g or 1¾ tbsp) of flaxseed oil daily, whereas the effects observed during the fish oil period were achieved with EPA + DHA intakes equivalent to eating more than 210 g (2.5 servings) of fatty fish daily (see Table 1). The biologic effects of these n-3 fatty acids might be similar with a more realistic comparison of ALA intake (e.g. 2.6–8 g ALA/day, which is equivalent to 2–6½ tbsp canola oil or 1 tsp–1 tbsp flaxseed oil) vs. EPA + DHA intake (e.g. 0.47 g EPA + DHA/day, which represents the current intake amount plus 2 servings of fish and four n-3-enriched eggs/week).

Choice of tissue

Clinical studies of n-3 bioequivalence provide data on the fatty acid composition of plasma and cellular lipids, but they do not provide data on the compositional changes that occurred in major tissues such as the heart and brain. There is a practical reason for this methodology: a procedure such as phlebotomy is used to obtain biologic material for analysis because few institutional research review committees would approve brain, heart, liver or eye biopsies to measure tissue EPA and DHA content in living subjects. Compared with biopsies, phlebotomy is relatively inexpensive and less invasive. This methodologic limitation probably cannot be overcome.

The significance of this methodologic limitation is seen in the findings of animal studies, which show that varying dietary levels of ALA, DHA, LA and arachidonic acid (AA, 20:4n-6) have profoundly different effects on the n-3 fatty acid content of the brain, retina, liver and heart [64–67]. For example, guinea pigs fed either canola oil, which provided ALA at a level of 7% of total fatty acids and no DHA, or a mixed fat diet that provided ALA (1% of total fatty acids) and DHA (0.7%), showed equivalent increases (1.4–1.7-fold, $p < 0.05$) in the DHA content of retinal phospholipids and brain phosphatidylethanolamine (PE). However, in guinea pig liver and heart phospholipids, the mixed fat diet increased the DHA content between 14- and 16-fold ($p < 0.05$), whereas the canola diet increased the DHA content 2-fold (significant at $p < 0.05$ for heart tissue only) [64].

These findings underscore how different animal tissues respond to different dietary levels of ALA and DHA. It may not be appropriate to extrapolate the results of animal studies to humans in terms of n-3 fatty acid metabolism [3]. There are few data on the response of human brain, retina, heart and liver to dietary n-3 fatty acids, other than several studies of the fatty acid content of the brain frontal cortex from term infants who died suddenly [68–70] and adipose tissue and segments of coronary arteries obtained from autopsied adults aged 52–90 years [71]. However, there are data obtained from asthenozoospermic and normozoospermic men suggesting that DHA, which is highly correlated with sperm motility [72], is not taken up by sperm [73]; rather, the high DHA content of sperm may depend on the availability of ALA as substrate [74].

Choice of lipid species

There is no consensus on which of the major lipid species (phospholipids, triacylglycerols, cholesteryl esters or non-esterified fatty acids) should be measured and, therefore, there is little consistency among the studies in the choice of lipid species. Several studies [3, 28, 30] reported the fatty acid composition of the same lipid species (phospholipids) in the same tissue (plasma). Few, if any, studies reported the n-3 fatty acid content of the non-esterified fatty acid fraction of plasma. This fraction can potentially supply the plasma membranes of cardiac myocytes with the free form of n-3 fatty acids needed to prevent lethal arrhythmias [75]. In dogs, pure intravenous infusions of ALA were as effective as pure EPA and DHA in preventing arrhythmias, possibly by a mechanism in which the n-3 fatty acids alter the electrophysiology of myocytes to make them resistant to arrhythmia [76, 77]. The availability of ALA from the non-esterified plasma fatty acid fraction may explain, in part, the cardiovascular protective effects of ALA reported in epidemiologic studies [75]. There are few data on how dietary n-3 fatty acids affect the fatty acid composition of this plasma fraction and whether changes in the lipid composition of this fraction have important health implications.

Moreover, certain differences between lipid species are seldom taken into account when designing clinical studies. For example, phospholipids tend to accumulate LA but virtually exclude ALA, whereas triacylglycerols accumulate both fatty acids [78, 79]. For example, Emken *et al.* [13] found that 60.8% of deuterated ALA in total plasma was incorporated into triacylglycerols, while only 14.5% was incorporated into the phosphatidylcholine (PC) fraction. The data from Layne *et al.* [9] also demonstrate these differences. In adult volunteers who consumed for 3 months a diet with either a low or high ratio of polyunsaturated fat to saturated fat, flaxseed oil consumption increased the ALA content of the VLDL fraction of plasma triacylglycerols between 5- and 6-fold ($p < 0.05$), whereas it increased the ALA content of the VLDL fraction of plasma phospholipids 2–3-fold ($p < 0.05$). At the end of the flaxseed oil supplementation period, the ALA content of triacyglycerol VLDL was 16–18-fold greater than that of phospholipid VLDL, whereas the ALA content of triacylglycerol LDL was only about 2-fold greater than that of phospholipid LDL. Even though flaxseed oil consumption resulted in a significant increase in the ALA content of nearly all lipoprotein fractions in plasma triacylglycerols and phospholipids, the magnitude of the increase in ALA content was greater in triacylglycerols than in phospholipids. The significance of these differences is not known, but limiting the fatty acid compositional analysis to only one or two lipid species may not provide a complete picture of n-3 fatty acid metabolism.

Finally, another confounding design factor is the analysis of the fatty acid composition of total plasma or platelet lipids or phospholipids rather than the fatty acid composition of individual lipid or phospholipid species (e.g. phosphatidylcholine, phosphatidylethanolamine). Important but subtle changes in fatty acid composition may be masked when lipid species are not analyzed separately [46].

Method of reporting fatty acid data

There is no consistency across studies in how the fatty acid composition of tissue lipids is reported. In some studies [3, 30–32, 39, 40], the data were reported as a percentage of total fatty acids, while in others [9, 28], the concentration of the fatty acid per mL or L of plasma was reported. The lack of agreement on how best to report fatty acid composition makes it difficult to compare findings across studies and to relate changes in fatty acid composition back to the dietary intervention.

Conclusions

The methodologic issues described in this paper present some complex challenges for the n-3 fatty acid research community. For example, most clinical studies employ dietary EPA + DHA enrichment levels that are not realistic and cannot be achieved by eating foods within the context of a typical diet. These high dietary EPA + DHA enrichment levels may have biologic effects that can only be achieved by taking fish oil supplements. It is unrealistic to expect most North Americans to take fish oil supplements regularly as a means of meeting their daily EPA + DHA needs, especially as consumers may object to the cost and fishy aftertaste of some fish oil supplements and the demands of a regimen that requires them to take several large capsules daily. Research based mainly on dietary supplements, not foods, ignores common eating patterns, and it does not, in this case, provide a real-world assessment of differences in the biologic effects of the various n-3 fatty acids.

Some standardization of clinical study design, especially in the choices of lipid species and fatty acids for analysis and in the reporting of data, would allow for greater comparisons across studies. To the extent that a more comprehensive analysis can be justified on a cost basis, the standardization of design would ensure a more complete analysis of how individual n-3 fatty acids differ in their biologic effects in humans.

One suggested course of action for addressing these issues is to organize national or international forums for discussing the appropriate n-3 dietary enrichment levels and the extent to which clinical studies can be standardized to provide a more discriminating picture of the biologic effects of individual n-3 fatty acids. Other topics that should be addressed include defining the bioequivalence of fatty acids, determining which indicators of biologic action should be used to measure bioequivalence, interpreting the literature regarding ALA conversion, and determining the relative importance of biologic endpoints as indicators of health for special populations (e.g. infants and children vs. adults at risk of chronic disease). The ultimate goal of improving the design of clinical studies is to provide data that helps researchers, health professionals and policy makers better understand how the n-3 fatty acids contribute to the growth and development of infants and children and how they help prevent or ameliorate many chronic diseases common in Western adult populations.

Acknowledgements

Presentation at the XVII World Congress was sponsored by the Canola Council of Canada and the Flax Council of Canada.

The author thanks the Canola Council and the Flax Council of Canada for their support in preparing this manuscript.

References

1. Gerster H: Can adults adequately convert α-linolenic acid (18:3n-3) to eicosapentaenoic acid (20:5n-3) and docosahexaenoic acid (22:6n-3)? Internat J Vit Nutr Res 68: 159–173, 1998
2. Adam O, Wolfram G, Zöllner N: Effect of α-linolenic acid in the human diet on linoleic acid metabolism and prostaglandin biosynthesis. J Lipid Res 27: 421–426, 1986
3. Dyerberg J, Bang HO, Aagaard O: α-Linolenic acid and eicosapentaenoic acid (letter). Lancet i: 199, 1980
4. McKeigue P: Diets for secondary prevention of coronary heart disease: Can linolenic acid substitute for oily fish (commentary)? Lancet 343: 1445, 1994
5. DeLany JP, Windhauser MM, Champagne CM, Bray GA: Differential oxidation of individual dietary fatty acids in humans. Am J Clin Nutr 72: 905–911, 2000
6. Vermunt SHF, Mensink RP, Simonis MMG, Hornstra G: Effects of dietary α-linolenic acid on the conversion and oxidation of ^{13}C-α-linolenic acid. Lipids 35: 137–142, 2000
7. Cunnane SC: The Canadian Society for Nutritional Sciences 1995 Young Scientist Award Lecture: Recent studies on the synthesis, β-oxidation, and deficiency of linoleate and α-linolenate: Are essential fatty acids more aptly named indispensable or conditionally dispensable fatty acids? Can J Physiol Pharmacol 74: 629–639, 1996
8. Garg ML, Thomson ABR, Clandinin MT: Interactions of saturated, n-6 and n-3 polyunsaturated fatty acids to modulate arachidonic acid metabolism. J Lipid Res 31: 271–277, 1990
9. Layne KS, Goh YK, Jumpsen JA, Ryan EA, Chow P, Clandinin MT: Normal subjects consuming physiological levels of 18:3(n-3) and 20:5 (n-3) from flaxseed or fish oils have characteristic differences in plasma lipid and lipoprotein fatty acid levels. J Nutr 126: 2130–2140, 1996
10. Garg ML, Thomson ABR, Clandinin MT: Effect of dietary cholesterol and/or ω3 fatty acids on lipid composition and Δ^5-desaturase activity of rat liver microsomes. J Nutr 118: 661–668, 1988
11. Leikin AI, Brenner RR: Cholesterol-induced microsomal changes modulate desaturase activities. Biochim Biophys Acta 922: 294–303, 1987
12. van Houwelingen AC, Hornstra G: *Trans* fatty acids in early human development. World Rev Nutr Diet 75: 175–178, 1994
13. Emken EA, Adlof RO, Gulley RM: Dietary linoleic acid influences desaturation and acylation of deuterium-labeled linoleic and linolenic acids in young adult males. Biochim Biophys Acta 1213: 277–288, 1994
14. Emken EA, Adlof RO, Duval SM, Nelson GJ: Effect of dietary docosahexaenoic acid on desaturation and uptake *in vivo* of isotope-labeled oleic, linoleic and linolenic acids by male subjects. Lipids 34: 785–791, 1999
15. Vermunt SHF, Mensink RP, Simonis AMG, Hornstra G: Effects of age and dietary n-3 fatty acids on the metabolism of [^{13}C]-α-linolenic acid. Lipids 34: S127, 1999
16. Cunnane SC, Ganguli S, Menard C, Liede AC, Hamadeh MJ, Chen Z-Y, Wolever TMS, Jenkins DJA: High α-linolenic acid flaxseed (*Linum usitatissimum*): Some nutritional properties in humans. Br J Nutr 69: 443–453, 1993
17. Pawlosky RJ, Hibbeln JR, Novotny JA, Salem N Jr: Physiological compartmental analysis of α-linolenic acid metabolism in adult humans. J Lipid Res 42: 1257–1265, 2001
18. Simopoulos AP, Leaf A, Salem N Jr: Workshop on the essentiality of and recommended dietary intakes for omega-6 and omega-3 fatty acids. Nutr Today 35: 166–167, 2000
19. de Lorgeril M, Salen P, Martin J-L, Monjaud I, Boucher P, Mamelle N: Mediterranean dietary pattern in a randomized trial: Prolonged survival and possible reduced cancer rate. Arch Intern Med 158: 1181–1187, 1998
20. Ascherio A, Rimm EB, Giovannucci EL, Spiegelman D, Stampfer M, Willett WC: Dietary fat and risk of coronary heart disease in men: Cohort follow up study in the United States. Br Med J 313: 84–90, 1996
21. Dolecek TA: Epidemiological evidence of relationships between dietary polyunsaturated fatty acids and mortality in the Multiple Risk Factor Intervention Trial. Proc Soc Exp Biol Med 200: 177–182, 1992
22. Hu FB, Stampfer MJ, Manson JE, Rimm EB, Wolk A, Colditz GA, Hennekens CH, Willett WC: Dietary intake of α-linolenic acid and risk of fatal ischemic heart disease among women. Am J Clin Nutr 69: 890–897, 1999
23. de Lorgeril M, Renaud S, Mamelle N, Salen P, Martin J-L, Monjaud I, Guidollet J, Touboul P, Delaye J: Mediterranean alpha-linolenic acid-rich diet in secondary prevention of coronary heart disease. Lancet 343: 1454–1459, 1994
24. de Lorgeril M, Salen P, Martin J-L, Monjaud I, Delaye J, Mamelle N: Mediterranean diet, traditional risk factors, and the rate of cardiovascular complications after myocardial infarction: Final report of the Lyon Diet Heart Study. Circulation 99: 779–785, 1999
25. Simon JA, Fong J, Bernert JT Jr, Browner WS: Serum fatty acids and the risk of stroke. Stroke 26: 778–782, 1995
26. Chan JK, McDonald BE, Gerrard JM, Bruce VM, Weaver BJ, Holub BJ: Effect of dietary α-linolenic acid and its ratio to linoleic acid on platelet and plasma fatty acids and thrombogenesis. Lipids 28: 811–817, 1993
27. Goh YK, Jumpsen JA, Ryan EA, Clandinin MT: Effect of ω3 fatty acid on plasma lipids, cholesterol and lipoprotein fatty acid content in NIDDM patients. Diabetologia 40: 45–52, 1997
28. Indu M, Ghafoorunissa: n-3 Fatty acids in Indian diets: Comparison of the effects of precursor (alpha-linolenic acid) vs product (long chain n-3 polyunsaturated fatty acids). Nutr Res 12: 569–582, 1992
29. Kwon J-S, Snook JT, Wardlaw GM, Hwang DH: Effects of diets high in saturated fatty acids, canola oil, or safflower oil on platelet function, thromboxane B_2 formation, and fatty acid composition of platelet phospholipids. Am J Clin Nutr 54: 351–358, 1991
30. Mantzioris E, James MJ, Gibson RA, Cleland LG: Dietary substitution with an α-linolenic acid-rich vegetable oil increases eicosapentaenoic acid concentrations in tissues. Am J Clin Nutr 59: 1304–1309, 1994
31. Sanders TAB, Roshanai F: The influence of different types of ω3 polyunsaturated fatty acids on blood lipids and platelet function in healthy volunteers. Clin Sci 64: 91–99, 1983
32. Caughey GE, Mantzioris E, Gibson RA, Cleland LG, James MJ: The effect on human tumor necrosis factor α and interleukin 1 β production of diets enriched in n-3 fatty acids from vegetable oil or fish oil. Am J Clin Nutr 63: 116–122, 1996
33. McDonald BE, Gerrard JM, Bruce VM, Corner EJ: Comparison of the effect of canola oil and sunflower oil on plasma lipids and lipoproteins and on *in vivo* thromboxane A_2 and prostacyclin production in healthy young men. Am J Clin Nutr 50: 1382–1388, 1989
34. Abbey M, Clifton P, Kestin M, Belling B, Nestel P: Effect of fish oil on lipoproteins, lecithin:cholesterol acyltransferase, and lipid transfer protein activity in humans. Arteriosclerosis 10: 85–94, 1990
35. Bierenbaum ML, Reichstein R, Watkins TR: Reducing atherogenic risk in hyperlipidemic humans with flax seed supplementation: A preliminary report. J Am Coll Nutr 12: 501–504, 1993

36. Chan JK, Bruce VM, McDonald BE: Dietary α-linolenic acid is as effective as oleic acid and linoleic acid in lowering blood cholesterol in normolipidemic men. Am J Clin Nutr 53: 1230–1234, 1991
37. Clandinin MT, Foxwell A, Goh YK, Layne K, Jumpsen JA: Omega-3 fatty acid intake results in a relationship between the fatty acid composition of LDL cholesterol ester and LDL cholesterol content in humans. Biochim Biophys Acta 1346: 247–252, 1997
38. Kelley DS, Branch LB, Love JE, Taylor PC, Rivera YM, Iacono JM: Dietary α-linolenic acid and immunocompetence in humans. Am J Clin Nutr 53: 40–46, 1991
39. Freese R, Mutanen M: α-Linolenic acid and marine long-chain n-3 fatty acids differ only slightly in their effects on hemostatic factors in healthy subjects. Am J Clin Nutr 66: 591–598, 1997
40. Kestin M, Clifton P, Belling GB, Nestel PJ: n-3 Fatty acids of marine origin lower systolic blood pressure and triglycerides but raise LDL cholesterol compared with n-3 and n-6 fatty acids from plants. Am J Clin Nutr 51: 1028–1034, 1990
41. Nestel PJ, Pomeroy SE, Sasahara T, Yamashita T, Liang YL, Dart AM, Jennings GL, Abbey M, Cameron JD: Arterial compliance in obese subjects is improved with dietary plant n-3 fatty acid from flaxseed oil despite increased LDL oxidizability. Arterioscler Thromb Vasc Biol 17: 1163–1170, 1997
42. Singer P, Wirth M, Berger I: A possible contribution of decrease in free fatty acids to low serum triglyceride levels after diets supplemented with n-6 and n-3 polyunsaturated fatty acids. Atherosclerosis 83: 167–175, 1990
43. Uauy R, Hoffman DR: Essential fat requirements of preterm infants. Am J Clin Nutr 71(suppl): 245S–250S, 2000
44. Gibson RA, Makrides M: n-3 Polyunsaturated fatty acid requirements of term infants. Am J Clin Nutr 71(suppl): 251S–255S, 2000
45. Allman-Farinelli MA, Hall D, Kingham K, Pang D, Petocz P, Favaloro EJ: Comparison of the effects of two low fat diets with different α-linolenic:linoleic acid ratios on coagulation and fibrinolysis. Atherosclerosis 142: 159–168, 1999
46. Corner EJ, Bruce VM, McDonald BE: Accumulation of eicosapentaenoic acid in plasma phospholipids of subjects fed canola oil. Lipids 25: 598–601, 1990
47. Freese R, Mutanen M, Valsta LM, Salminen I: Comparison of the effects of two diets rich in monounsaturated fatty acids differing in their linoleic/α-linolenic acid ratio on platelet aggregation. Thromb Haemostas 71: 73–77, 1994
48. Li D, Mann NJ, Sinclair AJ: Comparison of n-3 polyunsaturated fatty acids from vegetable oils, meat, and fish in raising platelet eicosapentaenoic acid levels in humans. Lipids 34: S309, 1999
49. Kelley DS, Nelson GJ, Love JE, Branch LB, Taylor PC, Schmidt PC, Mackey BE, Iacono JM: Dietary α-linolenic acid alters tissue fatty acid composition, but not blood lipids, lipoproteins or coagulation status in humans. Lipids 28: 533–537, 1993
50. Jones PJH, Lichtenstein AH, Schaefer EJ, Namchuk GL: Effect of dietary fat selection on plasma cholesterol synthesis in older, moderately hypercholesterolemic humans. Arterioscler Thromb 14: 542–548, 1994
51. Lichtenstein AH, Ausman LM, Carrasco W, Jenner JL, Gualtieri LJ, Goldin BR, Ordovas JM, Schaefer EJ: Effects of canola, corn, and olive oils on fasting and postprandial plasma lipoproteins in humans as part of a National Cholesterol Education Program step 2 diet. Arterioscler Thromb 13: 1533–1542, 1993
52. Cunnane SC, Hamadeh MJ, Liede AC, Thompson LU, Wolever TMS, Jenkins DJA: Nutritional attributes of traditional flaxseed in healthy young adults. Am J Clin Nutr 61: 62–68, 1995
53. Matheson B, Walker KZ, Taylor DM, Peterkin R, Lugg D, O'Dea K: Effect on serum lipids of monounsaturated oil and margarine in the diet of an Antarctic Expedition. Am J Clin Nutr 63: 933–938, 1996
54. Bierenbaum ML, Reichstein RP, Watkins TR, Maginnis WP, Geller M: Effects of canola oil on serum lipids in humans. J Am Coll Nutr 10: 228–233, 1991
55. Mantzioris E, Cleland LG, Gibson RA, Neumann MA, Demasi M, James MJ: Biochemical effects of a diet containing foods enriched with n-3 fatty acids. Am J Clin Nutr 72: 42–48, 2000
56. Connor WE, Bendich A, eds: Highly unsaturated fatty acids in nutrition and disease prevention. Am J Clin Nutr 71 (suppl): 169S–398S, 2000
57. Harris WS: n-3 Fatty acids and serum lipoproteins: Human studies. Am J Clin Nutr 65 (suppl): 1645S–1654S, 1997
58. Blok WL, Katan MB, van der Meer JWM: Modulation of inflammation and cytokine production by dietary (n-3) fatty acids. J Nutr 126: 1515–1533, 1996
59. GISSI-Prevenzione Investigators: Dietary supplementation with n-3 polyunsaturated fatty acids and vitamin E after myocardial infarction: Results of the GISSI-Prevenzione trial. Lancet 354: 447–455, 1999
60. Foulon T, Richard M-J, Payen N, Bourrain J-L, Beani J-C, Laporte F, Hadjian A: Effects of fish oil fatty acids on plasma lipids and lipoproteins and oxidant-antioxidant imbalance in healthy subjects. Scand J Clin Lab Invest 59: 239–248, 1999
61. US Department of Agriculture, Agricultural Research Service: Nutrient database for standard reference, release 14, 2001. Available online at www.nal.usda.gov/fnic
62. Food Surveys Research Group, Beltsville Human Nutrition Research Center, Agricultural Research Service, US Department of Agriculture: Intakes of 19 individual fatty acids: Results from the 1994–96 Continuing Survey of Food Intakes by Individuals. Available online at www.barc.usda.gov/bhnrc/foodsurvey/home.htm
63. Krauss RM, Eckel RH, Howard B, Appel LJ, Daniels SR, Deckelbaum RJ, Erdman JW Jr, Kris-Etherton P, Goldberg IJ, Kotchen TA, Lichtenstein AH, Mitch WE, Mullis R, Robinson K, Wylie-Rosett J, St Jeor S, Suttie J, Tribble DL, Bazzarre TL: AHA dietary guidelines: Revision 2000: A statement for healthcare professionals from the Nutrition Committee of the American Heart Association. Circulation 102: 2296–2311, 2000
64. Abedin L, Lien EL, Vingrys AJ, Sinclair AJ: The effects of dietary α-linolenic acid compared with docosahexaenoic acid on brain, retina, liver, and heart in the guinea pig. Lipids 34: 475–482, 1999
65. Berger A, Gershwin ME, German JB: Effects of various dietary fats on cardiolipin acyl composition during ontogeny of mice. Lipids 27: 605–612, 1992
66. Cleland LG, Gibson RA, Hawkes JS, James MJ: Comparison of cell membrane phospholipid fatty acids in five rat strains fed four test diets. Lipids 25: 559–564, 1990
67. Greiner RCS, Winter J, Nathanielsz PW, Brenna JT: Brain docosahexaenoate accretion in fetal baboons: Bioequivalence of dietary α-linolenic and docosahexaenoic acids. Pediatr Res 42: 826–834, 1997
68. Makrides M, Neumann MA, Gibson RA: Is dietary docosahexaenoic acid essential for term infants? Lipids 31: 115–119, 1996
69. Farquharson J, Jamieson EC, Abbasi KA, Patrick WJA, Logan RW, Cockburn F: Effect of diet on the fatty acid composition of the major phospholopids of infant cerebral cortex. Arch Dis Child 72: 198–203, 1995
70. Farquharson J, Cockburn F, Patrick WA, Jamieson EC, Logan RW: Infant cerebral cortex phospholipid fatty-acid composition and diet. Lancet 340: 810–813, 1992

71. Seidelin KN, Myrup B, Fischer-Hansen B: n-3 Fatty acids in adipose tissue and coronary artery disease are inversely related. Am J Clin Nutr 55: 1117–1119, 1992
72. Conquer JA, Martin JB, Tummon I, Watson L, Tekpetey F: Fatty acid analysis of blood serum, seminal plasma, and spermatozoa of normozoospermic vs. asthenozoospermic males. Lipids 34: 793–799, 1999
73. Conquer JA, Martin JB, Tummon I, Watson L, Tekpetey F: Effect of DHA supplementation on DHA status and sperm motility in asthenozoospermic males. Lipids 35: 149–154, 2000
74. Christophe A, Zalata A, Mahmoud A, Combaire F: Fatty acid composition of sperm phospholipids and its nutritional implications. Middle East Fertil Soc J 3: 46–53, 1998
75. Kang JX, Leaf A: Antiarrhythmic effects of polyunsaturated fatty acids: Recent studies. Circulation 94: 1774–1780, 1996
76. Billman GE, Kang JX, Leaf A: Prevention of ischemia-induced cardiac sudden death by n-3 polyunsaturated fatty acids in dogs. Lipids 32: 1161–1168, 1997
77. Billman GE, Kang JX, Leaf A: Prevention of sudden cardiac death by dietary pure ω-3 polyunsaturated fatty acids in dogs. Circulation 99: 2452–2457, 1999
78. Lands WEM, Libelt B, Morris A, Kramer NC, Prewitt TE, Bowen P, Schmeisser D, Davidson MH, Burns JH: Maintenance of lower proportions of (n-6) eicosanoid precursors in phospholipids of human plasma in response to added dietary (n-3) fatty acids. Biochim Biophys Acta 1180: 147–162, 1992
79. Lands WEM: Review: Stories about acyl chains. Biochim Biophys Acta 1483: 1–15, 2000

Molecular and Cellular Biochemistry **246:** 91–95, 2003.

Dissemination of *Chlamydia pneumoniae* to the vessel wall in atherosclerosis

Satoru Hirono and Grant N. Pierce

Cell Biology Laboratory, Division of Stroke and Vascular Disease, St. Boniface General Hospital Research Centre and Department of Physiology, Faculty of Medicine, University of Manitoba, Winnipeg, Canada

Abstract

Heart disease and stroke are the result of atherosclerotic vascular lesions. It is becoming increasingly clear that an infection may be an important initiating component within the atherogenic process. However, in order for the infection to contribute to atherosclerosis, it must first be capable of disseminating to the vessel wall. *Chlamydia pneumoniae* is an example of an infectious atherogenic stimulus. The present treatise reviews our knowledge concerning dissemination of infectious agents like *C. pneumoniae*. Three factors can be identified that modulate the severity of the infection in the vascular wall. First, although all vascular cell types appear to be infected with agents like *C. pneumoniae*, there are differences in the sensitivity to infection amongst these cell types. Second, the lipid environment is important in defining the effects of *C. pneumoniae* on atherosclerotic disease. Third, the inflammatory/atherosclerotic interaction is influenced by the specific infectious stimuli employed. The *in situ* atherogenic effects of *C. pneumoniae* may be specific to this organism and may not occur with related infectious agents like *C. trachomatis*. Despite the identification of these three factors, controversy exists surrounding specific characteristics of these effects. This may be the result of a plethora of differing experimental conditions (different labs, different lipids, different cell types or lines, and different *C. pneumoniae* characteristics (infection, dosage, duration, etc.)). Further study of these important phenomena is clearly warranted in view of the potential importance of infection to the atherosclerotic disease. (Mol Cell Biochem **246**: 91–95, 2003)

Key words: infection, arteriosclerosis, inflammation, bacteria, oxidized low-density lipoprotein

Introduction

Atherosclerotic arterial disease is the major cause of mortality today. Established cardiovascular risk factors do not fully explain the presence of atherosclerosis in a substantial number of patients. It is now recognized that an important inciting stimulus for atherosclerosis is an immunologically mediated inflammatory response [1]. Recent studies suggest that specific infectious pathways like bacterial, parasitic, or viral infection may initiate the inflammatory process [2, 3]. One agent that has attracted particular attention is the obligate intracellular prokaryotic human pathogen, *Chlamydia pneumoniae* [4, 5]. Epidemiological [6–8], pathological [9–12], and experimental studies [13–16] have linked *C. pneumoniae* infection to atherosclerosis and heart disease, however, the precise role of *C. pneumoniae* in the development and progression of atherosclerosis is still not clear.

The infectious pathway in the vasculature

The successful infection of the vascular wall with *C. pneumoniae* is the first critical step in the initiation of the atherogenic process. It is now indisputable that respiratory tract infection with *C. pneumoniae* can result in systemic dissemination to other organs [17] including aortic tissue [13–15]. The precise route through which *C. pneumoniae* travels from the respiratory tract to ultimately infect the vasculature is not entirely clear. One hypothesis is that the respiratory tract inoculation with *C. pneumoniae* results in an intracellular infection of alveolar macrophages that subsequently enter the vascular circulation [18]. The infected circulating monocytes are thought to transmit *C. pneumoniae* infection to endothelial cells via cell-to-cell contact, or otherwise migrate and

Address for offprints: G.N. Pierce, Division of Stroke and Vascular Disease, St. Boniface General Hospital Research Centre, 351 Tache Avenue, Winnipeg, Canada, R2H 2A6 (E-mail: gpierce@sbrc.ca)

reside within the arterial wall. It was shown *in vitro* that U-937 human macrophages infected with *C. pneumoniae* were capable of transmitting the infection to coronary artery endothelial cells [19]. Surprisingly, even uninfected monocytes increase the susceptibility of endothelial cells to *C. pneumoniae* infection when they are co-cultured together [20, 21]. This complex interplay between cells in the vascular wall has functional implications beyond dissemination of the infection. Coombes and Mahony [22] showed that conditioned medium from human endothelial cells infected with *C. pneumoniae* stimulated the replication of smooth muscle cells in a time- and dose-dependent manner.

Dissemination of the *C. pneumoniae* infection to the vasculature appears to be a specific process that can be modified by a number of factors. Moazed *et al.* [14] demonstrated that *C. pneumoniae* can disseminate to, but does not persist in the aorta following a single intranasal inoculation of normocholesterolemic C57BL/6J mice. The same group has reported that repeated inoculations of C57BL/6J mice on a normal chow diet result in longer persistence of *C. pneumoniae* and inflammatory changes in the aorta [23].

Even more important are the findings of an association between the lipid environment and the dissemination of the *C. pneumoniae* to the vessel wall. Hu *et al.* [15] found that chlamydial LPS antigen was not detected in the aorta from low-density lipoprotein receptor deficient (LDLR–/–) mice fed a regular chow regardless of chlamydial infection. However, when the mice were fed a cholesterol-supplemented diet, the antigen was detected in the aorta. This strongly suggested that the lipid environment was critical for disseminating the *C. pneumoniae* to the vessel wall. These findings were not inconsequential. This persistence of the organism in aortic tissue was associated with more advanced atherosclerotic lesions as well. The atherogenic effects of *C. pneumoniae* were also dependent upon the presence of a hypercholesterolemic environment. Infection with *C. pneumoniae* did not stimulate atherogenesis if the animals did not have elevated circulating cholesterol levels. Others [24, 25] have confirmed these observations. These data strongly suggested that cholesterol was associated with the atherogenic effects of *C. pneumoniae* but did not specifically identify how this occurred at a cellular level.

There are indications that the cellular lipid content and composition may be important. Knoebel *et al.* [26] found that rabbit and human smooth muscle cells pre-incubated with cationized LDL were more susceptible to *C. pneumoniae* infection. Cationized LDL enters cells in greater quantities than native LDL. Conversely, it has been suggested that exposure to oxidized LDL may be important for *C. pneumoniae* infection. Activation of endothelial cells by oxidized LDL within the developing lesions appeared critical for infected circulating monocytes to establish infection [27, 28]. However, others [29] have found that macrophages pre-incubated with modified LDL are less susceptible to *C. pneumoniae* infection. This may suggest that once transformed into foam cells, macrophages lose their ability to support and transmit *C. pneumoniae* infection. The divergence of findings regarding the effects of lipids on *C. pneumoniae* infection may be the result of a plethora of differing experimental conditions (different labs, different lipids, different cell types or lines, different *C. pneumoniae* characteristics (infection, dosage, duration, etc.)). Unfortunately, there still remains a clear need to understand how cholesterol (and its physiological forms: LDL, oxidized LDL, etc.) influences *C. pneumoniae* infection.

Dissemination may also be altered as a function of the strain of the infectious agent itself. *C. trachomatis* mouse pneumonitis (MoPn) is a mouse tropic strain of *C. trachomatis* that causes mouse pneumonia [30, 31] and has been used to study the respiratory pathology induced by chlamydial infection in mice. It has also been compared to *C. pneumoniae* in its capacity to stimulate atherosclerosis. For example, LDLR –/– mice inoculated intranasally with *C. pneumoniae* AR-39 strain at 0.5–1 × 10^7 IFU significantly enhanced the development of atherosclerotic lesions. However, inoculation of the same species of mice with the *C. trachomatis* MoPn strain at 0.5–1 × 10^4 IFU had no effect on atherosclerosis [15]. Interestingly, chlamydial LPS antigen was detected in the aorta from mice infected with either AR39 or MoPn strains [15]. Others have confirmed this lack of the atherogenic effect of *C. trachomatis* using a different mouse model [32]. The human biovar *C. trachomatis* E/UW-5/OT disseminated to the aorta of apolipoprotein E-deficient mice in a transient manner, but did not establish persistent infection whereas *C. pneumoniae* (AR39 strain) did [33]. Clearly, the different chlamydia species generate different effects although some aspects of their specific actions remain controversial.

The cellular effects of the different chlamydia species may be very different as well. *C. pneumoniae* can stimulate the migration of neutrophils and monocytes across infected endothelial cell monolayers *in vitro* [34]. The extent of neutrophil and monocyte transendothelial migration correlated with the expression level of IL-8 and MCP-1, respectively, induced by *C. pneumoniae* infection. In contrast, *C. trachomatis* caused no significant increase in IL-8 production or transendothelial migration of neutrophils. However, substantial levels of monocyte migration were induced by *C. trachomatis*, despite a lack of measurable MCP-1 stimulation [34]. These findings suggest that the up-regulation of IL-8 and MCP-1 in endothelial cells may be unique to *C. pneumoniae*. It also suggests that the cellular signaling pathways stimulated by the two species of chlamydia are very different.

The cell type exposed to the infectious agent may also modify dissemination of the infection. It has been established by a number of independent laboratories [26, 35–37] that *C.*

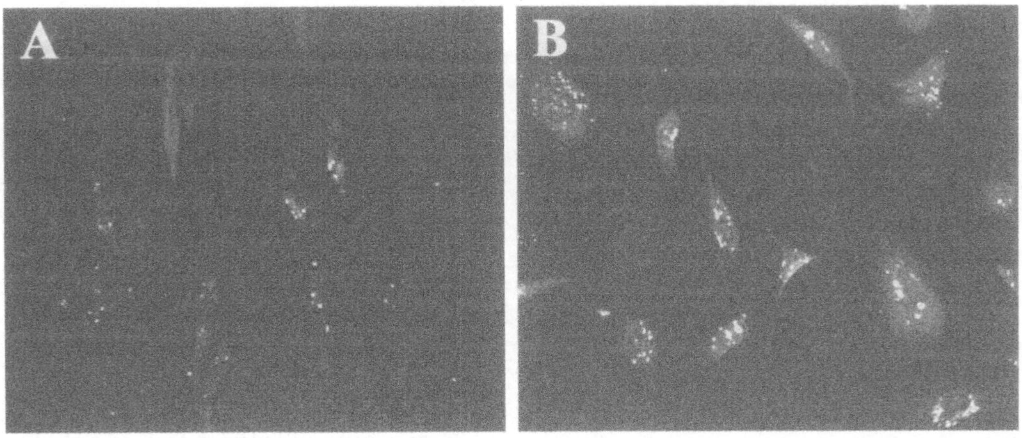

Fig. 1. Chlamydia pneumoniae infection of vascular smooth muscle cells and endothelial cells. (A) rabbit vascular smooth muscle cells and (B) human umbilical vein endothelial cells were infected with *C. pneumoniae* AR39 strain. At 72 h postinfection, monolayers were fixed with methanol and incubated with anti-chlamydial monoclonal antibody (Chemicon International Inc.). Multiple inclusion bodies were visualized by staining with FITC-conjugated anti-mouse IgG (Sigma). Magnification, ×400 (confocal fluorescence microscopy).

pneumoniae can infect the three principal cell types that are involved in the atherogenic process: endothelial cells, vascular smooth muscle cells, and monocytes/macrophages (Fig. 1). However, a large difference in the sensitivity of these cell types to infection is apparent when all the experimental factors are carefully controlled [36, 37]. Endothelial cells are strongly resistant to infection from *C. pneumoniae* whereas smooth muscle cells are not. Other evidence exists in support of the hypothesis that different vascular cell types possess different sensitivities to *C. pneumoniae* infection. For example, in contrast to vascular smooth muscle cells [26], RAW 264.7 macrophages pre-incubated with modified LDL became less susceptible to subsequent infection with *C. pneumoniae* [29].

The infection and cell signaling

It is not entirely clear how the interaction of *C. pneumoniae* with vascular cells achieves the physiological and pathological affects it produces. For example, it is still controversial whether active infection with a live organism is required to initiate its actions within the cells [19, 22, 38–40]. Molestina *et al.* [40] showed that heat-inactivation of *C. pneumoniae* completely abolished the ability of the organism to induce chemokine production in endothelial cells, whereas UV-inactivation resulted in only partial inhibition. This suggests that a potentially heat-labile substance of *C. pneumoniae* is responsible for endothelial cell activation [41] but further work is required. Others [22] have demonstrated that heat-inactivation of *C. pneumoniae* does not affect its ability to induce mitogenic activity in endothelial cells. This significant issue remains to be resolved. Another important question concerns the need for *C. pneumoniae* to be internalized in order to generate its cellular effects. Exposure of endothelial cells to *C. pneumoniae* induced an immediate (within 10–15 min) increase of protein tyrosine phosphorylation, p42/p44 MAPK stimulation and NF-κB activation/translocation [42]. These data suggest that attachment of *C. pneumoniae* is sufficient to trigger the signaling pathways in endothelial cells and internalization of the organism may not be necessary for the process. However, again, further work is necessary before definitive conclusions may be reached.

In summary, it is clear that an infectious agent can disseminate from the respiratory tract to the vasculature in sufficient quantities to stimulate arterial diseases like atherosclerosis. This dissemination appears to depend upon the concomitant presence of a cholesterol-enriched cellular environment but details of the characteristics of this environment are unavailable at present. Both the vascular cell types and the specific infectious stimuli can also modify the severity of the infection and the degree of progression of the atherosclerosis. However, again, detailed information concerning these factors is lacking. An interaction of the infectious agent with the cell surface appears to be critical in order to induce a cellular response but internalization of the organism cannot be completely ruled out yet as an important contributing factor. These deficiencies in our understanding may be the result of a plethora of differing experimental conditions (different labs, different lipids, different cell types or lines, and different *C. pneumoniae* characteristics (infection, dosage, duration, etc.)). Clearly, further work on all of these important issues needs to be initiated in order to delineate the mechanisms whereby a respiratory infection results in the stimulation of an atherogenic event.

Acknowledgements

This work was supported by a grant from the Canadian Institutes for Health Research. G.N. Pierce is a CIHR Senior Scientist. A Postdoctoral Fellowship from the Faculty of Medicine at the University of Manitoba, Canada supported S. Hirono. S. Hirono received a grant-in-aid from the Tsukada Medical Foundation, Japan.

References

1. Ross R: Atherosclerosis – an inflammatory disease. N Engl J Med 340: 115–126, 1999
2. Buja LM: Does atherosclerosis have an infectious etiology? Circulation 94: 872–873, 1996
3. Libby P, Egan D, Skarlatos S: Roles of infectious agents in atherosclerosis and restenosis – an assessment of the evidence and need for future research. Circulation 96: 4095–4103, 1997
4. Gupta S, Camm AJ: *Chlamydia pneumoniae* and coronary heart disease – coincidence, association, or causation? BMJ 314: 1778–1779, 1997
5. Gaydos CA, Quinn TC: The role of *Chlamydia pneumoniae* in cardiovascular disease. Adv Int Med 45: 139–173, 2000
6. Saikku P, Mattila K, Nieminen MS, Huttunen JK: Serological evidence of an association of a novel chlamydia, TWAR, with chronic coronary heart disease and acute myocardial infarction. Lancet ii: 983–986, 1988
7. Thom DH, Wang SP, Grayston JT, Siscovick DS, Stewart DK, Kronmal RA, Weiss NS: *Chlamydia pneumoniae* strain *TWAR* antibody and angiographically demonstrated coronary artery disease. Arterioscler Thromb 11: 547–551, 1991
8. Danesh J, Collins R, Peto R: Chronic infections and coronary heart disease: Is there a link? Lancet 350: 430–436, 1997
9. Kuo CC, Grayston JT, Campbell LA, Goo YA, Wissler RW, Benditt EP: *Chlamydia pneumoniae* (TWAR) in coronary arteries of young adults (15–34 years old). Proc Natl Acad Sci USA 92: 6911–6914, 1995
10. Jackson LA, Campbell LA, Schmidt RA, Kuo CC, Cappuccio AL, Lee MJ, Grayston JT: Specificity of detection of *Chlamydia pneumoniae* in cardiovascular atheroma – evaluation of the innocent bystander hypothesis. Am J Pathol 150: 1785–1790, 1997
11. Yamashita K, Ouchi K, Shirai M, Gondo T, Nakazawa T, Ito H: Distribution of *Chlamydia pneumoniae* infection in the atherosclerotic carotid artery. Stroke 29: 773–778, 1998
12. Taylor-Robinson D: *Chlamydia pneumoniae* in vascular tissue. Atherosclerosis 140(suppl 1): S21–S24, 1998
13. Fong IW, Chiu B, Viira E, Fong MW, Jang D, Mahony J: Rabbit model for *Chlamydia pneumoniae* infection. J Clin Microbiol 35: 48–52, 1997
14. Moazed TC, Kuo CC, Grayston JT, Campbell LA: Murine model of *C. pneumoniae* infection and atherosclerosis. J Infect Dis 175: 883–890, 1997
15. Hu H, Pierce GN, Zhong G: The atherogenic effects of chlamydia are dependent on serum cholesterol and specific to *Chlamydia pneumoniae*. J Clin Invest 103: 747–753, 1999
16. Summersgill JT, Molestina RE, Miller RD, Ramirez JA: Interactions of *Chlamydia pneumoniae* with human endothelial cells. J Infect Dis 181(suppl 3): S479–S482, 2000
17. Yang ZP, Kuo CC, Grayston JT: Systemic dissemination of *Chlamydia pneumoniae* following intranasal inoculation in mice. J Infect Dis 171: 736–738, 1995
18. Moazed TC, Kuo CC, Grayston JT, Campbell LA: Evidence of systemic dissemination of *C. pneumoniae* via macrophages in the mouse. J Infect Dis 177: 1322–1325, 1998
19. Quinn TC, Gaydos CA: *In vitro* infection and pathogenesis of *Chlamydia pneumoniae* in endovascular cells. Am Heart J 138: S507–S511, 1999
20. Lin TM, Campbell LA, Rosenfeld ME, Kuo CC: Monocyte-endothelial cell co-culture enhances infection of endothelial cells with *Chlamydia pneumoniae*. J Infect Dis 181: 1096–1100, 2000
21. Lin TM, Campbell LA, Rosenfeld ME, Kuo CC: Human monocyte-derived insulin-like growth factor-2 enhances the infection of human arterial endothelial cells by *Chlamydia pneumoniae*. J Infect Dis 183: 1368–1372, 2001
22. Coombes BK, Mahony JB: *Chlamydia pneumoniae* infection of human endothelial cells induces proliferation of smooth muscle cells via an endothelial cell-derived soluble factor(s). Infect Immun 67: 2909–2915, 1999
23. Campbell LA, Blessing E, Rosenfeld M, Lin TM, Kuo CC: Mouse model of *C. pneumoniae* infection and atherosclerosis. J Infect Dis 181(suppl 3): S508–S513, 2000
24. Blessing E, Lin T, Campbell LA, Rosenfeld ME, Lloyd D, Kuo CC: *Chlamydia pneumoniae* induces inflammatory changes in the heart and aorta of normocholesterolemic C57BL/6J mice. Infect Immun 68: 4765–4768, 2000
25. Blessing E, Campbell LA, Rosenfeld ME, Chough N, Kuo CC: *Chlamydia pneumoniae* infection accelerates hyperlipidemia induced atherosclerotic lesion development in C57BL/6J mice. Atherosclerosis 158: 13–17, 2001
26. Knoebel E, Vijayagopal P, Figueroa JE II, Martin DH: *In vitro* infection of smooth muscle cells by *Chlamydia pneumoniae*. Infect Immun 65: 503–506, 1997
27. Quinn MT, Parthasarathy S, Fong LG, Steinberg D: Oxidatively modified low density proteins: A potential role in recruitment and retention of monocyte/macrophages during atherogenesis. Proc Natl Acad Sci USA 84: 2995–2998, 1987
28. Cushing SD, Berliner JA, Valente AJ, Territo MC, Navab M, Parhami F, Gerrity R, Schwartz CJ, Fogelman A: Minimally modified low-density lipoprotein induces monocyte chemotactic protein 1 in human endothelial cells and smooth muscle cells. Proc Natl Acad Sci USA 87: 5134–5138, 1990
29. Rosenfeld ME, Blessing E, Lin TM, Moazed TC, Campbell LA, Kuo CC: Chlamydia, inflammation, and atherogenesis. J Infect Dis 181(suppl 3): S492–S497, 2000
30. Nigg C, Eaton MD: Isolation from normal mice of a pneumotropic virus which forms elementary bodies. J Exp Med 79: 497–510, 1944
31. Read TD, Brunham RC, Shen C, Gill SR, Heidelberg JF, White O, Hickey EK, Peterson J, Utterback T, Berry K *et al.*: Genome sequences of *Chlamydia trachomatis* MoPn and *Chlamydia pneumoniae* AR39. Nucleic Acids Res 28: 1397–1406, 2000
32. Blessing E, Nagano S, Campbell LA, Rosenfeld ME, Kuo CC: Effect of *Chlamydia trachomatis* infection on atherosclerosis in apolipoprotein E-deficient mice. Infect Immun 68: 7195–7197, 2000
33. Moazed TC, Campbell LA, Rosenfeld ME, Grayston JT, Kuo CC: *Chlamydia pneumoniae* infection accelerates the progression of atherosclerosis in apolipoprotein E-deficient mice. J Infect Dis 180: 238–241, 1999
34. Molestina RE, Miller RD, Ramirez JA, Summersgill JT: Infection of human endothelial cells with *Chlamydia pneumoniae* stimulates transendothelial migration of neutrophils and monocytes. Infect Immun 67: 1323–1330, 1999
35. Kaukoranta-Tolvanen SS, Laitinen K, Saikku P, Leinonen M: *Chlamydia pneumoniae* multiplies in human endothelial cells *in vitro*. Microb Pathog 16: 313–319, 1994
36. Godzik KL, O'Brien ER, Wang SK, Kuo CC: *In vitro* susceptibility of human vascular wall cells to infection with *Chlamydia pneumoniae*. J Clin Microbiol 33: 2411–2414, 1995
37. Gaydos CA, Summersgill JT, Sahney NN, Ramirez JA, Quinn TC:

Replication of *Chlamydia pneumoniae in vitro* in human macrophages, endothelial cells, and aortic artery smooth muscle cells. Infect Immun 64: 1614–1620, 1996

38. Kol A, Sukhova GK, Lichtman AH, Libby P: Chlamydial heat shock protein 60 localizes in human atheroma and regulates macrophage tumor necrosis factor-alpha and matrix metalloproteinase expression. Circulation 98: 300–307, 1998
39. Sasu S, LaVerda D, Qureshi N, Golenbock DT, Beasley D: *Chlamydia pneumoniae* and chlamydial heat shock protein 60 stimulate proliferation of human vascular smooth muscle cells via Toll-like receptor 4 and p44/p42 mitogen-activated protein kinase activation. Circ Res 89: 244–250, 2001
40. Molestina RE, Dean D, Miller RD, Ramirez JA, Summersgill JT: Characterization of a strain of *Chlamydia pneumoniae* isolated from coronary atheroma by analysis of the *omp1* gene and biological activity in human endothelial cells. Infect Immun 66: 1370–1376, 1998
41. Kol A, Bourcier T, Lichtman AH, Libby P: Chlamydial and human heat shock protein 60s activate human vascular endothelium, smooth muscle cells, and macrophages. J Clin Invest 103: 571–577, 1999
42. Krull M, Klucken AC, Wuppermann FK, Fuhrmann O, Magerl C, Seybold J, Hippenstiel S, Hegemann JH, Jantos CA, Suttorp N: Signal transduction pathways activated in endothelial cells following infection with *Chlamydia pneumoniae*. J Immunol 162: 4834–4841, 1999

Molecular and Cellular Biochemistry **246**: 97–103, 2003.

Bcl10 protein can act as a transcription activator in yeast

Ming Chen, Ling Yun Li and Yi-Peng Qi
Wu Han University, School of Life Sciences, Molecular Virology Laboratory, Wu Han, P.R. China

Abstract

The Bcl10 gene has recently been cloned from the chromosomal translocation t(1:14) (p22; q32) in a low-grade mucosa-associated lymphoid tissue (MALT) lymphoma, and was implicated in the pathogenesis of this and several other tumor types. In yeast two hybrid systems, when it was fused to Gal4 DNA-binding domain of pGBT9, this fusion protein can activate the expression of reporter genes without Gal4-AD domain. Through deletion assay and secondary structure prediction, we found that the N-terminal of Bcl10 contributed more to activation than the C-terminal. It has been recently reported that Bcl10 expression is correlated with the activation of NF-κB in eukaryotic cells , so it may represent an important role in protein expression. Our research is the first to demonstrate a new function of Bcl10: transcription activation in yeast. (Mol Cell Biochem **246**: 97–103, 2003)

Key words: Bcl10, transcription activator, yeast hybrid system

Introduction

Many genes related to cell apoptosis have been discovered, and the studies on them have greatly helped to cancer research. In 1999, a novel apoptotic signal gene was identified [1] through the cloning of chromosome breakpoints in extranodal (MALT-type) marginal zone lymphomas carrying the t(1:14) (p22; q32) translocation [1]. The full-length coding sequence of this gene was predicted to encode a 234 amino acid protein with a molecular weight of 26.2 kDa.

Bcl10 contains an amino-terminal caspase recruit domain (CARD) homologous to that found in several apoptotic molecules. Wild-type Bcl10 promoted apoptosis and suppressed malignant transformation *in vitro*, whereas truncated mutants lost the pro-apoptotic activity and exhibited gain of function enhancement of transformation [2]. However, the mechanism of regulation of apoptosis mediated by Bcl10 has not been demonstrated [3].

It has been testified overexpressed Bcl10 reduced c-Jun N-terminal kinase activation and induced NF-κB activation, instead antigen-driven NF-κB signals and proliferation were found to be defective in Bcl10–/– lymphocytes [4]. CARD10 [5], a novel Bcl10 interactor, functions as a molecular scaffold for the assembly of a Bcl10 signaling complex that activate NF-κB. All of these corroborate the role of Bcl10 as an important component of the NF-κB signaling pathway.

Recent research on Bcl10 clarified the important role it plays in apoptosis and transcription activation, but the definitive mechanism is still unknown. Whether the close relation of Bcl10 and NF-κB reveals Bcl10 itself is a transcription related gene? In our research work, we had hoped to find some proteins that interact with Bcl10 in yeast hybrid system, but the results were beyond our expectations. For Bcl10 protein, it can activate transcription of reporter genes when it was fused to the Gal4-BD without the Gal4-AD. The fact that Bcl10 can activate transcription in yeast suggest a possibility that Bcl10 protein may act as a transcriptional activator under certain conditions. It emphasizes that Bcl10 is an important component in gene expression, and it would further broaden the known function of Bcl10 protein.

Address for offprints: Y.-P. Qi, Wu Han University, School of Life Sciences, Molecular Virology Laboratory, Wu Han, PR China 430072
(E-mail: qiyipeng@whu.edu.cn)

Materials and methods

Materials

Yeast strain HF7z (it's chromosome is integrated with His and LacZ reporter gene), plasmid pGBT9 (5.4 kb cloning vector; used to generate fusions of the bait protein with the Gal4 DNA-BD), pACT2-AD (8.1 kb cloning vector; used to generate fusions of a known protein from cDNA library with the Gal4 AD), and many materials for yeast hybrid system were bought from the CLONTECH Company. Primers were synthesized by SAIBAISHENG Company. Human fiber cDNA library was kindly provided by Doctor Qi Bing (Wu Han University, China). Restriction enzymes, T_4 DNA ligase, plasmid SH14063, pCL1 and pBluscript were bought from HUA MEI Company (China).

Methods

Proliferation of plasmid library and extraction of plasmid DNA

Plasmid library were proliferated according to CLONTECH User Manual [6] and plasmids were extracted according to Molecular Clone [7].

PCR amplification

The primers p1:5′-AA**G AAT TC**C CAT CTC CCG CCT CCA CCA TG-3′ and p2:5′-ATG ACA AAG TGC TGT TAC TGT **GAC GTC** TC-3′ (black word stands for restriction enzyme site: PstI and EcoRI) were designed according to published Bcl10 gene sequence. Human fiber cDNA was used as template and the requirements for PCR was: denaturation temperature 95°C, 5 min; 30 cycles (95°C, 1 min → 55°C, 2 min → 72°C, 3 min). The products were purified by using low melting point agarose and then was used as template in the second round PCR to get resulted Bcl10 gene. The oligonucleotides primers used to amplify the desired Bcl10 subfragments which were later subcloned into pGBT9 and pACT2 are shown below:

P3: 5′-AAG AAT TCG ACT ACT TAC AGG AAA AC-3′
P4: 5′-CTA GAA CAG GCA AAT TCA CTG CAG AA-3′
P4′: 5′-CTG AAT TCT CTC TGA ATT TGC CTG TTC TA-3′
P5: 5′-CTG AAT TCA AGA AGG ACG CCT TAG AAA AT-3′
P6: 5′-CCC TGC AGT GCG AAG TGT AGT TGA AGA GA-3′

The olignucleotide primer pairs used to amplify Bcl10 subfragments were: p1 and p4 to get $Bcl10_{1-172}$, p2 and p4′ to get $Bcl10_{170-234}$, p3 and p6 to get $Bcl10_{68-194}$, p1 and p6 to get $Bcl10_{1-194}$,p2 and p5 to get $Bcl10_{15-234}$.

Construction of vectors

The resultant PCR products containing a 5′-EcoRI site and a 3′-PstI site were ligated in-frame with the DNA encoding the Gal4 DNA binding domain of pGBT9 for six recombinant plasmids pG/Bcl10, pG/$Bcl10_{1-172}$, pG/$Bcl10_{172-234}$, pG/$Bcl10_{68-194}$, pG/$Bcl10_{1-194}$ and pG/$Bcl10_{15-234}$. With the same methods, the resultant Bcl10 gene was cloned to the sequencing vector pBluescript for another recombinant plasmid pB/Bcl10.

In addition, through two-time subcloned, Bcl10 gene was also inserted into pACT2 in frame controlled by alcohol dehydrogenase promotor for recombinant plasmid pA/Bcl10. In this plasmid, the Gal4 active domain of pACT2 was substituted by Bcl10 (Fig. 2).

LiAc-mediated yeast transformation

In the LiAc transformation method, yeast competent cells are prepared and suspended in a LiAc solution with certain plasmid DNA to be transformed, along with the herring test DNA. Plasmids pG/Bcl10, pA/Bcl10 and five Bcl10 defective mutants fused to Gal4-BD in the plasmid pGBT9 are separately introduced to HF7Z yeast strain with small-scale LiAc yeast transformation procedure [6]. Trp and His minus plates were used as selective plates.

β-Galactosidas assays

The colony-lift filter assay was used to screen large numbers of transformants that survive the His growth selection or in one-hybrid library screening. The first yeast transformants, grown well in Trp-minus plates are selected by using Trp and His minus plates for their second growth. The activation of transcription of defective mutants was measured by the growth of yeast on Trp and His minus plates and the activity of β-galactosidas assays. The procedures for the colony-lift filter and liquid culture using ONPG as substrate were described in CLONTECH yeast protocols. The calculation formula of activity in β-galactosidas assays is:

$$\beta\text{-Galactosidas unites} = 1{,}000 \times OD420/(t \times V \times OD_{600})$$

Analysis with software ANTHEPROT 4.3

Protein sequence analytical software ANTHEPROT 4.3 was created by Institute of Biology and Chemistry of Proteins in France. Hydrophobicity profile and Secondary structure prediction of Bcl10 were made with ANTHEPROT 4.3, which are showed in Fig. 7 and 8.

Results

Amplification and cloning of Bcl10 gene and it's subfragments

By using designed p1 and p2 as primers and human fiber cDNA as template, Bcl10 gene with length of 730 bp was obtained by two rounds of PCR (Fig. 1a). Five Bcl10 subfragments also were acquired by PCR with different primer

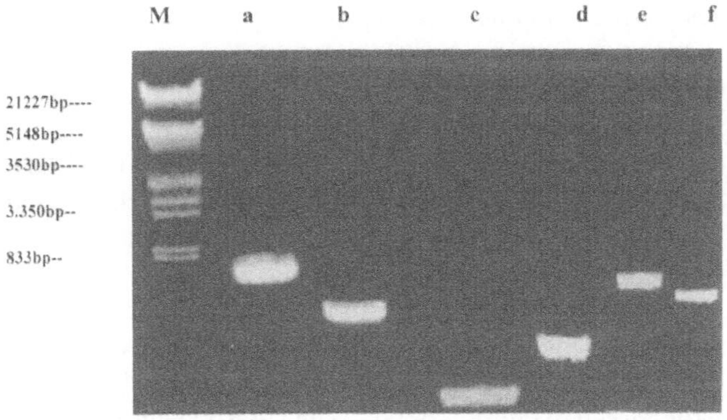

Fig. 1. PCR Amplification of Bcl10 and it's subfragments from human fiber cDNA library. M: lambda DNA/EcoRI + HindIII marker; a: Bcl10 gene (730 bp); b: $Bcl10_{1-172}$ (516 bp); c: $Bcl10_{172-234}$ (186 bp); d: $Bcl10_{68-194}$ (378 bp); e: $Bcl10_{15-234}$ (657 bp); f: $Bcl10_{1-194}$ (582 bp).

pairs (Figs 1b–1f). Bcl10 gene and those subfragments were inserted into various vectors to construct expressing vectors and sequencing vectors. All the structures of these recombinant plasmids are showed in Fig. 2. Bcl10 gene was confirmed by sequencing and the result was consistent with the reports.

Recombinant plasmids were identified by PCR and restriction digestion. P1 and p2 as primers, Bcl10 gene with 730 bp band was amplified from pG/Bcl10 and pB/Bcl10 (Figs 3a and 3b). By using restriction digestion, both fragments of 5.5 kb and 730 bp were cut from pG/Bcl10, two bands of 2.93 kb and 730 bp were also released from pB/Bcl10 (Fig. 3d), two bands of 7.34 kb and 730 bp were digested from pA/Bcl10 (Fig. 3e). The data demonstrated that Bcl10 has been cloned in pGBT9, pACT2 and pBluscript DNA. Identification of five defective mutants by PCR (the results are not displayed) was in accordance with the PCR amplification results in Fig. 1. The inserts of all the recombinants were further verified by sequencing (the data are not shown).

Transcriptional activation of Bcl10 protein

Transcriptional activation by the Gal4-Bcl10 fusion protein in the yeast interaction trap system is shown in Figs 4 and 5. When the yeast was cotransformed with pACT2-cDNA (which contains the Gal4-AD domain) and pG/Bcl10, both reporter genes His and lacZ were activated. All of the thousands of second transformants growing in His minus plates turned blue after half an hour in the colony-lift filter assay without exceptions (Fig. 4) which exclude the possibility that some interaction protein were found. If some interaction proteins were found, only limited number of colonies turn blue in colony-lift filter assay.

Since Bcl10 is a acidic protein (pI = 5.4), a feature shared by many transcription activators , and Bcl10 was related to

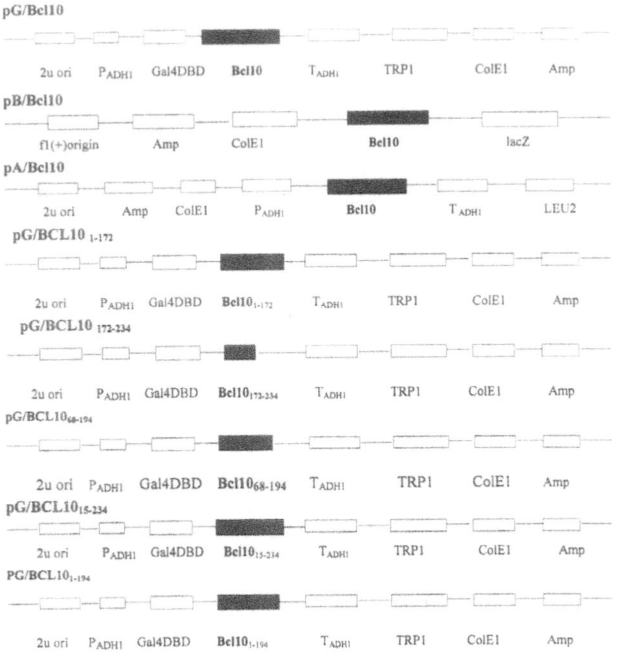

Fig. 2. Structure of eight recombinant plasmids.

the activation of NF-κB, we explored the possibility that Bcl10 protein may activate gene expression. We testified that Bcl10 directly fused to DNA-biding domain could also activate transcription (Fig. 5–2). However, when pA/Bcl10 (AD

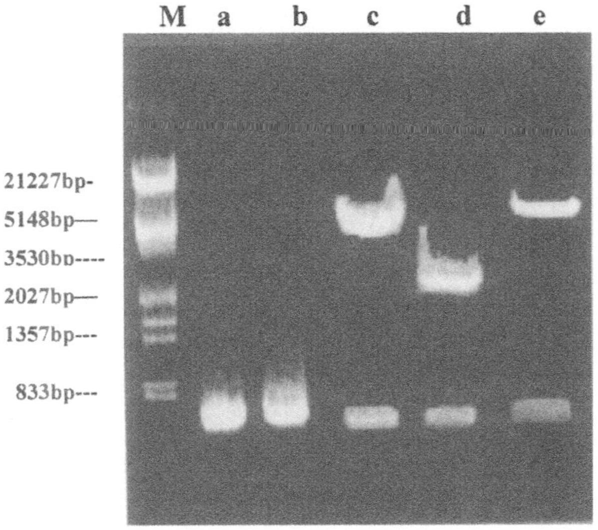

Fig. 3. Restriction digestion and PCR amplification of recombinant plasmids. M: lambda DNA/EcoRI + HindIII marker; a: pG/Bcl10 by PCR; b: pB/Bcl10 by PCR; c: pG/Bcl10 digested by PstI and EcoRI; d: pB/Bcl10 digested by PstI and EcoRI; e: pC/Bcl10 digested by PstI and EcoRI.

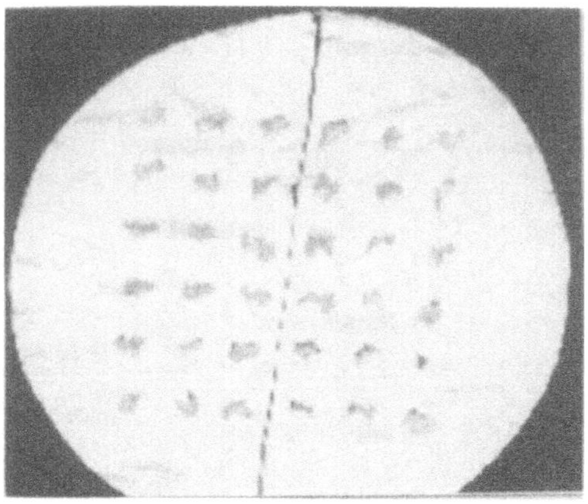

Fig. 4. Bcl10 protein activated the expression of reporter gene His and lacZ. His minus plates used as selective plates. Yeast transformed with pG/Bel10 + pACT2-cDNA were detected in colony-lift filter assay.

was substituted by Bcl10) alone were testified, the activation almost was lost (Fig. 5–5). Although pG/Bcl10 + pACT2-cDNA could activate the expression of reporter genes (Fig. 5–3), pG/Bcl10 alone also could activate transcription, so the expression of His and lacZ were activated not because Bcl10 interacted with some protein from cDNA library, it is presumably when Bcl10 is brought into UAS domain of yeast DNA, it triggers transcription activation.

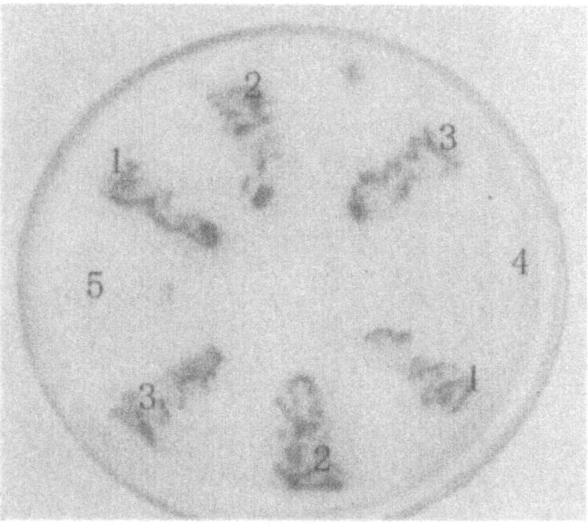

Fig. 5. Yeast transformed with recombinant plasmids growing in His minus plate and in colony-lift filter assay. 1. Positive control: pCL1; 2. pG/Bel10; 3. pG/BCL10 + pACT2-cDNA; 4. Negative control: SH14063; 5. pA/Bel10.

The activation domain in Bcl10

We constructed five fusion protein expression vectors by using five Bcl10 subfragments cloned with pGBT9 (Fig. 2). They respectively expressed Gal4-BD/Bcl10$_{1-172}$, Gal4-BD/Bcl10$_{172-234}$, Gal4-BD/Bcl10$_{68-194}$, Gal4-BD/ Bcl10$_{15-234}$ and Gal4-BD/Bcl10$_{1-194}$. Those recombinant plasmids and two control plasmids (pG/Bcl10, pA/Bcl10) containing the whole length Bcl10 gene were transformed into yeast separately.

Through identification with selective SD medium and analysis by β-galactosidas assays, we confirmed the transcription activity of Bcl10 and it's function domain. Observed from His minus plates, the yeast transformed with pG/Bcl10 grow normally in His minus plates, otherwise the yeast transformed with pA/Bcl10 did not grow (Fig. 6A). It demonstrated this activation must be generated through the binding domain of Gal4 bringing Bcl10 to UAS region of yeast genome. Three fusion proteins pG/Bcl10$_{172-234}$, pG/Bcl10$_{68-194}$, pG/Bcl10$_{15-234}$, lost the function of activating transcription of downstream gene. The results illustrated that as far as pG/Bcl10$_{172-234}$ and pG/Bcl10$_{15-234}$ are concerned, the c-terminal

A

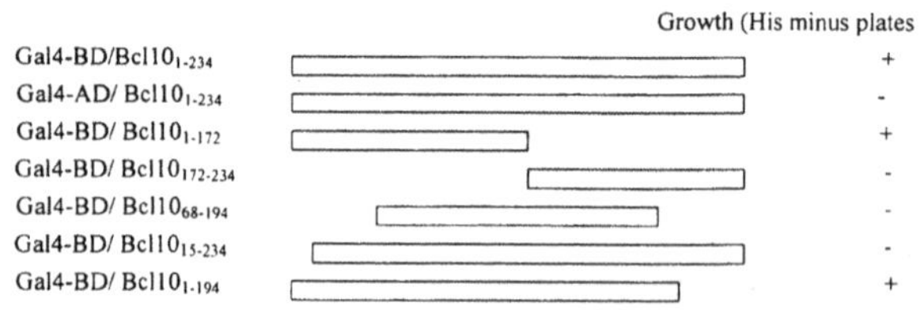

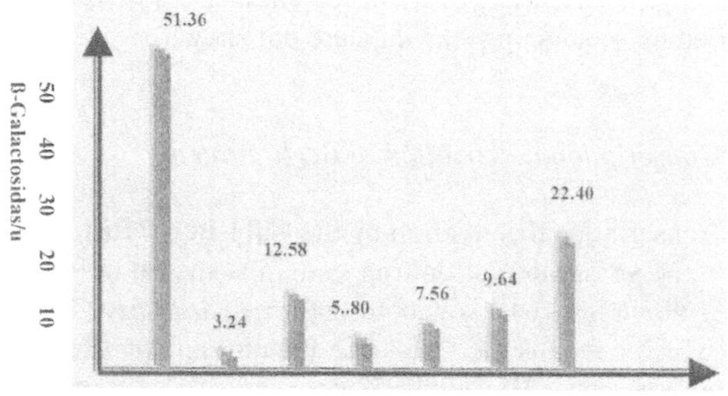

Fig. 6. Transcriptional activation by Bcl10 and it's defective mutants. The effects of various fusion proteins were monitored by the growth of yeast in the absence of His and quantified by measuring the activity of β-galactodase. A: Bcl10 and its defective mutants fused to the DNA biding domain were introduced into the yeast strain HF7Z. B: The transcription activity of Bcl10 and its defective mutants were quantified by measuring the activity of β-galactodase. (a: Gal4-BD/1–234; b: Gal4-AD/1–234; c: Gal4-BD/1–172; d: Gal4-BD/172–234; e: Gal4-BD/68–194; f: Gal4-BD/15–234; g: Gal4-BD/1–194).

of Bcl10 is integrated, but their n-terminal were deleted by 171 and 14 amino acids respectively; In other hand, defective mutant pG/Bcl10$_{68-194}$ also was deleted 67 amino in n-terminal, so even if these three defective mutants were fused with Gal4-DBD, they can't maintain the yeast growth in His minus plates and can't activate transcription of His gene. In addition, the other two defective mutants pG/Bcl10$_{1-172}$, pG/Bcl10$_{1-194}$ both reserved n-terminal sequence, they can activate the expression of His gene. The conclusion is that the Bcl10 protein possesses the function of transcription activation, it's activation domain probably is located in 1 ~ 15 amino acids of n-terminal.

Through β-galactosidas assays, the same conclusion also can be reached. From Fig. 6, we can see the function of activation of Bcl10 is tightly correlated with n-terminal. If n-terminal is deleted, transcription activation was lost. The more it's n-terminal is preserved, the less it's activation decrease. When the n-terminal was deleted 15,67 and 171 amino, the activation were 18.77, 14.72 and 11.29% of the full length gene (Fig. 6B, d–f). However if only n-terminal is integrated, but c-terminal is deleted (Fig. 6B, c and g), their activation retained 24.50 and 43.61% of the full length. So the conclusion should be that although as far as the transcription activation of Bcl10 is concerned, n-terminal is more important than c-terminal, the activation of the whole length molecule is the highest. Thus the transcription activation of Bcl10 is relevant to the integrity and structure of the whole molecule.

Hydrophobicity profile and secondary structure prediction

With protein analytical software ANTHEPROT 4.3, hydrophobicity profile of Bcl10 is showed in Fig. 7. We can see the Bcl10 protein has several hydrophobicity function domains. Three fragments: Bcl10$_{1-172}$, Bcl10$_{172-234}$, Bcl10$_{68-194}$ separately were deleted some hydrophobicity domains, so their transcription activation dropped largely. For Bcl10$_{1-194}$, it was deleted one small hydrophobicity domain in c-terminal, it's activation remains about 50% compared to that of the whole length. Bcl10$_{15-234}$ conserve all the important hydrophobicity domains, however the activation still greatly decreased. Thus, n-terminal has more influence on transcription activation than c-terminal. Accordingly, this activation is not only related to hydrophobicity, but also to the whole molecule structure. With the same protein analytical software, secondary structure prediction of Bcl10 was illustrated in Fig. 8. As we can see, there is a big α-helix in n-terminal, especially the peak is located in the fifteenth amino acid residue. So it may account for that n-terminal is essential for transcription activity. In the meantime, some little α-helix and β-turn

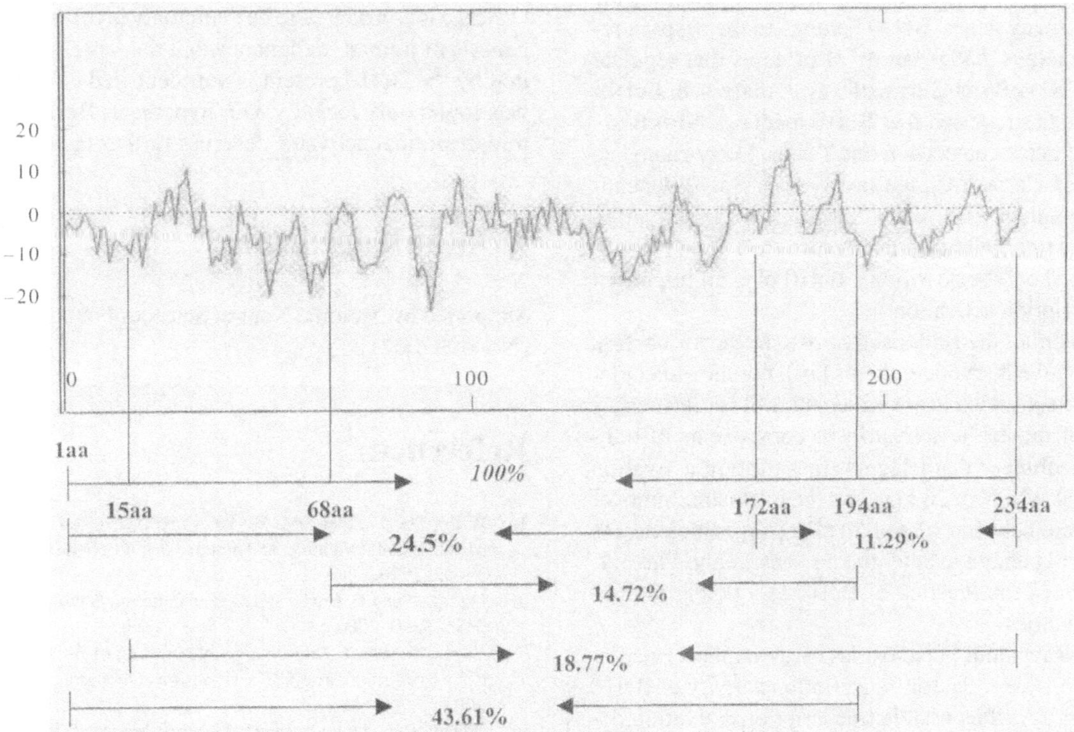

Fig. 7. The relation between hydrophobicity and transcription activity of Bcl10.

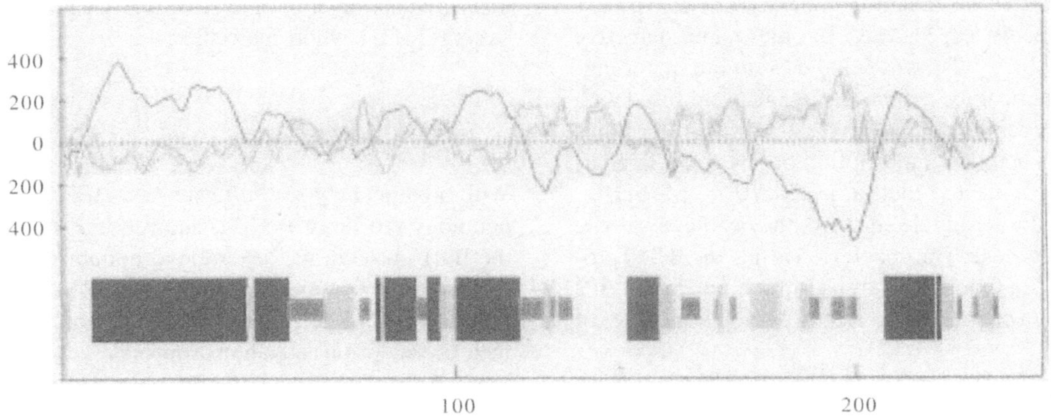

Fig. 8. Second structure of Bcl10 protein with GOR 1 method.

are in the middle or c-terminal, if they are deleted, both have some effects on the activation function of the whole protein.

Discussion

Fusion protein Bcl10-Gal4BD activated the expression of reporter gene, the fact which identified Bcl10 can activate transcription in yeast. Transcription activity of Bcl10 may be due to its close relation with NF-κB. NF-κB [8], an oxidative stress -sensitive nuclear transcription factor, controls the expression of many genes. Bcl10 belongs to the caspase recruitment domain (CARD) family of proteins that regulate apoptosis and NF-κB signaling pathways. Analysis of Bcl10-deficient mice has revealed that Bcl10 mediates NF-κB activation by antigen receptors in B and T cells. Many guanylate kinase, such as Carmal [9], are involved in Bcl10-dependent signaling pathway in T-cells that leads to the activation of NF-κB. The tight relationship between Bcl10 and NF-κB, some kinase , show the possibility Bcl10 play an important way in transcription activation.

Bcl10 are acidic, the acidic activators appear to function universally in all eukaryotic systems [10], but the molecular mechanism of activation is not understood. It is presumably the binding of the acidic activators to components of transcriptional machinery that triggers transcription activation [11]. Bcl10 has a MW of 26 kDa and the exclusion limits of nuclear pore are between 40 and 60 kDa [12], which means Bcl10 are small enough to enter the nucleus freely. This further strengthened the function of Bcl10 as a potential transcription activator.

Deletion assay and β-galactosidas assay demonstrated n-terminal is more essential for transcription activity of Bcl10 than c-terminal. As the possible transcription activation domain, n-terminal possesses a big α-helix and contains CARD, these secondary structure contributed to its important role in transcription activity. However, the activity must be closely related to the construction and conformation of the whole molecule.

At least for yeast UAS, there is no evidence that Bcl10 protein can bind to DNA, however, this is not unusual because several transcription factors including Vp1 [13] and Vp16 [14], contain activation domains but are not able to bind to DNA yet they often contain protein-protein interaction domains. On the other hand, Bcl10 perhaps can bind to upstream regulation sequence of some other genes, which is still a question needed to be solved.

Bcl10 is considered to be commonly involved in the pathogenesis of human malignancy, but the large function exhibited by the Bcl10 protein is not identified completely for it was found only recently. Our hypothesis Bcl10 can act as a transcriptional activator deserves further testing.

Acknowledgements

Supported by National Natural Sciences Foundation of China (No. 3997002).

References

1. Willis TG, Jadayel DM: Bcl10 is involved in t(1:14) (p22;q32) of MALT B cell lymphoma and mutated in multiple tumor types. Cell 96: 35–45, 1999
2. Du MQ, Peng H, Liu H: Bcl10 gene mutation in lymphoma. Blood 15: 3885–3890, 2000
3. Yui D, Yoneda T: Interchangeable binding of Bcl10 to TRAF2 and cIAPs regulates apoptosis signaling. Oncogene 20: 4317–4323, 2001
4. Yoneda T, Imaizumi K, Maeda M, Yui D, Manabe T: Regulatory mechanisms of TRAF2-mediated signal transduction by Bcl10, a MALT lymphoma-associated protein. J Biol Chem 275: 11114–11122, 2000

5. Wang L, Guo Y, Huang WJ: CARD10 is a novel caspase recruitment domain/membrane-associated guanylate kinase family member that interacts with BCL10 and activates NF-κB. J Biol Chem 276: 21405–21409, 2001
6. Clontech Matchmaker Gal4 Two-Hybrid User Manual. 1992, pp 28–35
7. Samblank J, Fuliqi EF: Extraction of plasmids. Mol Clone 24–25, 1998
8. Yang F, de Villiers WJ, McClain CJ: Green tea polyphenols block endotoxin-induced tumor necrosis factor-production and lethality in a murine model. J Nutr 128: 2334–2340, 1998
9. Gaide O, Martinon F, Micheau O, Bonnet D, Thome M, Tschopp JC: A CARD-containing binding partner of Bcl10, induces Bcl10 phosphorylation and NF-κB activation. FEBS Lett 496: 121–127, 2001
10. Wang J, Goodman HM, Zhang H: An Arabidopis 14-3-3 protein can act as a transcriptional activator in yeast. FEBS Lett 443: 282–284, 1999
11. Ptashne M, Gann A: Transcriptional activation by recruitment. Nature 386: 569–577, 1997
12. Rhee Y, Gurel F, Gafni Y, Dingwall C, Citovsky V: A genetic system for detection of protein nuclear import and export. Nature Biotech 18: 433–437, 2002
13. McCarty DR, Hattori T, Carson CB, Vasil V, Lazar M, Vasil IK: The Viviparous-1 developmental gene of maize encodes a novel transcriptional activator. Cell 54: 541–555, 1991
14. Cress WD, Triezenberg SJ: Critical structural elements of the VP16 transcriptional activation domain. Science 251: 87–90, 1991

Molecular and Cellular Biochemistry **246**: 105–110, 2003.

Activation of peroxisome proliferator-activated receptors α and γ1 inhibits human smooth muscle cell proliferation

Peter Zahradka,[1,4] Natalia Yurkova,[1] Brenda Litchie,[1] Michael C. Moon,[1,3,4] Dario F. Del Rizzo[3] and Carla G. Taylor[2]

[1]*Institute of Cardiovascular Sciences, St. Boniface General Hospital Research Centre; Departments of* [2]*Human Nutritional Sciences,* [3]*Surgery,* [4]*Physiology, University of Manitoba, Winnipeg, Canada*

Abstract

Atherosclerotic lesions occur as a result of excess lipid deposition within the vascular tissues. The peroxisome proliferator-activated receptors (PPARs) present in adipose and hepatic tissues have been shown to promote fatty acid oxidation and lipid storage. An immunohistochemical assessment of PPARα and PPARγ revealed both proteins were also present in the medial and intimal layers of human arteries, predominately in regions containing smooth muscle cells. In agreement with this observation, smooth muscle cells isolated from these vessels were found by RT-PCR to express both PPARα and PPARγ1. The functionality of these receptors was tested with selective PPAR agonists. Mitogenic stimulation of smooth muscle cell proliferation was blocked by 15d-PGJ_2, a PPARγ agonist, as well as by WY14643, a PPARα agonist. These data indicate PPAR activation by selective agonists could influence lesion progression directly, as well as indirectly through reductions in serum lipoprotein and triglyceride levels. (Mol Cell Biochem **246**: 105–110, 2003)

Key words: peroxisome proliferator-activated receptor, human smooth muscle, DNA synthesis, WY14643, PGJ_2

Introduction

The peroxisome proliferator-activated receptors (PPARs) belong to the nuclear receptor superfamily of ligand-activated transcription factors that include the steroid, retinoid and thyroid hormone receptors (reviewed in [1, 4, 5]). Three distinct PPARs, designated PPARα, PPARγ and PPARβ/δ, have been described. In addition, PPARγ has been found to consist of two variants, PPARγ1 and PPARγ2. Each PPAR is differentially expressed in many tissues, and location is closely related to function.

PPARs are primarily involved in the regulation of lipid metabolism and homeostasis. They function by inducing gene expression through the PPRE, a promoter element with the consensus sequence AGGTCAGCTGTCA. While few endogenous PPAR ligands have been identified, *in vitro* reporter systems have shown that PPARs can be activated by a diverse group of lipid or lipid-like compounds such as long chain fatty acids, eicosanoids and synthetic peroxisome proliferators [1, 2]. In addition to their structural differences, these agonists are also classified by their ability to selectively activate either PPARα or PPARγ, or to non-selectively activate both. In this context, it has been proposed that omega-3 fatty acids influence fat storage by activating PPARs [3], and the established roles of PPARγ in adipocyte differentiation and PPARα in the regulation of fatty acid oxidation [4] lend support for this premise. Consequently, PPAR agonists may protect against atherogenesis.

In vascular tissues, PPARα has been detected in both endothelial (EC) and smooth muscle cells (SMC), and has been found in infiltrating monocytes/macrophages [5–7]. In the presence of PPARα agonists, there is a marked reduction in the production of inflammatory mediators, specifically IL-6 by SMCs and tissue factor by macrophages [5, 8]. PPARγ is

Address for offprints: P. Zahradka, Institute of Cardiovascular Sciences, St. Boniface Research Centre, 351 Tache Avenue, Winnipeg, MB, Canada R2H 2A6 (E-mail: peterz@sbrc.ca)

also expressed in ECs and SMCs [9, 10]. In contrast, PPARγ expression in monocytes is low, but is induced during differentiation into macrophages [7]. PPARγ agonists have been shown to reduce matrix metalloproteinase expression, particularly MMP-9, in both macrophages and SMCs, and decrease chemokine secretion by ECs [8, 11]. Furthermore, the PPARγ agonists 15-deoxy-$\Delta^{12,14}$-prostaglandin J_2 (15d-PGJ_2) and troglitazone have both been shown to inhibit SMC migration and proliferation [12, 13]. Based on this evidence, PPAR agonists likely have a more direct effect on the atherosclerosis than was earlier thought.

In a recent study, we found that SMCs isolated from vessels exhibiting significant atherosclerotic lesions were unable to enter a quiescent state [14]. While this property was correlated with elevated levels of *c-fos* expression, other factors are also expected to contribute. This investigation was therefore initiated to determine whether changes in PPAR expression or response to ligands was altered. We observed that PPARα and PPARγ were expressed in the smooth muscle cells of both arterial and venous conduits. As well, we show for the first time that agonists of PPARα, in addition to PPARγ, are capable of inhibiting the proliferation of human SMCs.

Materials and methods

Cell culture

Primary cultures of human SMCs were prepared by migration from free-floating explants of human internal mammary artery, radial artery and saphenous vein segments as described in Del Rizzo *et al.* [14]. The vessel segments were obtained from excess bypass graft conduits in accordance with ethical guidelines and a protocol approved by the University of Manitoba. Cells were propagated in Dulbecco's MEM (DMEM; Gibco-BRL) containing 20% FBS (Gibco-BRL). When 75% confluent, the growth medium was replaced with DMEM supplemented with 5 μg/mL transferrin, 1 nmol/L selenium, 20 mmol/L ascorbate and 10 nmol/L insulin for 96 h. Quiescent cells were used for all growth assays.

Immunohistochemistry

Artery segments were flash frozen with a dry ice/ethanol bath in O.C.T. and stored at −80°C. Sections of 7 μm were prepared with a cryostat, placed onto glass slides and immersed in Streck tissue fixative for 10 min. After extensive washing with PBS, the sections were incubated sequentially with blocking buffer, primary antibody to PPAR (polyclonal antibodies for both PPARα and PPARγ were from Santa Cruz Biotechnology, Inc.) and, finally, with Cy3-coupled secondary antibody (diluted 1:1000). Images visualized by epifluorescence microscopy (Olympus BH-2) were digitally captured with a DAGE-MTI CCD camera.

DNA Synthesis

Triplicate sets of quiescent cells, prepared in 24-well dishes, were treated with mitogen ± PPAR agonist in the presence of 1 μCi [^{3}H]-thymidine (Perkin Elmer), which was added to the cultures 24 h after stimulation, for 72 h. Incorporation of radiolabel into DNA was monitored by trichloroacetic acid precipitation according to Zahradka *et al.* [15].

Reverse transcription-polymerase chain reaction amplification

Total RNA was isolated from cells in 6-well culture dishes using TRIzol. The RNA was resuspended in RNase-free water and concentration determined by spectrophotometic absorbance at 260-nm. Possible genomic DNA contamination of the samples was removed with a 10 min DNase I (Gibco-BRL) incubation. Reverse transcription of 1 μg of RNA was conducted according to the protocol (25 cycles of amplification and 62°C annealing temperature) recommended for the Access RT-PCR System (Promega). The specific forward and reverse oligodeoxynucleotide primers employed were: GAPDH (s) 5′-CGCTGTGAACGGATTTGGCCGTAT-3′, GAPDH (as) 5′-AGCCTTCTCCATGGTGGTGAAGAC-3′; PPARα (s) 5′-AGATTTCGCAATCCATCGGC-3′, PPARα (as) 5′-GCGTGG-ACTCCGTAATGATA-3′, PPARγ1 (s) 5′-CTCGAGGACACCGGA GAG-3′, PPARγ1 (as) 5′-GTCATTTCTGCGGCCACG-3′; PPARγ2 (s) 5′-GGTGAAACTCTGGGAGATTCT-3′, PPARγ2 (as) 5′-TGT-AATCTGCAACCACTGGAT-3′ [16–18]. Amplification products were analyzed by electrophoresis in 2% agarose gels and visualized with ethidium bromide. Band intensity captured on Polaroid 667 black and white instant film was quantified by scanning densitometry. Control reactions (minus RNA, minus RT and minus primers) were used to demonstrate the specificity of the PCR reaction.

Data measurement and statistical analysis

Radiotracer assay data were quantified and plotted as means ± S.E.M. of individual experiments (n = 3). Student's *t*-test was used to compare treatment means vs. controls. Statistical significance was set at $p < 0.05$.

Results

Distribution of PPARs in human vessels

Human arterial (radial, internal mammary) and venous (saphenous) segments were sectioned and examined by indirect immunofluorescence staining employing PPAR-specific primary antibodies. Both PPARα and PPARγ were detected in these vessels (Fig. 1). Staining was strongest in the media, implicating SMCs as the source. Staining was also evident in regions exhibiting an intima. No PPAR staining was present in the adventitia. Finally, the cytosolic distribution of PPAR staining indicates these receptors were not activated. These data suggest both PPAR isoforms are distributed throughout the vessels, and that there is a strong correlation with SMCs.

PPARs in human SMCs

SMCs prepared from explants of human vessels have been shown to retain the characteristics exhibited *in situ* [14]. It is therefore assumed that our results are representative of cells present in the vessel. Total mRNA was isolated from the cells and subjected to reverse transcriptase polymerase chain reaction (RT-PCR) amplification. Primers were designed to selectively amplify PPARα, PPARγ1 and PPARγ2. It was observed that both PPARα and PPARγ1 were expressed by SMCs from each of the three human conduits employed in this study (Fig. 2). In contrast, PPARγ2 was not detected. These results confirm that both PPARα and PPARγ1 are present in the SMCs of human arteries and veins.

Effect of PPAR agonists on SMC proliferation

The presence of PPARα and PPARγ1 in human SMCs suggest these cells should be responsive to PPAR agonists. Since PPARγ agonists have been shown to restrict SMC proliferation, we tested the ability of 15d-PGJ_2 to block DNA synthesis by human saphenous vein and internal mammary artery SMCs. This agent was highly effective, exhibiting an EC_{50} of 3.5 μM (Figs 3A and 3B). Interestingly, stimulation of DNA synthesis (1.4–2 fold) was observed with 1 μM 15d-PGJ_2 in all replicate experiments. Additionally, WY14643, a PPARα agonist, was also effective in blocking DNA synthesis (Figs 3C and 3D), displaying an EC_{50} of 90–140 μM depending on the source of the cells. This is the first indication that activation of PPARα is coupled to cell proliferation and may represent a potentially novel application of PPARα agonists.

Discussion

The results obtained over the course of this study have provided several lines of evidence for PPARS as important elements of SMCs that contribute to vascular function. First, we have shown by immunological and PCR amplification methods that PPARα and PPARγ are expressed by human SMCs

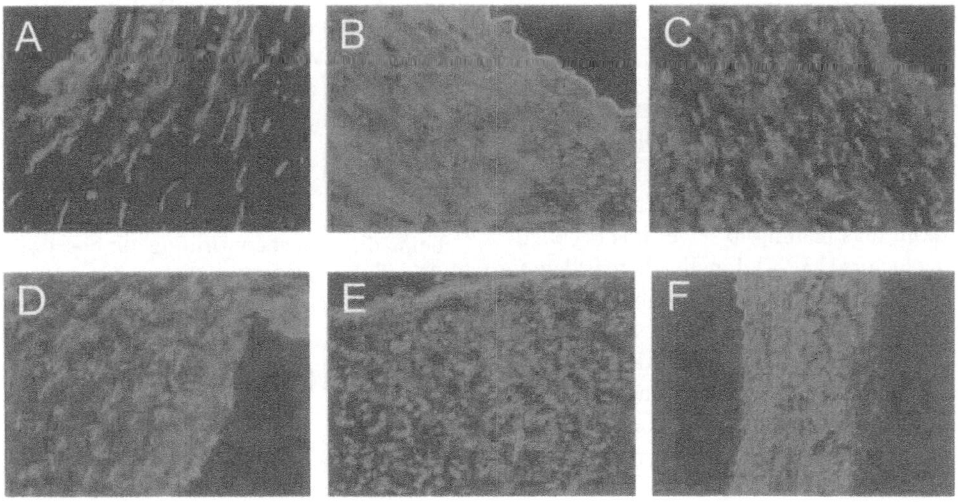

Fig. 1. Expression of PPARα and PPARγ in human vascular tissues. Segments of saphenous vein (A, D), radial artery (B, E) and internal mammary artery (C, F) were frozen and sectioned as described in 'Materials and methods'. Individual sections were incubated with antibody (Santa Cruz Biotechnology, 1:200 dilution) to either PPARα (A–C) or PPARγ (D–F), and after processing with Cy3-conjugated secondary antibody, the images were digitally captured.

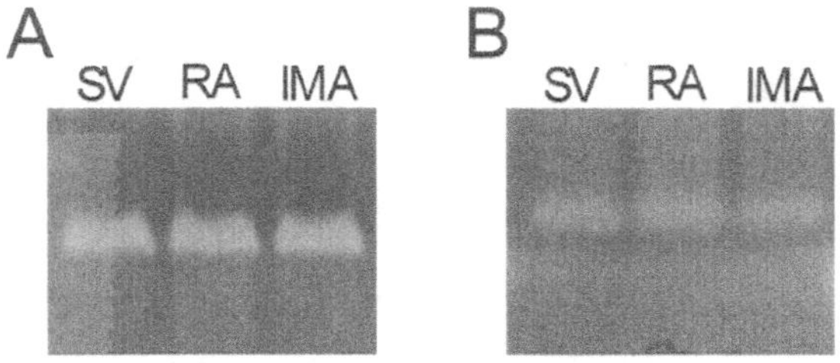

Fig. 2. Human smooth muscle cells express both PPARα and PPARγ1 mRNA, but not PPARγ2 mRNA. Total cell RNA was extracted with TRIzol and 1 μg was amplified by reverse transcriptase-polymerase chain reaction as described in 'Materials and methods'. Amplification products were stained with ethidium bromide staining after agarose gel electrophoresis, and photographed. The figure illustrates representative data for PPARα (panel A) and PPARγ1 (panel B). SMCs were derived from saphenous vein (SV), radial artery (RA) and internal mammary artery (IMA) as described in 'Materials and methods'.

of arterial and venous conduits, thus extending the earlier observations of Law *et al.* [19]. Second, RT-PCR revealed that PPARγ1 is the isoform present in these cells. Third, we have demonstrated a novel connection between SMC proliferation and activation of PPARα. The latter finding may be particularly significant with respect to how PPARα activation influences the formation of atherosclerotic lesions.

While the functional significance of PPARs to vascular tissue still remains to be clarified, PPARα and PPARγ are known to influence serum lipid levels through their roles in fatty acid oxidation and in adipogenesis, respectively [4]. For this reason, fibrates (PPARα agonists) and glitazones (PPARγ agonists) indirectly correlate with a reduction or inhibition of atherogenesis [20, 21]. Nevertheless, there is also strong evidence to suggest that activation of smooth muscle PPARγ can repress the SMC properties that promote the development of atherosclerotic lesions. The most direct effect of PPARγ agonists, including both glitazones and 15d-PGJ_2 is the inhibition of SMC proliferation and migration [12, 13, 17], key events in atherogenesis. These effects presumably contribute to the positive action of troglitazone in preventing in-stent restenosis [22]. Additionally, a reduction in chemokine gene expression resulting from 15d-PGJ_2 treatment decreases monocyte infiltration, thus leading to fewer macrophage-associated paracrine factors [23, 24]. Finally, there is data to suggest that 15d-PGJ_2 promotes the differentiation of SMCs [12] and in this way prevents the phenotypic modulation that leads to foam cell formation. Alternatively, PPARγ activation may decrease foam cell formation by suppressing expression of LOX-1, the oxidized LDL receptor [25]. However, it is also worth noting that both 15d-PGJ_2 and troglitazone have been reported to inhibit cell proliferation via PPARγ-independent mechanisms [26, 27].

Unlike PPARγ, a direct effect of PPARα on cell proliferation in vascular tissues has not been identified. Nevertheless, it has been shown that PPARα activators such as the fibrates promote the regression of atheromas by reducing their lipid content [28]. Also, the PPARα agonist WY14643 has been shown to induce expression of the ABCA1 gene which mediates cholesterol efflux from foam cells [29]. However, while activation of PPARα has been shown to decrease the release of inflammatory cytokines [5], possibly by antagonizing NF-κB [5, 30], there is no mechanism currently identified by which WY14643 can interfere with SMC proliferation. At best, activation of PPARα may induce apoptosis of SMCs [31].

In this investigation, a direct link between PPARα activation and SMC proliferation was observed (Fig. 3B). This finding provides novel insight into a new mechanism by which PPARα may influence vascular function. Since we found no evidence of apoptosis, how can these data be explained, especially in relation to the report of Miwa *et al.* [12] which stated that WY14643 does not inhibit the proliferation of SMCs from human umbilical arteries? First, Miwa *et al.* [12] did not employ concentrations as high as those used in this study. They report that 100 μM did not inhibit DNA synthesis in response to serum stimulation, although there was a non-significant decline of 15% in thymidine incorporation. In contrast, this study demonstrates that the EC_{50} for WY14643 is approximately 120 μM, suggesting that the WY14643 concentration employed by Miwa *et al.* [12] was insufficient to produce an inhibitory effect. In a separate study, it was shown that WY14643 suppresses expression of cdk1 and $p21^{cip1}$ in hepatic tissue, which results in a marked decline in cellular PCNA content [32]. Independently, we have shown that NF-κB promotes SMC proliferation by repressing $p21^{cip1}$ expression [33]. Since a relationship between PPARα and NF-κB exists, and NF-κB modulates $p21^{cip1}$ expression, a mechanism by which PPARα could directly control SMC proliferation may be postulated.

The relationship between PPARs and SMCs suggests PPAR activators may influence SMC proliferation through multiple mechanisms. At present, we propose that PPARα agonists, by inhibiting SMC proliferation, may have greater potency in preventing atherosclerosis than therapies solely restricted to lowering serum lipid levels. Future studies are being directed at confirming our observations and extending this work to identify the mechanism of action for WY14643.

Acknowledgements

This investigation was funded by grants to the Canadian Institutes of Health Research Group in Experimental Cardiology (to P.Z.) and the Natural Sciences and Engineering Research Council (to C.G.T.). M.M. was supported by a fellowship from the Manitoba Health Research Council.

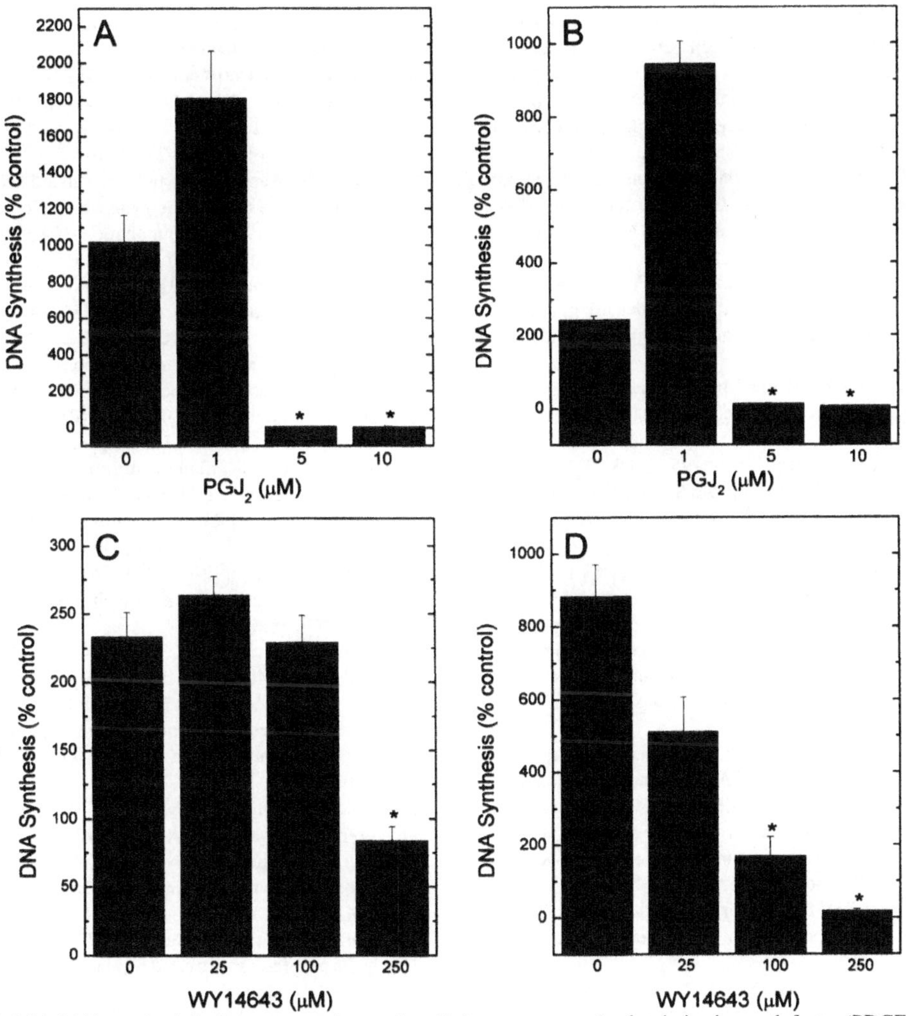

Fig. 3. PPAR agonists inhibit DNA synthesis by human smooth muscle cells in response to platelet-derived growth factor (PDGF) stimulation. Quiescent SMCs were prepared by explant culture from human saphenous vein (A, C) and human internal mammary artery (B, D) as described in 'Materials and methods'. The cells were subsequently treated with PDGF (0.1 μg/ml) in the presence of PGJ_2 (A, B) or WY14643 (C, D). DNA synthesis was monitored by incorporation of $[^3H]$thymidine as described in 'Materials and methods'. The data are presented as means ± S.E. (n = 3). Statistically significant differences from control ($p < 0.05$) are indicated (*).

References

1. Torra IP, Chinetti G, Duval C, Fruchart JC, Staels B: Peroxisome proliferator-activated receptors: From transcriptional control to clinical practice. Curr Opin Lipidol 12: 245–254, 2001
2. Qi C, Zhu Y, Reddy JK: Peroxisome proliferator-activated receptors, coactivators, and downstream targets. Cell Biochem Biophys 32:187–204, 2000
3. Clarke SD: Polyunsaturated fatty acid regulation of gene transcription: A mechanism to improve energy balance and insulin resistance. Br J Nutr 83(suppl 1): S59–S66, 2000
4. Schoonjans K, Staels B, Auwerx J: The peroxisome proliferator activated receptors (PPARs) and their effects on lipid metabolism and adipocyte differentiation. Biochim Biophys Acta 1302: 93–109, 1996
5. Staels B, Koenig W, Habib A, Merval R, Lebret M, Torra IP, Delerive P, Fadel A, Chinetti G, Fruchart JC, Najib J, Maclouf J, Tedgui A: Activation of human aortic smooth-muscle cells is inhibited by PPARalpha but not by PPARgamma activators. Nature 393: 790–793, 1998
6. Inoue I, Shino K, Noji S, Awata T, Katayama S: Expression of peroxisome proliferator-activated receptor alpha (PPAR alpha) in primary cultures of human vascular endothelial cells. Biochem Biophys Res Commun 246: 370–374, 1998
7. Chinetti G, Griglio S, Antonucci M, Torra IP, Delerive P, Majd Z, Fruchart JC, Chapman J, Najib J, Staels B: Activation of proliferator-activated receptors alpha and gamma induces apoptosis of human monocyte-derived macrophages. J Biol Chem 273: 25573–25580, 1998
8. Marx N, Mackman N, Schonbeck U, Yilmaz N, Hombach VV, Libby P, Plutzky J: PPARalpha activators inhibit tissue factor expression and activity in human monocytes. Circulation 103: 213–219, 2001
9. Marx N, Bourcier T, Sukhova GK, Libby P, Plutzky J: PPARgamma activation in human endothelial cells increases plasminogen activator inhibitor type-1 expression: PPARgamma as a potential mediator in vascular disease. Arterioscler Thromb Vasc Biol 19: 546–551, 1999

10. Marx N, Sukhova GK, Collins T, Libby P, Plutzky J: PPARalpha activators inhibit cytokine-induced vascular cell adhesion molecule-1 expression in human endothelial cells. Circulation 99: 3125–3131, 1999
11. Chinetti G, Fruchart JC, Staels B: Peroxisome proliferator-activated receptors (PPARs): Nuclear receptors with functions in the vascular wall. Z Kardiol 90(suppl 3): 125–132, 2001
12. Miwa Y, Sasaguri T, Inoue H, Taba Y, Ishida A, Abumiya T: 15-Deoxy-delta(12,14)-prostaglandin J(2) induces G(1) arrest and differentiation marker expression in vascular smooth muscle cells. Mol Pharmacol 58: 837–844, 2000
13. Gouni-Berthold I, Berthold HK, Weber AA, Seul C, Vetter H, Sachinidis A: Troglitazone and rosiglitazone inhibit the low density lipoprotein-induced vascular smooth muscle cell growth. Exp Clin Endocrinol Diabetes 109: 203–209, 2001
14. Del Rizzo DF, Yurkova N, Moon MC, Litchie B, Zahradka P: Platelet-derived growth factor-induced expression of c-fos in human vascular smooth muscle cells: Implications for long-term graft patency. Ann Thorac Surg 74: 90–95, 2002
15. Zahradka P, Elliot T, Hovland K, Larson DE, Saward L: Repression of histone gene transcription in quiescent 3T6 fibroblasts. Eur J Biochem 217: 683–690, 1993
16. Zahradka P, Harris KD, Triggs-Raine B, Lamontagne G, Leblanc N: PCR-based analysis of voltage-gated K+ channels in vascular smooth muscle. Mol Cell Biochem 145: 39–44, 1995
17. Marx N, Schonbeck U, Lazar MA, Libby P, Plutzky J: Peroxisome proliferator-activated receptor gamma activators inhibit gene expression and migration in human vascular smooth muscle cells. Circ Res 83: 1097–1103, 1998
18. Benson S, Wu J, Padmanabhan S, Kurtz TW, Pershadsingh HA: Peroxisome proliferator-activated receptor (PPAR)-gamma expression in human vascular smooth muscle cells: Inhibition of growth, migration, and c-fos expression by the peroxisome proliferator-activated receptor (PPAR)-gamma activator troglitazone. Am J Hypertens 13: 74–82, 2000
19. Law RE, Goetze S, Xi XP, Jackson S, Kawano Y, Demer L, Fishbein MC, Meehan WP, Hsueh WA: Expression and function of PPARgamma in rat and human vascular smooth muscle cells. Circulation 101: 1311–1318, 2000
20. Rubins HB, Robins SJ, Collins D, Fye CL, Anderson JW, Elam MB, Faas FH, Linares E, Schaefer EJ, Schectman G, Wilt TJ, Wittes J: Gemfibrozil for the secondary prevention of coronary heart disease in men with low levels of high-density lipoprotein cholesterol. Veterans Affairs High-Density Lipoprotein Cholesterol Intervention Trial Study Group. N Engl J Med 341: 410–418, 1999
21. Li AC, Brown KK, Silvestre MJ, Willson TM, Palinski W, Glass CK: Peroxisome proliferator-activated receptor gamma ligands inhibit development of atherosclerosis in LDL receptor-deficient mice. J Clin Invest 106: 523–531, 2000
22. Takagi T, Akasaka T, Yamamuro A, Honda Y, Hozumi T, Morioka S, Yoshida K: Troglitazone reduces neointimal tissue proliferation after coronary stent implantation in patients with non-insulin dependent diabetes mellitus: A serial intravascular ultrasound study. J Am Coll Cardiol 36: 1529–1535, 2000
23. Elangbam CS, Tyler RD, Lightfoot RM: Peroxisome proliferator-activated receptors in atherosclerosis and inflammation – an update. Toxicol Pathol 29: 224–231, 2001
24. Zhang X, Wang JM, Gong WH, Mukaida N, Young HA: Differential regulation of chemokine gene expression by 15-deoxy-delta 12,14 prostaglandin J2. J Immunol 166: 7104–7111, 2001
25. Chiba Y, Ogita T, Ando K, Fujita T: PPARgamma ligands inhibit TNF-alpha-induced LOX-1 expression in cultured endothelial cells. Biochem Biophys Res Commun 286: 541–546, 2001
26. Palakurthi SS, Aktas H, Grubissich LM, Mortensen RM, Halperin JA: Anticancer effects of thiazolidinediones are independent of peroxisome proliferator-activated receptor gamma and mediated by inhibition of translation initiation. Cancer Res 61: 6213–6218, 2001
27. Maggi LB Jr, Sadeghi H, Weigand C, Scarim AL, Heitmeier MR, Corbett JA: Anti-inflammatory actions of 15-deoxy-delta 12,14-prostaglandin J2 and troglitazone: Evidence for heat shock-dependent and -independent inhibition of cytokine-induced inducible nitric oxide synthase expression. Diabetes 49: 346–355, 2000
28. Betz E, Quack G: Effect of etofibrate on the development and the regression of atheromas in a rabbit model of atherosclerosis. Vasa 19: 157–160, 1990
29. Chinetti G, Lestavel S, Bocher V, Remaley AT, Neve B, Torra IP, Teissier E, Minnich A, Jaye M, Duverger N, Brewer HB, Fruchart JC, Clavey V, Staels B: PPAR-alpha and PPAR-gamma activators induce cholesterol removal from human macrophage foam cells through stimulation of the ABCA1 pathway. Nat Med 7: 53–58, 2001
30. Jones DC, Ding X, Daynes RA: Nuclear receptor peroxisome proliferator-activated receptor alpha (PPARalpha) is expressed in resting murine lymphocytes. The PPARalpha in T and B lymphocytes is both transactivation and transrepression competent. J Biol Chem 277: 6838–6845, 2002
31. Diep QN, Touyz RM, Schiffrin EL: Docosahexaenoic acid, a peroxisome proliferator-activated receptor-alpha ligand, induces apoptosis in vascular smooth muscle cells by stimulation of p38 mitogen-activated protein kinase. Hypertension 36: 851–855, 2000
32. Ma X, Stoffregen DA, Wheelock GD, Rininger JA, Babish JG: Discordant hepatic expression of the cell division control enzyme p34cdc2 kinase, proliferating cell nuclear antigen, p53 tumor suppressor protein, and p21Waf1 cyclin-dependent kinase inhibitory protein after WY14,643 ([4-chloro-6-(2,3-xylidino)-2-pyrimidinylthio]acetic acid) dosing to rats. Mol Pharmacol 51: 69–78, 1997
33. Zahradka P, Werner JP, Buhay S, Litchie B, Thomas S: NF-κB activation is essential for angiotensin II-dependent proliferation and migration of vascular smooth muscle cells. J Mol Cell Cardiol 34: 1609–1621, 2002

Molecular and Cellular Biochemistry **246:** 111–116, 2003.

High levels of CUG-initiated FGF-2 expression cause chromatin compaction, decreased cardiomyocyte mitosis, and cell death

Cheryl J.A. Hirst,[1,2] Meenhard Herlyn,[4] Peter A. Cattini[1] and Elissavet Kardami[1,2,3]

Departments of [1]Physiology; [3]Human Anatomy and Cell Science; [2]Institute of Cardiovascular Sciences, St. Boniface General Hospital Research Centre, University of Manitoba, Winnipeg, Manitoba, Canada; [4]The Wistar Institute, Philadelphia, PA, USA

Abstract

Fibroblast growth factor 2 (FGF-2) is a multifunctional mitogen present in CUG-and AUG-initiated forms, referred to as 'hi' and 'lo' FGF-2, respectively. We have used an adenoviral vector to express the predominantly nuclear human 'hi' FGF-2 and examined the relationship between expression levels, mitotic entry, cell number and chromatin compaction of cardiac myocytes, over 1–3 days in culture. At a multiplicity of infection (m.o.i.) of 50, levels of 'hi' FGF-2 (assessed by Western blotting) and mitotic index (fraction of myocyte nuclei staining positive for phosphorylated histone H3) paralleled each other, becoming maximal at 2 days. At 200 m.o.i., maximal expression of 'hi' FGF-2 (approximately double that at 50 m.o.i) was achieved at 2 days and coincided with decreased mitotic index and increased chromatin compaction. At 3 days compaction was maximal, mitotic index was minimal, and cell numbers decreased, accompanied by the appearance of DNA laddering, an indicator of apoptosis. Overall, the lower dose of 'hi' FGF-2 and early time points favored a proliferative phenotype while the higher dose, and later time points, promoted chromatin compaction, inhibition of proliferation and cell death. (Mol Cell Biochem **246**: 111–116, 2003)

Key words: CUG FGF-2, cardiomyocytes, chromatin compaction, mitosis, apoptosis

Introduction

FGF-2 is a member of a large family of heparin-binding proteins [1]. It exists as different molecular weight isoforms, originating from alternative initiation codons [2]. CUG-initiated forms produce proteins of 22–24 and 34 kDa ('hi' FGF-2); they are localized predominantly to the nucleus [2]. The AUG-initiated 18 kDa FGF-2 ('lo' FGF-2) is found in the nucleus, cytosol and in association with the cell surface [2]. 'The N-terminal extension of 'hi' FGF-2 contains nuclear localization signal-like sequences that target it to the nucleus [3].

Most studies have explored the role of the 18 kDa FGF-2. Nevertheless, isoform- and cell type-specific functions have been described [4]. For example, both 'lo' and 'hi' isoforms are reported to promote cell proliferation, however, only 'lo' FGF-2 stimulates cell migration, while 'hi' FGF-2 is reported as either inhibitory [5] or having no effect on migration [1]. In addition, the growth stimulatory effects of transfected 'lo' (but not 'hi') FGF-2 on NIH 3T3 fibroblasts are mediated by the tyrosine kinase receptor [6]. Thus, it has been suggested that 'lo' FGF-2 operates in an autocrine/paracrine manner, while 'hi' FGF-2 exerts its effects via an intracrine pathway [4]. Preferential expression of 'hi' FGF-2 is brought about by stress or cell transformation [7].

In the heart, 'hi' or 'lo' FGF-2 predominate in immature or adult myocardium, respectively, and hypothyroidism promotes accumulation of 'hi' FGF-2 [8]. 'Hi' FGF-2 levels are transiently elevated following cardiac injury [9]. Previously,

Address for offprints: E. Kardami, Institute of Cardiovascular Sciences, St. Boniface General Hospital Research Centre, 351 Tache Avenue, Winnipeg, Manitoba, Canada, R2H 2A6 (E-mail: ekardami@sbrc.ca)

using calcium phosphate-mediated transient gene transfer, we showed that overexpression of 'hi' or 'lo' FGF-2, stimulated myocyte proliferation via an autocrine/paracrine route [10, 11]. In addition, 'hi' FGF-2 caused an unusual form of nuclear disruption; characterized by formation of multiple chromatin 'clumps' surrounded by intact nuclear envelope [10–12]. We have now used adenoviral vectors to express human 'hi' FGF-2 and examine the relationship between 'hi' FGF-2 expression levels, cell proliferation and development of chromatin compaction in cardiac myocytes maintained 1–3 days in culture.

Materials and methods

Cell culture

Primary cultures of rat ventricular myocytes were obtained using one-day-old Sprague-Dawley rat pups, as described [11]. Myocytes were plated at 0.7×10^6/35 mm dish on collagen-coated plates or coverslips in the presence of growth medium consisting of 10% fetal bovine serum (FBS; HyClone), in Dulbeco's modified Eagle's medium (DMEM; Gibco). After 1 day in culture cells were transfected with the adenoviral vectors in DMEM medium. Cells were analyzed 24–72 h after transfection. Less than 5% non-muscle cells (α-actinin-negative) were present in these cultures at any time point (data not shown).

Adenoviral vectors and gene transfer

Construction of adenoviral vectors expressing CUG- (Ad.'hi' FGF-2) and/or AUG-('lo' FGF-2) initiated species of FGF-2 has been described previously [13]. Cardiac myocytes were transfected with adenoviral vectors expressing mainly 'hi' FGF-2 or control virus at m.o.i. (multiplicity of infection) of 50 or 200. Transfection efficiency was assessed by immunostaining for FGF-2, using high dilutions of an anti-FGF-2 serum (1:5000 [14]); over 90% of cells in 'hi' FGF-2 transfected cultures presented strong nuclear anti-'hi'-FGF-2 staining, compared to 0% in vector-transfected cultures (data not shown).

DNA analysis

DNA was extracted from myocyte cultures and analysed in 1.5% agarose gels as described [12]. To obtain positive controls for DNA ladder formation, myocytes were treated with staurosporine (Sigma), a protein kinase C inhibitor that has been shown to cause apoptosis [15], as described [12].

Immunofluorescence

A previously characterized site-specific antibody highly specific for the phosphorylated form of histone H3 (ser10) was used as described [15, 16]. Myocytes were processed for immunofluorescence as described previously [12]. All coverslips were counter-stained with Hoechst 33342 and mounted on slides. A Nikon Diaphot, as well as a Zeiss Axiovert 200 epifluorescence microscopes, equipped with appropriate filters, were used to view the coverslips and take photographs.

Western blotting

Myocytes were lysed in 1% SDS buffer containing 50 mmol/L Tris-HCl (ph 7.4), as well as protease and phosphatase inhibitors. Lysates were boiled for 5 min, sonicated and centrifuged for 10 min in a microcentrifuge. Nuclei were obtained using the Nuclei-EZ kit (Sigma) as per manufacturer's instructions. The BCA assay (Pierce) was used to determine protein concentration of the supernatants. Twenty μg of protein were loaded onto 12% SDS-polyacrylamide gels. Electrophoretic transfer and western blotting was done as described [12]. After rinsing, membranes were incubated with anti-rabbit secondary antibody linked to horseradish peroxidase. Antigen antibody complexes were visualized by enhanced chemiluminescence (SuperSignal Chemiluminescent Substrate; Pierce). The bands were quantitated using a Bio-Rad Imaging densitometer.

Determination of mitotic index and cell number

Phosphorylated histone H3 (phospho H3) is used as a marker associated with mitotic chromatin condensation [12,16]. It was used in our experiments to determine the mitotic index (MI = fraction of cells in mitosis, or about to enter mitosis) of myocytes transfected with the 'hi' FGF-2 adenovirus. Phospho H3-positive nuclei as well as total number of nuclei were counted in each field. A total of 32 fields (in 4 coverslips per treatment) were examined. An estimate for relative cell numbers was obtained by averaging number of nuclei per field.

Differences between vector-transfected or Ad.'hi' FGF-2-transfected cultures, or between Ad.'hi' FGF-2-transfected cultures at different time points and dosages were examined for statistical significance by analysis of variance, and the Tukey's *post hoc* test, using the InStat program, where $p < 0.05$ was considered significant.

Results

Cardiomyocytes transfected with adenoviral vectors expressing either exclusively 'lo' (18 kDa) or predominantly 'hi' (22–

24 kDa) FGF-2 were analyzed by Western blotting: As shown in Fig. 1, Ad.'lo' FGF-2 produced an 18 kDa immunoreactive band. Ad.'hi' FGF-2 produced bands at 22–24 kDa (corresponding to CUG-initiated 'hi' FGF-2) as well as the 18 kDa band corresponding to the AUG-initiated 'lo' FGF-2, as described previously for expression in melanocytes [13]. Nuclear protein (5 μg/lane) obtained from Ad.'hi' FGF-2 (but not vector-only) transfected cultures contained exclusively the 22–24 kDa bands of human 'hi' FGF-2, as expected.

To assess the effects of viral vector dosage and time in culture on 'hi' FGF-2 expression, myocytes were transfected with two different doses (m.o.i. 50 and m.o.i. 200) of 'hi' FGF-2 carrying adenovirus and examined 24, 48 and 72 h later for relative abundance of the 22–24 kDa FGF-2 bands. Parallel cultures were transfected with 50 and 200 m.o.i. of vector alone, to control for non-specific effects of the virus.

At 50 m.o.i., relative levels of 'hi' FGF-2 were maximal at 48 h, declining slightly at 72 h (Fig. 2A). At 200 m.o.i., relative levels of 'hi' FGF-2 were higher than those at 50 m.o.i., at all time points, as expected (Fig. 2A); they reached a maximum at 48 h, decreasing at 72 h.

To obtain an estimate of the fraction of myocytes in mitosis (i.e. mitotic index, MI), we counted numbers of nuclei staining positive for phospho-H3, as well as total number of nuclei in several randomly selected fields; both Ad.'hi' FGF-2 as well as vector-only infected cultures were assessed. To compensate for any effects of the viral infection per se, we obtained a 'normalized' index, by dividing the MI from Ad.'hi' FGF-2 infected cultures at a specific m.o.i. and time point with the corresponding MI values from vector -infected cultures. Normalized index values therefore represent the changes in MI due to expression of 'hi' FGF-2, irrespectively of effects of the vector and are shown in Fig. 2B. 'Hi' FGF-2, at the 50 m.o.i. dose, induced maximal increase in mitosis at 48 h (MI = 0.047 ± 0.04, S.E.M.), significantly increased from either its earlier time point (24 h) or the corresponding vector (48 h; MI = 0.032 ± 0.003). This increase in mitosis at 48 h coincides with increased levels of 'hi' FGF-2 (Fig. 2A).

At 200 m.o.i. and 24–48 h, mitotic fraction is not significantly different than corresponding vector values. At 72 h, however, normalized index illustrates that the mitotic fraction (MI = 0.022

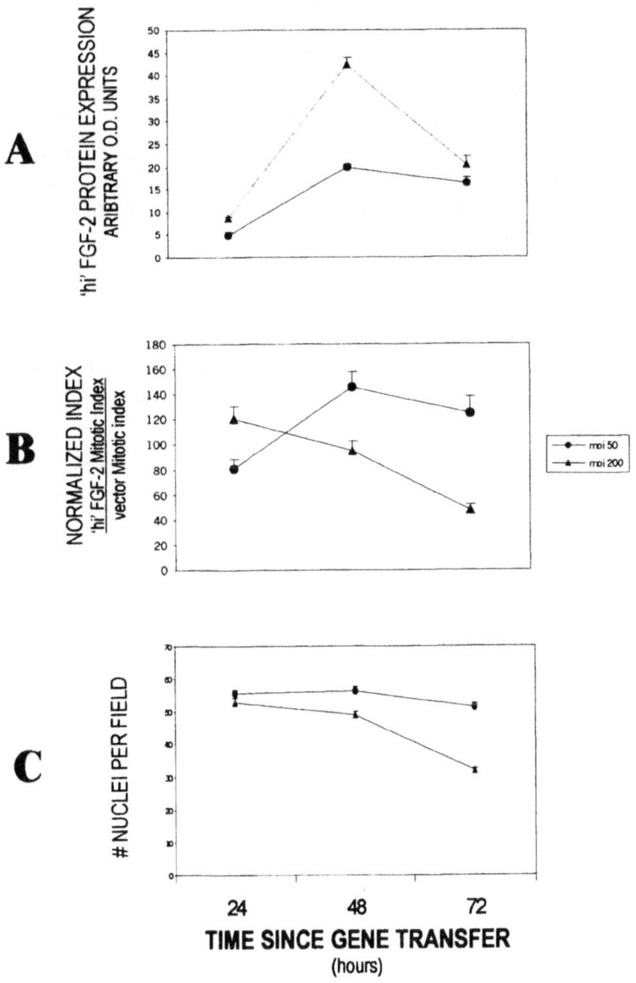

Fig. 2. (A) Relative levels of 'hi' FGF-2 (in arbitrary densitometry O.D. units) depicted as a function of dosage (m.o.i.) and time in culture, as indicated. The data are depicted as the mean ± S.E.M., n = 3. (B) Normalized index (MI in Ad.'hi' FGF-2 transfected cultures/MI in vector only transfected cultures) as a function of dosage (m.o.i.) and time in culture, as indicated. MI is the fraction of nuclei expressing phospho- H3. The data are depicted as the mean ± S.E.M., n = 4. (C) Average number of nuclei counted per random visual field (mean ± S.E.M., n = 32) as a function of dosage (m.o.i.) and time in culture, as indicated (n = 32).

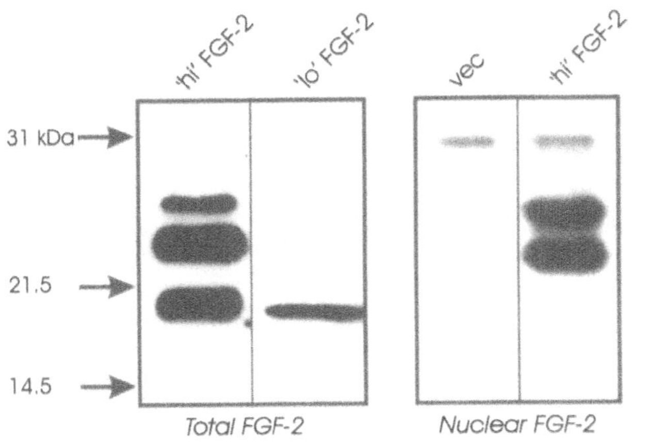

Fig. 1. Expression of human 'hi' or 'lo' FGF-2 after adenoviral transfection of cardiomyocytes, detected by Western blotting, as indicated. Total FGF-2 represents immunoreactive FGF-2 present in total myocyte culture lysates (20 μg/lane). Nuclear FGF-2 represents immunoreactive FGF-2 detectable in nuclear protein obtained from myocyte cultures (5 μg/lane) transfected with vector or 'hi' FGF-2, as indicated. The ~ 30 kDa band present in both nuclear FGF-2 lanes is likely an endogenous FGF-2-like protein.

± 0.002) is significantly reduced compared to either corresponding vector values (MI = 0.045 ± 0.004), previous time points at same m.o.i, or any of the values at 50 m.o.i.

To obtain an estimate of cell numbers at the various time points, we compared average numbers of total nuclei counted per field (n = 32). As seen in Fig. 2C, at 50 m.o.i., nuclear numbers remained unchanged at 24–48 h, decreasing slightly at 72 h (Fig. 2C). At 200 m.o.i., nuclear numbers decreased significantly (Fig. 2C) compared to earlier time points at the same m.o.i., or to any values of the 50 m.o.i groups.

We also determined the fraction of myocyte nuclei presenting the 'compacted chromatin' phenotype. This is characterized by the appearance of multiple condensed 'clumps' of chromatin within the nucleus (Fig. 3A). At 24 h, neither 'hi' FGF-2 dosage was associated with the appearance of nuclear disruption. Nuclear compaction was detectable at 48 h at the 200 m.o.i (but not the 50 m.o.i) dose, and reached high levels (nearly 50 % of total nuclei) at 72 h and at 200 m.o.i. At this time point, even the 50 m.o.i dose elicited a degree of chromatin compaction (17%); this was, nevertheless, significantly smaller compared to the higher dose at the same time point ($p < 0.01$) (Fig. 3B).

We examined the distribution patterns of clumped chromatin and 'hi' FGF-2 in the nucleus, under high magnification: as shown in Fig. 3C, 'hi' FGF-2 shows a homogeneous pattern of localization, while chromatin has formed distinct 'clumps'. At this level of analysis there was no evidence of specific co-staining for FGF-2 and chromatin 'clumps'.

Cells transfected with vector only (50 and 200 m.o.i.) did not display chromatin clumping nor the decline in cell numbers/mitotic index induced by the high dose of 'hi' FGF-2 at 72 h (data not shown).

Genomic DNA was analyzed for presence of DNA laddering, a hallmark of apoptotic cell death (Fig. 4). Cells treated with staurosporine, an inducer of apoptosis, provided a positive control for ladder formation. DNA laddering was not detectable at 24 or 48 h, at either 50 or 200 m.o.i. At 72 h, DNA laddering was detectable at both 50 and 200 m.o.i. of Ad.'hi' FGF-2- but not vector-transfected cells.

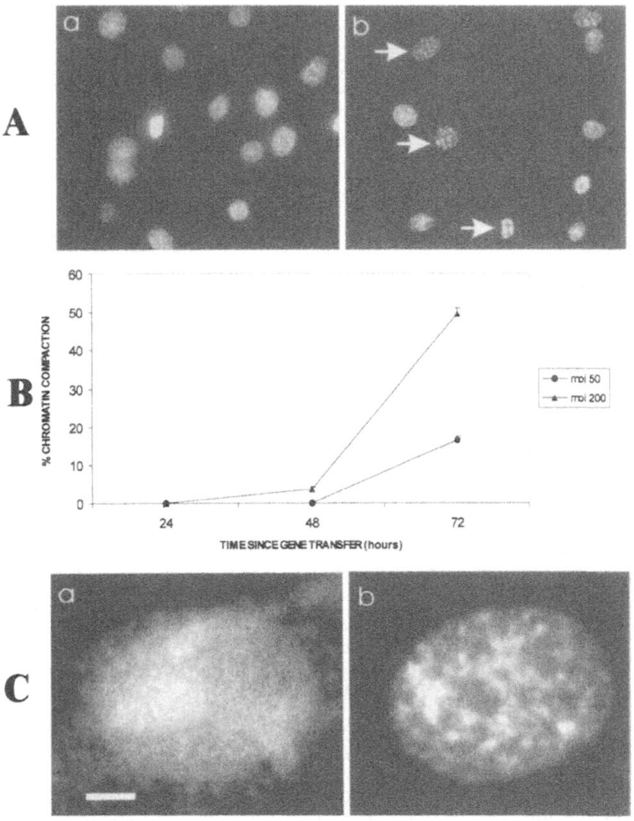

Fig. 3. Effect of 'hi' FGF-2 overexpression on chromatin compaction. (A) (a) Nuclear staining of cultures transfected with 200 m.o.i. of vector or (b) Ad. 'hi' FGF-2 for 72 h. Arrows indicate the compacted chromatin. (B) Percentage of nuclei displaying chromatin compaction as a function of dosage (m.o.i.) and time in culture, as indicated. Data are shown as the mean ± S.E.M. (n = 4). (C) Simultaneous fluorescence labeling for (a) FGF-2 and (b) DNA. Anti-FGF-2 antibodies were used at a dilution detecting overexpressed 'hi' FGF-2, but not endogenous FGF-2. Bar in (a) corresponds to 5 μM.

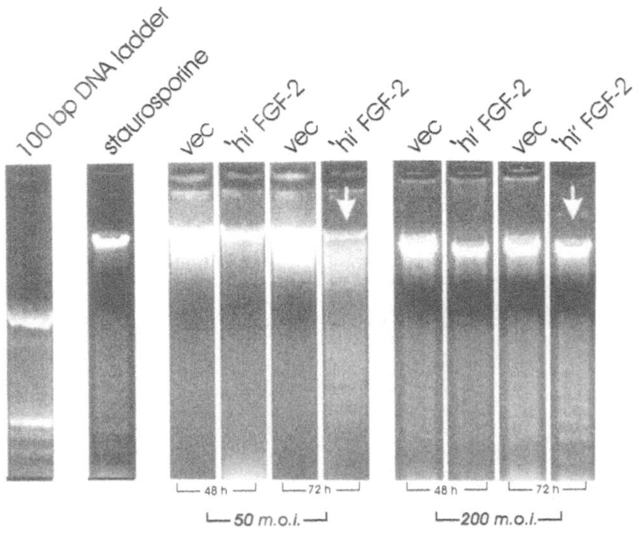

Fig. 4. Effect of 'hi' FGF-2 overexpression on DNA ladder formation. Electrophoretic analysis of genomic DNA from control, vector-transfected cells and 'hi' FGF-2 infected cells, at 48 and 72 h, and 50–200 m.o.i. as indicated. Arrow shows DNA ladder. DNA from cells treated with staurosporine (apoptosis inducer) is used as a positive control for DNA ladder formation.

Discussion

We have shown that the appearance of 'hi' FGF-2 induced chromatin disruption is both dose and time dependent, being maximal at higher dosage and three days in culture. In addition, higher levels of 'hi' FGF-2 expression were associated with decreased mitotic index, cell loss and appearance of DNA ladder.

Relative levels of 'hi' FGF-2 were a function of both dosage and time in culture at least until the 48 h time point. Until then the introduced gene likely continued to be expressed, resulting in increased protein accumulation. The decline in accumulated 'hi' FGF-2 at 72 h, especially at the 200 m.o.i., could be a consequence of decreased expression at this time point and/or increased degradation, coinciding with decreases in cell number and DNA ladder formation.

'Hi' FGF-2 can affect myocytes both in an autocrine/paracrine manner, stimulating DNA synthesis, but also in an intracrine manner, causing the compacted nuclear phenotype [11]. Our data indicate that higher levels of expression, especially at later time points, may 'tip-the-balance' towards disruption of chromatin, and loss of ability for cell division. Myocytes were plated at relatively high densities, reaching confluence at 24 h from plating. This, combined with a degree of ongoing cell loss/cell detachment, may be the reason we did not observe increases in cell number, despite the increases in mitotic index observed at 48 h with the low FGF-2 dose.

The compacted chromatin phenotype was only detectable at 48 h, at the high 'hi' FGF-2 dose. It is intriguing that we did not see a simple relationship between 'hi' FGF-2 levels and compaction: similar levels of 'hi' FGF-2 are observed at 50 m.o.i. and 48 h, and at 72 h for both 50 or 200 m.o.i., corresponding to significantly different levels of compaction (0, 17 and 50%, respectively). These findings suggest that development of the disrupted nuclear phenotype is not solely a consequence of increased 'hi' FGF-2, and that myocytes need to be exposed to this factor for a certain time (> 48 h in our experimental system) before they develop nuclear disruption. A natural consequence of this reasoning is that 'hi' FGF-2 triggers a series of events leading to chromatin disruption and eventually to cell death.

Previously we showed that pure 'hi' FGF-2 is capable of increasing histone-reconstituted chromatin compaction in the test tube [12] and thus we suggested that nuclear 'hi' FGF-2 may cause 'clumping' by direct interaction with chromatin. Our present findings do not support this premise. In addition, the distinct patterns of localization of compacted chromatin and 'hi' FGF-2 within the nucleus might also argue against a direct interaction. There is, however, the possibility (currently under investigation) that an interaction occurs at earlier time points, before effects on chromatin are detectable.

We have reported that the chromatin phenotype caused by 'hi' FGF-2, at least at the early stages of multiple clumps, did not present certain features of apoptotic cell death such as disruption of nuclear lamina, staining for 'TUNNEL', or P17. Furthermore, clumping was still observed in the presence of caspase inhibitors, Bcl-2, and 'lo' FGF-2 [12]. Use of adenoviral vectors to express 'hi' FGF-2 has enabled us to increase transfection efficiency considerably (over 90%, compared to 10–20% of previous studies) and conduct time and dose dependent studies. Thus we were able to disrupt chromatin in the majority of transfected cells (50% at 72 h and 200 m.o.i), and to obtain evidence of DNA laddering at this time point. We thus conclude that 'hi' FGF-2 expression eventually leads to cardiomyocyte cell death presenting some apoptotic features.

In conclusion, we have provided evidence that prolonged cardiomyocyte exposure to the intracrine action of 'hi' FGF-2 leads to mitotic arrest, chromatin disruption and finally cell loss, presumably through some form of apoptotic cell death. This contrasts to the autocrine/paracrine effects of FGF-2, which results in mitotic stimulation and protection from injury [1]. 'Hi' FGF-2, preferentially accumulated during stress, [7] may contribute to the development of cell death.

Acknowledgement

This research was funded by the Canadian Institutes for Health Research (EK). We thank Dr. J. Davie (Manitoba Institute of Cell Biology, University of Manitoba) for the gift of the phospho-H3 antibody.

References

1. Bikfalvi A, Klein S, Pintucci G, Rifkin DB: Biological roles of fibroblast growth factor-2. Endocr Rev 18: 26–45, 1997
2. Bugler B, Amalric F, Prats H: Alternative initiation of translation determines cytoplasmic or nuclear localization of basic fibroblast growth factor. Mol Cell Biol 11: 573–577, 1991
3. Quarto N, Finger FP, Rifkin DB: The NH2-terminal extension of high molecular weight bFGF is a nuclear targeting signal. J Cell Physiol 147: 311–318, 1991
4. Delrieu I: The high molecular weight isoforms of basic fibroblast growth factor (FGF-2): An insight into an intracrine mechanism. FEBS Lett 468: 6–10, 2000
5. Piotrowicz RS, Maher PA, Levin EG: Dual activities of 22–24 kDa basic fibroblast growth factor: Inhibition of migration and stimulation of proliferation. J Cell Physiol 178: 144–153, 1999
6. Bikfalvi A, Klein S, Pintucci G, Quarto N, Mignatti P, Rifkin DB: Differential modulation of cell phenotype by different molecular weight forms of basic fibroblast growth factor: Possible intracellular signaling by the high molecular weight forms. J Cell Biol 129: 233–243, 1995

7. Vagner S, Touriol C, Galy B, Audigier S, Gensac MC, Amalric F, Bayard F, Prats H, Prats AC: Translation of CUG- but not AUG-initiated forms of human fibroblast growth factor 2 is activated in transformed and stressed cells. J Cell Biol 135: 1391–1402, 1996
8. Liu L, Doble BW, Kardami E: Perinatal phenotype and hypothyroidism are associated with elevated levels of 21.5–22 kDa basic fibroblast growth factor in cardiac ventricles. Dev Biol 157: 507–516, 1993
9. Padua RR, Kardami E: Increased basic fibroblast growth factor (bFGF) accumulation and distinct patterns of localization in isoproterenol-induced cardiomyocyte injury. Growth Factors 8: 291–306, 1993
10. Pasumarthi KB, Doble BW, Kardami E, Cattini PA: Over-expression of CUG- or AUG-initiated forms of basic fibroblast growth factor in cardiac myocytes results in similar effects on mitosis and protein synthesis but distinct nuclear morphologies. J Mol Cell Cardiol 26: 1045–1060, 1994
11. Pasumarthi KB, Kardami E, Cattini PA: High and low molecular weight fibroblast growth factor-2 increase proliferation of neonatal rat cardiac myocytes but have differential effects on binucleation and nuclear morphology. Evidence for both paracrine and intracrine actions of fibroblast growth factor-2. Circ Res 78: 126–136, 1996
12. Sun G, Doble BW, Sun JM, Fandrich RR, Florkiewicz R, Kirshenbaum L, Davie JR, Cattini PA, Kardami E: CUG-initiated FGF-2 induces chromatin compaction in cultured cardiac myocytes and *in vitro*. J Cell Physiol 186: 457–467, 2001
13. Nesbit M, Nesbit HK, Bennett J, Andl T, Hsu MY, Dejesus E, McBrian M, Gupta AR, Eck SL, Herlyn M: Basic fibroblast growth factor induces a transformed phenotype in normal human melanocytes. Oncogene 18: 6469–6476, 1999
14. Kardami E, Murphy LJ, Liu L, Padua RR, Fandrich RR: Characterization of two preparations of antibodies to basic fibroblast growth factor which exhibit distinct patterns of immunolocalization. Growth Factors 4: 69–80, 1990
15. Yue TL, Wang C, Romanic AM, Kikly K, Keller P, DeWolf WE Jr, Hart TK, Thomas HC, Storer B, Gu JL, Wang X, Feuerstein GZ: Staurosporine-induced apoptosis in cardiomyocytes: A potential role of caspase-3. J Mol Cell Cardiol 30: 495–507, 1998
16. Chadee DN, Hendzel MJ, Tylipski CP, Allis CD, Bazett-Jones DP, Wright JA, Davie JR: Increased Ser-10 phosphorylation of histone H3 in mitogen-stimulated and oncogene-transformed mouse fibroblasts. J Biol Chem 274: 24914–24920, 1999

Molecular and Cellular Biochemistry **246:** 117–120, 2003.

In vitro effect of nicotine and cotinine on the susceptibility of LDL oxidation and hemoglobin glycosylation

S. Asgary,[1] G.H. Naderi,[2] N. Sarrafzadegan[3] and M. Gharypur[2]

Departments of [1]Pharmacognosy; [2]Biochemistry; [3]Cardiology, Isfahan Cardiovascular Research Center, Isfahan University of Medical Science, Isfahan; [3]Isfahan, Iran

Abstract

Nicotine, a major component of cigarette smoke, plays an important role in the development of cardiovascular disease and lung cancer in smokers. This study was designed to determine the *in vitro* effects of nicotine and its metabolite cotinine on the susceptibility of LDL to oxidation and hemoglobin glycosylation. Three different concentrations of each one (10, 15, 25 μg/ml) were used. Our data show that nicotine and cotinine are inhibitors for Cu^{2+}-induced LDL oxidation but also they increase the glycosylation degree of hemoglobin. Nicotine at final concentrations of (10, 15, 25 μg/ml) increases the rate of hemoglobin glycosylation 25, 32 and 47%, respectively, and cotinine at final concentrations increase the rate of glycosylation 8, 10 and 12%, respectively. Therefore promoting hemoglobin glycosylation is one of the alternations caused by smoking that increase risk of cardiovascular disease. (Mol Cell Biochem **246**: 117–120, 2003)

Key words: nicotine, cotinine, LDL oxidation, hemoglobin glycosylation

Introduction

Smoking is associated with an increased incidence of cardioactive vascular disease in numerous epidemiological studies [1–3]. The mechanisms underlying this association are only partly known. Some atherogenic mechanisms induced by smoking include an increase in serum triglyceride and a decrease in serum high-density lipoprotein cholesterol (HDL), vasospasm, hyperfibrinogenemia, increased platelet aggregation, impaired fibrinolysis, and insulin resistance [4–11], and it is increased level of hemoglobin A_{IC} found in hypertensive and normotensive smokers [12].

It is now generally accepted that oxidation of low density lipoprotein (LDL) plays a significant pathogenic role of atherosclerosis, it is, however controversial whether LDL prepared from smokers is more susceptible to oxidation than LDL from non smokers [13–17]. Of the thousands of chemicals in cigarette smoke, some of them are suspected to contribute to smoking induced vascular disease including nicotine, carbon monooxide, polycyclic aromatic hydrocarbons and tobacco glycoproteins.

Nicotine, one of the most pharmacologically tobacco components, has a wide range of cardiovascular effects [18–20]. The aim of the recent study was to find the *in vitro* effects of nicotine, a major component of cigarette smoke, and its metabolite cotinine on the susceptibility of LDL oxidation and hemoglobin glycosylation.

Materials and methods

Effect on hemoglobin glycosylation

The method is the same as our previous report [21, 22]. Blood from normal volunteers was drawn using EDTA as an anticoagulant. The red blood cells were washed thrice with 0.14 M NaCl solution. Then 1 vol. of red blood cells suspension was lysed with 2 vol. of 0.1 M phosphate buffer, pH 7.4,

Address for offprints: S. Asgary, Isfahan Cardiovascular Research Center, Isfahan University of Medical Science, P.O. Box 81465-1148, Isfahan, Iran (E-mail: isfcavasrc@hotmail.com)

and 0.5 vol. CCl_4. After lysing, the hemolysate was freed from debris by centrifugation. The upper layer was separated and the hemoglobin concentrations were measured by the Drabkin method [23, 24].

One ml of hemoglobin solution (5 g/100 ml) and 1 ml of solution containing glucose (2 g/100 ml) and gentamycin (20 mg/100 ml) in 0.1 M phosphate buffer, pH 7.4 were incubated in the dark at room temperature. Then the glycosylation degree of hemoglobin in the presence and absence of nicotine and cotinine were measured by the colorimetric method [23, 25].

Effect on susceptibility of LDL oxidation

The susceptibility of LDL to *in vitro* oxidation was assessed by the technique described by Regnstrom *et al.* [26]. The LDL preparation was dialyzed against a 100-fold volume of 0.02 M phosphate buffer (pH 7.4) 0.16 M NaCl (Dialysis buffer), then purged with nitrogen for 15 h at 4°C in darkness. The buffer was changed 3 times. EDTA-free LDL was diluted in dialysis buffer to a final concentration of 25 µg/ml and oxidation initiated by addition of a freshly prepared aqueous $CuSO_4$ solution to a final concentration of 1.66 µmol/l. LDL oxidation kinetics were monitored by the change in 234 nm absorbance at 30°C on a Shimadzu spectrophotometer uv-3100. The dienes formed during LDL oxidation produce an absorbance spectrum with a distinct peak at 234 nm with essentially no interindividual variation. The initial absorbance at 234 nm was taken as the baseline and the changes in absorbance in the presence of nicotine and cotinine and also in their absence (control) were recorded every 30 min for 4 h. The differences of absorbance in the test and control tubes were used as the measure of LDL susceptibility to oxidation.

Statistical analysis

Each examination was repeated 3 times and the means of value is reported. The analyses were performed by means of Students *t*-test for independent samples, using SPSS. A level of $p < 0.05$ was considered to be significant.

Results

Table 1 shows that both nicotine and cotinine have significant effects on hemoglobin glycosylation that was dose-dependent. Nicotine at final concentrations of 10, 15, 25 µg/ml increases the rate of hemoglobin glycosylation 25, 32 and 47%, respectively, while cotinine at the same concentrations increases the rate of hemoglobin glycosylation 8, 10 and 12%, respectively.

Table 1. In vitro effect of nicotine and cotinine on hemoglobin glycosylation

Compounds	Increasing of % glycosylation 10 (µg/ml)	15 (µg/ml)	25 (µg/ml)
Nicotine	25	32	47
Cotinine	8	10	12

Significantly different from control (*t*-test): $p < 0.05$.

Table 2 shows the effects of nicotine and cotinine on the susceptibility of LDL to Cu^{2+} induced oxidation. Nicotine and cotinine significantly increased lag time before the onset of conjugated deines formation ($p < 0.05$). Prolongation of the lag time was dose-dependent (Fig. 1).

Discussion

Pharmacological effect of nicotine and cotinine is not only important for smokers but these observations are also clinically important because of the widespread use of smokeless tobacco products [27] and the increased prescription of nicotine-containing chewing gum in smoking cessation therapy [28]. Studies show sustained levels of nicotine among young males using smokeless tobacco, causing concern for subsequent cardiovascular risk. Also, there is little information on the effects of nicotine replacement on cardiovascular risk in cessation programs. Nicotine increases HDL and LDL cholesterol and the serum levels of total cholesterol by increasing the synthesis and secretion of triglyceride-rich lipoprotein [29]. Some studies established that smoking increased oxidation of LDL cholesterol. In addition, it was demonstrated that cigarette smoking increased plasma levels of autoantibodies against oxidized LDL and that smoking is not associated with increased oxidizability of LDL [30].

Our data show that nicotine and cotinine are inhibitors for Cu^{2+}-induced LDL oxidation. Also, their chemical structures reflect their actions as anti-oxidant, it is currently believed that peroxidases may oxidize LDL *in vivo* and induce generating pro-oxidants from anti-oxidants. Therefore, the *in vivo* effect of these two substances may be different from their *in vitro* effects.

Table 2. In vitro effect of nicotine and cotinine on susceptibility of low-density lipoprotein to Cu^{2+}-induced LDL oxidation

Compounds	Lag time (min) 0 (µg/ml) (control)	10 (µg/ml)	15 (µg/ml)	25 (µg/ml)
Nicotine	55.2 ± 2.6	85 ± 4.5	93 ± 4.3	108.7 ± 12.4
Cotinine	58.1 ± 3.0	69.9 ± 4.9	76.3 ± 5.6	83 ± 4.3

X ± S.D. of 3 separate experiments. Significantly different from control (*t*-test): $p < 0.05$.

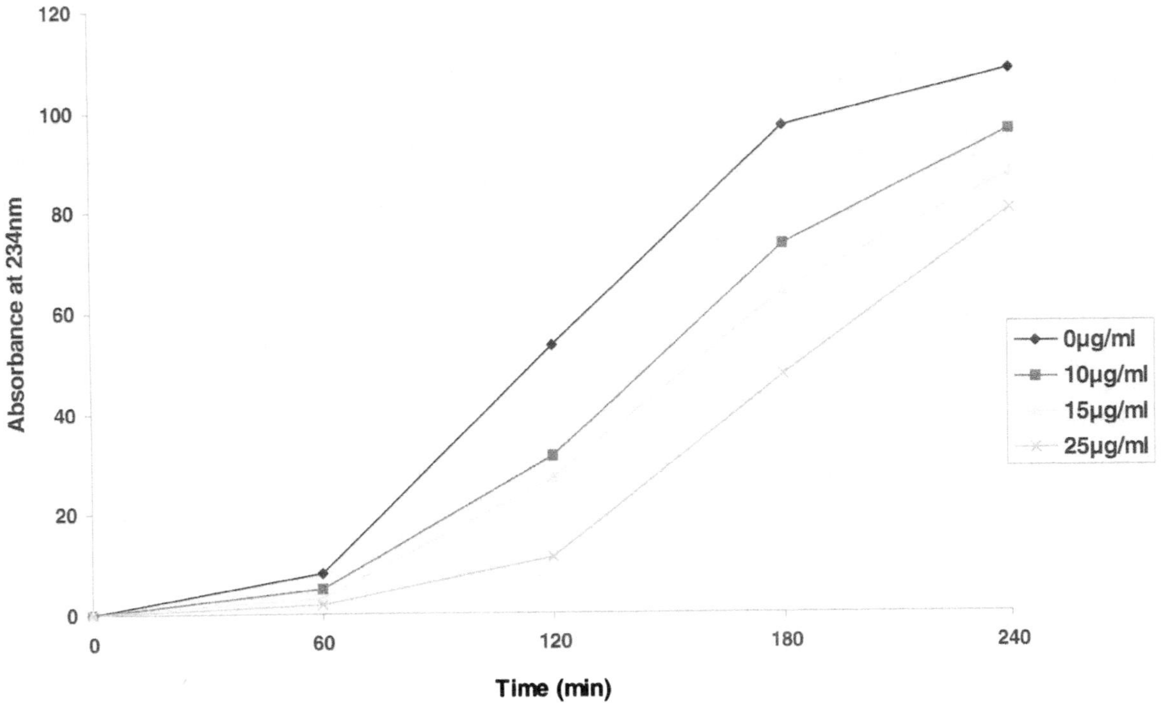

Fig. 1. Comparison of changes in absorbance at 234 nm due to Cu^{2+} induced LDL oxidation in control and test tubes containing nicotine.

Also, it may be possible to identify other components in cigarette smoke that increase LDL oxidation. There may be other mechanisms rather than LDL oxidation induced by nicotine and cotinine that may lead to development of atherosclerosis. Our finding nevertheless demonstrated that nicotine and cotinine increases non-enzymatic glycosylation of hemoglobin that is one of the atherogenic mechanisms induced by smoking. Since non-enzymatic glycosylation of body proteins, due to prolonged hyperglycemia, is an important consequence of diabetes therefore, smoking in particular in diabetics will have more complications.

References

1. Taylor AE, Johnson DC, Kazemi H: Environmental tobacco smoke and cardiovascular disease a position paper from the council on cardiopulmonary and critical care. Am Heart Assoc Circ 86: 699–702, 1992
2. Diez-Roux AV, Nieto FJ, Comstock GW, Howard G, Szklo M: The relationship of active and passive smoking to carotid atherosclerosis 12–14 years later. Prev Med 24: 48–55, 1995
3. Tell GS, Polak JF, Ward BJ, Kittner SJ, Savage PJ, Robbins J: For the cardiovascular health study (CHS) collaborative research group. Relation of smoking with carotid artery wall thicknesses and stenosis in older adults: The cardiovascular health study. Circulation 90: 2905–2908, 1994
4. Winnifred MD: Smoking and cardiovascular function. J Hypertens 8(suppl): 817–823, 1996
5. Hostmark AT, Bevy I, Brudal S, Berg SR, Kierulf P, Bjerkedal T: Coronary risk factors in middle-aged men as related to smoking, coffee intake and physical activity. Scand J Soc 20: 96–203, 1992
6. Krupski WC: The peripheral vascular consequences of smoking. Ann Vasc Surg 5: 291–304, 1991
7. Howard G, Wagenknecht LE, Burke G, Diez Roux A, Evans GW, Mcgovern P, Nieto J, Tell GS: Cigarette smoking and progression of atherosclerosis. JAMA 279: 119–124, 1998
8. Stachenko SJ, Reeder BA, Lindsay B: Smoking prevalence and associated risk factors in Canadian adults. CMAJ (suppl): 20–27, 1992
9. Quillen JE, Rossen JD, Oskarsson HJ, Minor RL Jr, Lopez JAG, Winford MD: Acute effect of cigarette smoking on the coronary circulation, constriction of epicardial and resistance vessels. J Am Coll Cardcol 22: 642–647, 1993
10. Zhu BQ, Parmley W: Hemodynamic and vascular effects of active and passive smoking. AM Heart J 130: 1270–1275, 1995
11. Attval S, Fowelin J, Lager I: Smoking induces insulin resistance a potential link with the insulin resistance syndrome. J Int Med 233: 327–332, 1993
12. Nilsson PM, Lind L, Pollare T, Berne C, Lithell HO: Increased level of hemoglobin A_{1c} but not impaired insulin sensitivity, found in hypertensive and normotensive smokers. Metabolism 44: 557–561, 1995
13. Scheffler E, Wiest E, Woehrle J, Otto I, Schulz I, Haber L: Smoking influences the atherogenic potential of low-density lipoprotein. Clin Invest 70: 263–268, 1992
14. Scheffler E, Hube L, Fruhbis J, Schulz I, Ziegler R, Dresel HA: Alteration of plasma low-density lipoprotein from smoker. Atherosclerosis 82: 261–265, 1990

15. Princen HMG, Van Poppel G, Vogelzang C, Buytenhek R, Kok FJ: Supplemention with vitamin E but not B-carotene *in vivo* protects low-density lipoproteins from lipid peroxidation *in vitro*. Arterioscler Thromb 12: 554–562, 1992
16. Harats D, Ben-Naim M, Dabach Y, Hollaader G, Stein O, Stein YM: Cigarette smoking renders LDL susceptible to peroxidative modification and enhanced metabolism by macrophages. Atherosclerosis 79: 245–252, 1989
17. Siekmeier R, Wulfroth P, Wieland H, Groß W, Marz W: Low-density lipoprotein susceptibility of *in vitro* oxidation in healthy smokers and non smokers. Clin Chem 42: 524–530, 1996
18. Lidia A, Peyton J III, Marc H, Neal LB: Divergent tolerance to metabolic and cardiovascular effects of nicotine in smokers with low and high levels of cigarette consumption. Clin Pharmacol Ther 56: 55–64, 1994
19. Jerry BJ, Xianobang Y: Mechanism of the hypertensive response to central injection of nicotine in conscious rats. Brain Res Bull 32: 35–41, 1993
20. Siegel D, Benowitz NL, Ernster VL, Grady DG, Hauck WW: Smokeless tobacco, cardiovascular risk factors, and nicotine and cotinine levels in professional baseball players. Am J Public Health 82: 417–421, 1992
21. Asgary S, Naderi GH, Sarrafzadegan N, Ghassemi N, Boshtam M, Rafiei M, Arefian A: Anti-oxidant effect of flavonoids on hemoglobin glycosylation. Pharm Acta Helv 73: 223–226, 1999
22. Asgary S, Naderi GH, Sarrafzadegan N, Vakili R: The inhibitory effects of pure flavonoids on *in vitro* protein glycosylation. Herbal Pharmacother 2: 47–57, 2002
23. Fairbanks VF, Klee GG: Biochemical aspects of hematology. In: C.A. Burtis, E.R. Ashwood (eds). Tietz Textbook of Clinical Chemistry, 2nd edn. W.B. Saunders Company, Philadelphia, 1994, pp 2020–2030
24. Van Kampen EJ, Zijlstra WG: Determination of hemoglobin and its derivatives. Adv Clin Chem 8: 1414–1487, 1965
25. Fluckiger R, Winter Halter KH: Biochemical and clinical aspects of hemoglobin abnormalities. Academic Press, New York, 1978, p 208
26. Regnstrom J, Nilsson J, Tornvall P, Landon C, Hamsten A: Susceptibility to low-density lipoprotein oxidation and coronary atherosclerosis in man. Lancet 339: 1183–1184, 1992
27. Maxwell JC: Maxwell report: Smokeless grows; cigars decline, tobacco reporter, 1993, p 56–57.
28. Russel MA, Wilson C, Feyerbend C, Cole PV: Effect of nicotine chewing gum on smoking behavior and as aid to cigarette withdrawal. Brit Med J 2: 391–393, 1976
29. Ashakumary L, Vijayammal PL: Effect of nicotine on lipoprotein metabolism in rats. Lipids 32: 311–315, 1997
30. Heitzer T, Yla-Herttuala S, Luoma J, Kurz S, Munzel T, Just H, Olschewski M, Drexler H: Cigarette smoking potentiates endothelial dysfunction of forearm resistance vessels in patients with hypercholesterolemia. Role of oxidized LDL. Circulation 93: 1346–1353, 1996

Molecular and Cellular Biochemistry **246:** 121–127, 2003.

Plasmid lipid and lipoprotein pattern in the Inuit of the Keewatin district of the Northwest territories

Michael J.B. Kutryk[1] and Bram Ramjiawan[2]

[1]*St. Michael's Hospital, University of Toronto, Toronto, Ontario;* [2]*Division of Stroke and Vascular Disease, St. Boniface General Hospital Research Centre and the Department of Physiology, University of Manitoba, Winnipeg, Manitoba, Canada*

Abstract

It has long been recognized that certain indigenous populations, such as the traditional living Inuit, have been relatively spared from ischemic heart disease despite their high fat diet. There is evidence to suggest that elevated serum levels of oxidated cholesterol may be a strong risk factor for atherogenesis. This study was performed to examine the plasma levels of lipid, triglyceride, lipoprotein cholesterol and oxidated cholesterol derivatives in the Inuit of the Keewatin district of Canada. Lipoprotein isolation from plasma was performed by serial centrifugation and apolipoprotein concentrations and native and esterified cholesterol content of the lipoprotein isolated were determined. Analysis of cholesterol and its oxidation products was performed using high pressure liquid chromatography. The Inuit showed a decreased concentration of both VLDL apolipoprotein and LDL apolipoprotein. The incidence of oxidatived cholesterol in Inuit VLDL and LDL was higher than control except for the 25 α-hydroxy derivative in VLDL. These data provide insight into the relative contributions of genetic and environmental influences on the development of ischemic heart disease. (Mol Cell Biochem **246**: 121–127, 2003)

Key words: cholesterol, indigenous populations, Inuit, oxidized cholesterol, apolipoprotein

Introduction

It has long been recognized that certain indigenous populations, such as the traditional living Inuit, have been spared the current epidemic of ischemic heart disease [1]. A review of the risk factors for underdeveloped populations suggests that environment rather than genetic factors is responsible [2]. However, in the Inuit, constitution may be involved as they exhibit low serum cholesterol values despite a high fat diet [3]. Elevated serum cholesterol is known to be a risk factor for ischemic heart disease. There is evidence, however, that elevated serum levels of oxidated cholesterol may be an even stronger risk factor for atherogenesis than serum cholesterol itself [4]. Both animal studies [4], and epidemiological data [5] have shown that diets high in oxidized derivatives of cholesterol results in markedly increased morbidity from coronary artery disease. In addition, *in vitro* studies shown that lipid laden foam cells similar to those seen in atheromatous plaques can be generated by treatment of vascular smooth muscle cells in culture with oxidatively modified derivatives of cholesterol (Kutryk and Pierce, unpublished data). Thus, it is becoming apparent that native cholesterol may not be a direct risk factor for ischemic heart disease, rather oxidized cholesterol may be responsible.

The Inuit of the Keewatin District are an ideal population in which to examine the contribution of oxidized derivatives of cholesterol in the development of ischemic heart disease. The Inuit of this region have been shown to have a decreased incidence of coronary artery disease compared with those of Caucasian descent. Oxidized derivatives of cholesterol have been detected in human sera and have been shown to be a component of lipoproteins [6–9]. Inuit from other regions also have been shown to be at reduced risk for coronary artery disease [10–12], and it has been suggested that lower cholesterol and triglyceride concentrations due to lowered B-lipoprotein levels are responsible for this lowered morbidity.

Address for offprints: M.J.B. Kutryk, 30 Bond Street, Toronto, Ontario, Canada, M5B 1W8 (E-mail: kutrykm@smh.toronto.on.ca)

To date there have not been any studies examining the lipid/lipoprotein profile or oxidated cholesterol content of the plasma of the Inuit of the Keewatin District of the NWT. This report presents data on plasma lipid, triglyceride, lipoprotein cholesterol and oxidated cholesterol derivatives in the Inuit of the Keewatin district. The epiemiolgic significance of these data is discussed.

Materials and methods

Materials

All chemical and reagents were purchased form Sigma Chemical Company.

Subjects

The study subjects were Inuit males (n = 13) of the Keewatin District of the North West Territories greater than 18 years of age who presented to the Churchill Health Centre between May 1, 1989 and June 13, 1989. Control subjects consisted of 12 healthy male Caucasian volunteers aged 18–52 years living in an urban centre. Informed consent was obtained from each patient before enrolment.

Examination of the patients included measurements of height and weight. Medical histories were obtained and individuals with documented coronary artery disease or diabetes were excluded from the study.

Lipid and lipoprotein measurements

Venous blood samples were collected in the presence and absence of EDTA. Plasma was separated immediately and collected by centrifugation at 1500 × g for 10 min at 4°C, quickly frozen and maintained at –30°C for later isolation of lipoproteins. Serum was separated from the EDTA-free aliquot by centrifugation at 1500 × g for 10 min at a 4°C. Serum samples were maintained at 4°C and sampled for total HDL and LDL cholesterol and triglycerides. None of the samples contained evidence of chylomicrons and all were sampled within one week. Triglyceride concentrations in serum were measured by the enzymatic method of Bucolo and Davis [13]. Total serum cholesterol was assayed by the enzymatic assay of Omodeo Sale *et al.* [14] after enzymatic conversion of esterified derivatives of cholesterol to free cholesterol. Beta lipoproteins (roughly equivalent to VLDL and LDL fractions) were precipitated from serum by the addition of phosphotungstate [15]. The remaining supernatant containing lipoproteins (roughly equivalent to HDL fractions) was assayed for total esterified plus free cholesterol. By using a correction factor, the lipoprotein fraction was corrected for high triglyceride containing particles which correlated with the VLDL fraction.

Lipoprotein isolation from plasma was performed by serial centrifugation based on the method described by Orr *et al.* [16]. Blood was collected into evacuated tubes such that the final concentration of EDTA was 1 mg/ml. This concentration of EDTA has been shown to act as an effective antioxidant by sequestering heavy metal catalysts. Plasma was isolated by centrifugation at 1500 × g at 4°C for 15 min. To the plasma 5,5′–dithiobis (2-nitrobenzoic acid) was added to a final concentration of 1.5 mM to inhibit lecithin cholesterol acyl transferase. Phenylmethyl-sulphonyl fluoride (2 mM final) was added to inhibit proteolysis. Thimerosol (final concentration 0.08 mg/ml) and sodium azide (0.13 mg/ml final) were added as bactericides. VLDL (< 1.006 g/ml) was isolated by flotation centrifugation for 24 h at 37,000 rpm in a Beckman ultracentrifuge fitted with a Type 42.1 rotor. After the VLDL was removed, the plasma was adjusted to d = 1.063 g/ml with solid NaCl (by interpolation of data from Baxter and Wallace [17]) and LDL was isolated by centrifugation for 24 h at 42,000 rpm in a Type 42.1 rotor.

Total apoprotein concentration of the LDL and VLDL solutions was determined by the method of Lowry *et al.* [18] using bovine serum albumin as a standard.

Native and esterified cholesterol content of the isolated lipoprotein fractions were determined as described previously [14] after lipid extraction from the lipoprotein [19].

Analysis of cholesterol oxidation products

Aliquots of isolated VLDL and LDL were subjected to lipid extraction as described above. The organic solvent was evaporated under a stream of dry N_2, the dried lipid suspended in 1 ml of methylene chloride, and filtered through 0.45 μm HPLC grade filters. The solvent was again evaporated as above, the dried lipid suspended in 0.1 ml methylene chloride of which 0.02 ml was used for HPLC analysis. HPLC analysis of cholesterol and its oxidation products was performed with a Beckman 116 programmable solvent module fitted with 5 μl particle size 156 × 4.6 mm column (Ultasphere SI) that was eluted with hexane/isopropanol (95:5 v/v) at a flow rate of 1 ml/min. Detection of peaks was accomplished with a Beckman model 166 programmable UV-visible detector. Identification of oxidated cholesterol derivatives in the isolated VLDL and LDL samples was made by comparison to a chromatogram of known standards. A representative standard chromatogram of cholesterol and several oxidation products is shown in Fig. 1.

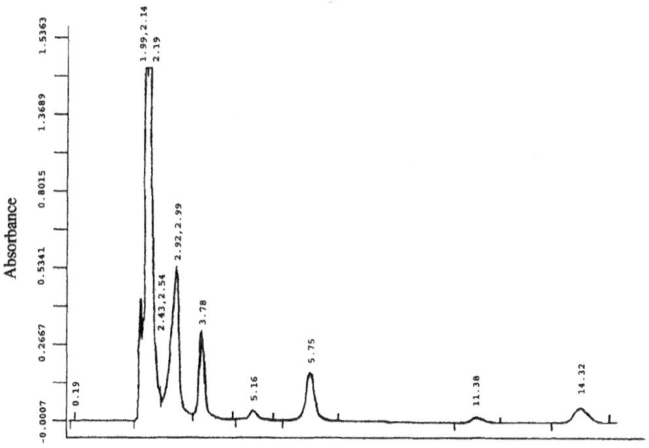

Fig. 1. A high pressure liquid chromatograph of cholesterol oxidation standards. The chromatographic conditions are described in the text. The oxidation products are indicated by their retention times and are as follows: 2.92, 2.99, 4-cholesten-3-one; 3.70, cholesterol; 5.16, 20-α-hydroxycholesterol; 6.75, 25-α-hydroxycholesterol; 11.38, 6-ketocholesterol; 14.32, 5-cholesten-3-ol-7-one.

Data analysis

The results were analyzed statistically with a Student's *t*-test and a p level of 0.05 was arbitrarily chosen as a level of significance.

Results

Characteristics of the study group

Table 1 shows the age and height/weight relationships of the two groups on this study. The Inuit subjects had a mean age that was significantly greater than the control group and exhibited significantly more weight per unit height that did the control subjects.

Table 1. Age and height/weight characteristics of study population

	Age	Ponderal index
Inuit	44 ± 6*	11.8 ± 0.4*
Control	32 ± 3	12.5 ± 0.2

Ponderal index is calculated by dividing height (in inches) by the cube root of weight (in pounds). Values represent means ± S.E. (n = 11–12). *$p < 0.05$ vs. control values.

Serum lipoprotein profile

Table 2 shows the lipoprotein profile of both control and Inuit serum. No significant difference was apparent in the total cholesterol (native plus esterified) or the triglyceride levels of the serum. The alpha-lipoprotein fraction (which roughly corresponds to HDL as obtained by flotation ultracentrifugation) was significantly higher in the Inuit population. No difference was observed between the groups in the beta-lipoprotein fraction (roughly corresponding to LDL). The Inuit also exhibited a significantly lower beta-lipoprotein to alpha-lipoprotein ratio.

Plasma lipoprotein profile

Table 3 shows the VLDL/LDL lipoprotein profile of both control and Inuit plasma. Inuit plasma exhibited significantly lower levels of both VLDL and LDL apolipoproteins. The decreased concentration of LDL apolipoprotein is reflected in the significantly higher value for LDL native cholesterol and LDL cholesterol esters. No significant differences were observed in the content of VLDL native cholesterol or VLDL cholesterol esters.

Quantification of oxidated derivatives of cholesterol

In order to determine the content of oxidated derivatives of cholesterol in the lipoprotein subfractions, aliquots of LDL and VLDL were subjected to HPLC analysis. Figure 2 shows typical chromatographs from lipoprotein fractions of control subjects. The VLDL (Fig. 2a) or the LDL (Fig. 2b) fractions did not demonstrate the presence of oxidated cholesterol. In contrast, typical chromatographs obtained from lipoproteins fractionated from Inuit plasma are presented in Fig. 3. Figure 3a shows a representative chromatograph of VLDL fractionated from Inuit plasma. Only a cholesterol peak is present. However, in the LDL sample from the Inuit (Fig. 3b)

Table 2. Serum lipoprotein profile

	Total cholesterol (mM/L)	Triglycerides (mM/L)	α-Lipoprotein cholesterol (mM/L)	β-Lipoprotein cholesterol (mM/L)	β/a
Inuit	5.9 ± 0.7	1.6 ± 0.7	1.8 ± 0.3*	3.4 ± 0.5	2.1 ± 0.4*
Control	5.2 ± 0.3	2.0 ± 0.4	1.2 ± 0.1	3.2 ± 0.2	2.8 ± 0.3

See 'Materials and methods' for assay procedures. α-Lipoprotein corresponds roughly to HDL. β-Lipoprotein corresponds roughly to LDL. Values represent means ± S.E. (n = 5–12). *$p < 0.05$ vs. control.

Table 3. Plasma lipoprotein profile as determined by enzymatic assay

VLDL apo-lipoprotein (mg/ml)	LDL apo-lipoprotein (mg/ml)	VLDL cholesterol (μmol/mg protein)	LDL cholesterol (μmol/mg protein)	LDL cholesterol (μmol/mg protein)	VLDL cholesterol esters (μmol/mg protein)	LDL cholesterol esters (μmol/mg protein)
Inuit	0.50 ± 0.05*	2.01 ± 0.14*	0.59 ± 0.13	0.92 ± 0.06*	0.63 ± 0.10	1.73 ± 0.13*
Control	0.71 ± 0.10	3.67 ± 0.14	0.79 ± 0.08	0.65 ± 0.03	0.49 ± 0.06	0.88 ± 0.07

See 'Materials and methods' for assay procedures. Values represent means ± S.E. *$p < 0.05$ vs. control.

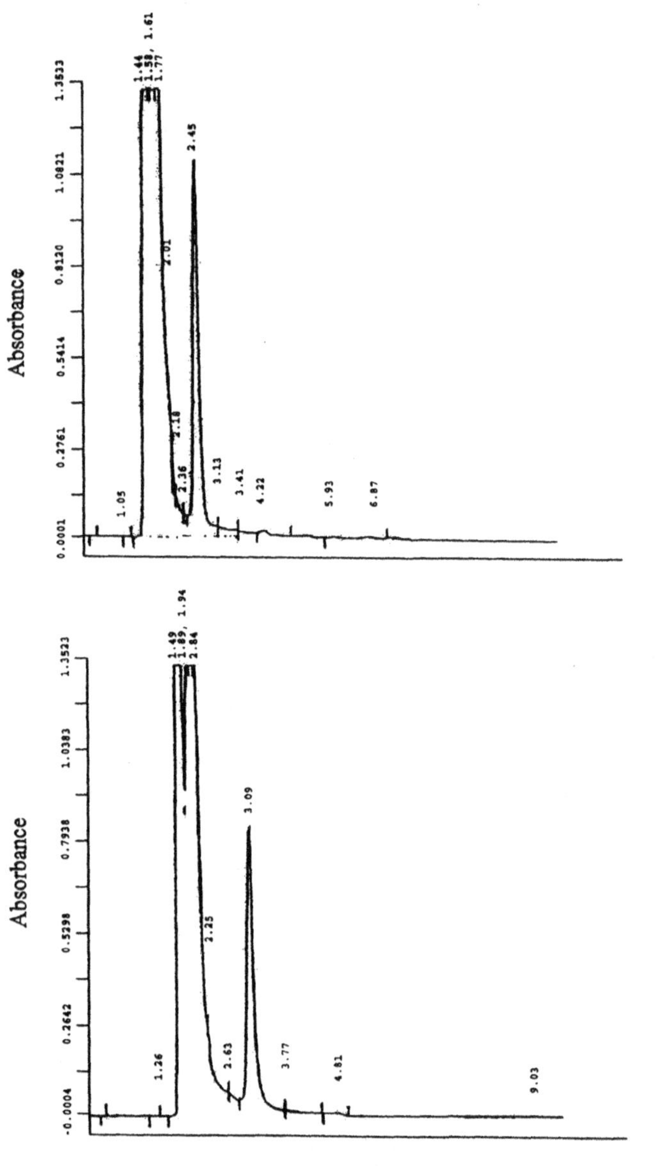

Fig. 2. Typical HPLC derived chromatographs of lipoprotein cholesterol of control subjects. The products are indicated by their retention times. (A) HPL chromatograph of VLDL. 2.65, cholesterol; 4.22 identified as 25-α-hydroxycholesterol by doping the sample as described in text (figure not shown). (B) HPL chromatograph of LDL. 3.18, cholesterol. Chromatographic conditions described in the text.

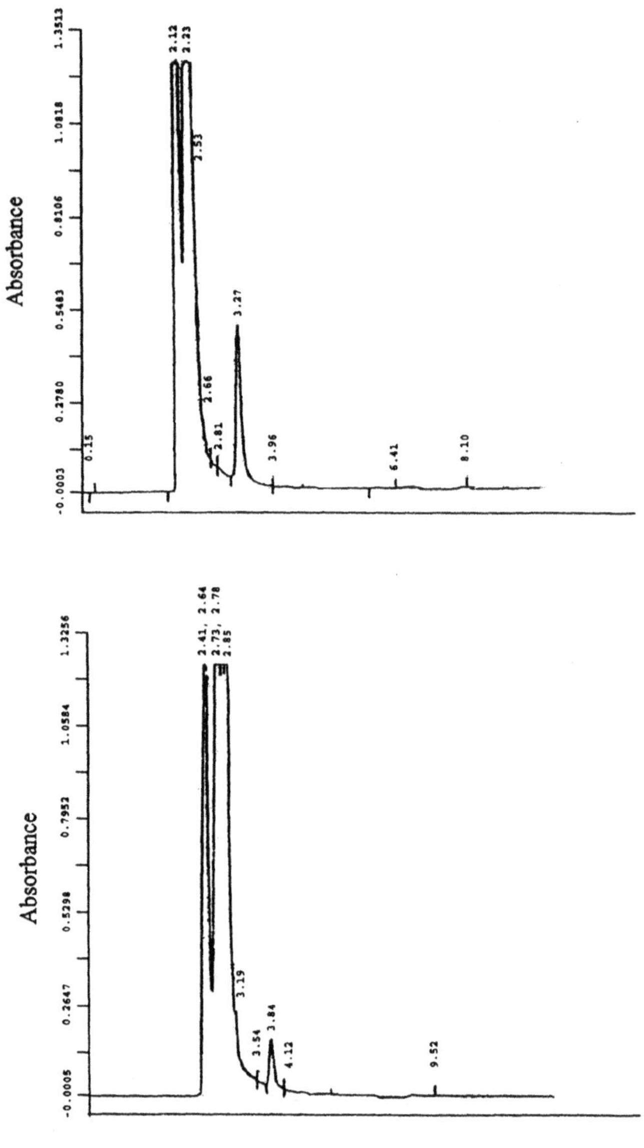

Fig. 3. Typical HPLC derived chromatographs of lipoprotein cholesterol of Inuit subjects. The products are indicated by their retention times. (A) HPL chromatograph of LDL. 3.27, cholesterol. (B) HPL chromatograph of LDL. 3.86, cholesterol. Chromatographic conditions described in the text.

a significant peak corresponding to 25-alpha-hydroxycholesterol is apparent in addition to the cholesterol peak. This oxidation product has been identified by adding purified 25-alpha-hydroxycholesterol to the Inuit sample on a second HPLC run.

Some of the subjects demonstrated relatively large amounts of oxidated cholesterol in their lipoprotein samples. Figure 4 is a chromatograph of the VLDL isolated from an 18 year old Inuit male who presented to the Churchill Health Centre complaining of loss of vision and pain in his right eye. Slit lamp examination confirmed a diagnosis of iridocyclitis and suitable topical treatment was initiated. Blood was drawn after initiation of therapy. The chromatograph in Fig. 4 clearly shows the presence of several oxidatively modified derivatives of cholesterol.

Table 4 shows the incidence of various oxidated derivatives in the isolated lipoprotein fractions of plasma. Only two of the oxidation products detected could be identified. 4-cholesten-3-one was observed in 46% of the VLDL isolates from the Inuit plasma. The range of concentrations detected was from 0.004–0.514 µmole/mg protein. Only 3 of 12 control VLDL isolates (25%) exhibited this derivative (the range was 0.108–0.199 µmole/mg protein). Only one of the Inuit

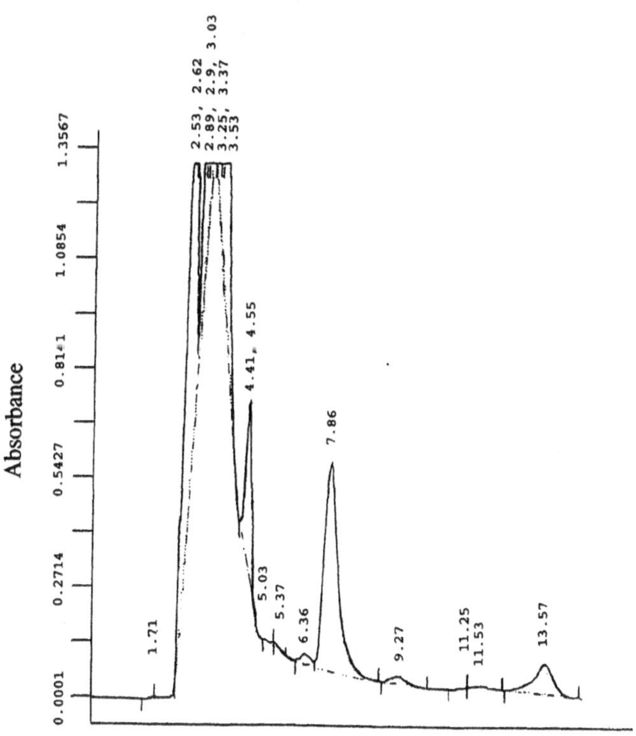

Fig. 4. HPL chromatograph of VLDL lipoprotein cholesterol of 18 year old Inuit male showing particularly high levels of oxidized derivatives of cholesterol. 6.36, 4-cholesten-3-one; 7.06, cholesterol; 9.17, 20-α-hydroxycholesterol; 13.57, 6-ketocholesterol. Chromatographic conditions described in the text.

Table 4. Incidence of identified oxidated derivatives of cholesterol in isolated lipoprotein fractions of plasma

	4-cholesten-3-one		25-α-hydroxycholesterol	
	VLDL	LDL	VLDL	LDL
Inuit	46	8	46	38
Control	25	0	58	17

Values represent % of total population.

subjects (8%) demonstrated the 3-keto-derivative in the LDL isolate and none of the control LDL isolates exhibited this oxidized form. The more prevalent oxidation product, 25-alpha-hydroxycholesterol, was present in 46% of the Inuit VLDL isolates (6 of 13), the levels ranging from 0.013–1.78 µmole/mg protein. Fifty-eight percent (7 of 12) of the control VLDL isolates exhibited the 25-hydroxy derivative, levels ranging from 0.002–0.353 µmole/mg protein. Five of thirteen (38%) of the Inuit subjects showed the 25-hydroxy derivative (0.002–0.242 µmole/mg) while only 2 of the control subjects (17%) exhibited this derivative (0.014 and 0.123 µmole/mg).

Table 5 shows the comparison between Inuit and control with respect to cholesterol and cholesterol oxidation products as determined by HPLC. There are no significant differences between the control and Inuit VLDL cholesterol contents whereas the native cholesterol content of LDL is significantly higher in the Inuit. This is consistent with the data obtained using the enzymatic assay (Table 3). When the oxidated cholesterol species are included, the Inuit VLDL content of total cholesterol becomes significantly different from its control. It is important to note that this information cannot be obtained with the more conventionally employed enzymatic determination of cholesterol since oxidated derivatives of cholesterol are not detectable with the enzymatic method. LDL values were not significantly different between the two groups.

Discussion

The Inuit included in this study showed characteristics different from the control population. The Inuit exhibited a mean age 12 years greater than the control population. As well, the

Table 5. Native and oxidized cholesterol content of plasma as determined by HPLC

	Native cholesterol µmol/mg protein		Total cholesterol µmol/mg protein	
	VLDL	LDL	VLDL	LDL
Inuit	0.58 ± 0.10	0.14 ± 0.05*	0.97 ± 0.22*	0.43 ± 0.06*
Control	0.42 ± 0.06	0.17 ± 0.02	0.53 ± 0.10	0.18 ± 0.02

Total cholesterol represents the sum of native and the identifiable oxidated derivatives. Values for total cholesterol represents mean ± S.E. *$p < 0.05$.

ponderal index (PI) was significantly lower in the Inuit group. Previous studies have shown the Inuit to have ponderal indices lower than those of the predominantly Caucasian North American population. Insurance statistics have shown significantly higher mortality rates for persons with a PI below 12.5 [20, 21]. The validity of using PI to assess obesity in the Inuit population has come under question [24]. Other assessments of obesity, such as skinfold measurements, have shown that the Inuit are more lean than their North American counterparts despite their low PI. A lower PI than other Canadians is thought to reflect a constitutive difference in stature [24].

All measurements in this study were performed on random draw blood samples which conforms to the guidelines established by the Manitoba Health Services Commission for screening for hypercholesterolemia. All blood was collected at least two hours postpradial to avoid confounding the results with chylomicra. Chylomicra contamination was not present in any of the samples taken.

Previous studies have shown that the Inuit population had significantly lower levels of serum cholesterol, triglycerides and β-lipoproteins compared with Caucasian controls [3, 11, 12, 22]. In contrast, the present study has shown no significant differences between Inuit and control with respect to these parameters (Table 2). This discrepancy may in part be due to acculturation of the Inuit, which has been particularly rapid over the past 15 years [2]. Similar to previous reports [3, 11, 12, 22], our data show the Inuit to have significantly higher levels of serum α-lipoprotein (HDL), which has been shown to have a strong negative relationship with coronary artery disease [23]. This may help explain the continued lower incidence of ischemic heart disease in the Inuit.

The Inuit showed a decreased concentration of both VLDL apolipoprotein and LDL apolipoprotein (Table 3). Since no significant difference was observed in the total amount of β-lipoprotein cholesterol, this suggests that the apoprotein in the Inuit is associated with a larger amount of both cholesterol and cholesterol esters. This difference would be expected to alter the physical properties of the LDL complex, however, the physiological significance of this difference has yet to be examined.

The concentration of VLDL associated cholesterol is significantly higher in the control group (data not shown), however the amount of VLDL cholesterol associated with the apoprotein moieties is constant between groups. That the VLDL cholesterol is significantly elevated in the control groups, yet total cholesterol is not significantly different between the Inuit and controls reflects the higher alpha-lipoprotein associated cholesterol content of the Inuit. It has long been known that elevated VLDL cholesterol is a positive risk marker for coronary artery disease while elevated HDL cholesterol is a negative marker [23]. Thus these two factors may be acting in concert to reduce the risk of ischemic heart disease in the Inuit population.

There is increasing evidence implicating oxidatively modified derivatives of cholesterol as the atherogenetic species in individuals at risk for developing ischemic heart disease. The incidence of oxidatived cholesterol in the Inuit VLDL and LDL was higher than control except for the 25 α-hydroxy derivative in VLDL (Table 4). Consistent with the enzymatic assay findings, HPLC data show the Inuit to have a significantly higher LDL cholesterol content when expressed per unit apoprotein (Table 5). When the oxidized derivatives are considered, however, the Inuit display significantly higher levels of total oxidized plus native cholesterol when compared with controls. These data must be analyzed with caution. The selection procedure for the Inuit participants may bias the data making interpretation difficult. Most of the Inuit subjects were ill and presented to the Churchill Health Centre seeking medical attention. All of these patients were placed on pharmocotherapy before samples were drawn. It is well known that illness can affect the plasma lipoprotein profile and it has been recently documented that various illnesses can affect the plasma content of oxidized species of cholesterol [8, 9]. It is interesting to note that the three Inuit that displayed the greatest amount of oxidized cholesterol and the greatest variety of derivatives were suffering from pneumonia and were being treated with antibiotics at the time of sampling. In contrast, the control subjects were all healthy volunteers, none of whom were taking any medications.

Sample size also confounds the interpretation of these data. It is difficult to draw firm conclusions concerning risk factors or risk markers for morbity from such a limited and internally biased population base. This study was undertaken as a pilot study and in this context proved very successful. It suggests that further, more extensive investigations of indigenous populations undergoing acculturation, like the Inuit, may provide insight into the relative importance of cultural and racial factors in the pathogenesis of ischemic heart disease.

References

1. Keys A: Coronary heart disease – the global picture. Atherosclerosis 22: 149–192, 1975
2. Shephard RJ: The influences of race and environment on ischemic heart disease. Can Med Assoc J 111: 1336–1340, 1974
3. de Knijft P, Johansen LG, Rosseneneu M, Franti RR, Jespersen J, Havekes LM: Lipoprotein profile of a Greenland Inuit population. Influence of anthropometric variables, Apo E and A4 polymorphism, and lifestyle. Arterioscler Thromb 12: 1371–1379, 1992
4. Quinn MT, Parthasarathy S, Fong LG, Steinberg D: Oxidatively modified low density lipoproteins: A potential role in recruitment and retention of monocyte/macrophages during atherogenesis. Proc Natl Acad Sci USA 84: 2995–2998, 1987
5. Jacobson MS: Cholesterol oxides in Indian ghee: Possible cause of unexplained high risk of atherosclerosis in Indian immigrant populations. Lancet 2: 656–658, 1987

6. Smith LL, Teng JI, Lin YY, Seitz PK, McGehee MF: Sterol metabolism – XLVII. Oxidized cholesterol esters in human tissues. J Steroid Biochem 14: 889–900, 1981
7. Javitt NB, Kok E, Burstein S, Cohen B, Kutscher J: 26-Hydroxycholesterol. Identification and quantitation in human serum. J Biol Chem 256:12644–12646, 1981
8. Gray MF, Lawrie TD, Brooks CJ: Isolation and identification of cholesterol-oxide and other minor sterols in human serum. Lipids 6: 836–843, 1971
9. Koopman BJ, van der Molen JC, Wolthers BG, Vanderpas JB: Determination of some hydroxycholesterols in human serum samples. J Chromatogr 416: 1–13, 1987
10. Schaefer O: Are Eskimos more or less obese than other Canadians? A comparison of skinfold thickness and ponderal index of Canadian Eskimos. Am J Clin Nutr 30: 1623–1628, 1977
11. Bang HO, Dyerberg J, Nielsen AB: Plasma lipid and lipoprotein pattern in Greenlandic West-coast Eskimos. Lancet 1: 1143–1145, 1971
12. Dyerberg J, Bang HO: Lipid metabolism, atherogenesis, and haemostasis in Eskimos: The role of the prostaglandin-3 family. Haemostasis 8: 227–233, 1979
13. Bucolo G, David H: Quantitative determination of serum triglycerides by the use of enzymes. Clin Chem 19: 476–482, 1973
14. Omodeo Sale F, Marchesini S, Fishman PH, Berra B: A sensitive enzymatic assay for determination of cholesterol in lipid extracts. Anal Biochem 142: 347–350, 1984
15. Burstein M, Scholnick HR, Morfin R: Rapid method for the isolation of lipoproteins from human serum by precipitation with polyanions. J Lipid Res 11: 583–595, 1970
16. Orr JR, Adamson GL, Lindgren FT: Preparative ultracentrifugation and analytic ultracentrifugation of plasma lipoproteins. In: E.G. Perkins (ed). Analysis of Fats, Oils and Lipoproteins. American Oil Chemists Society, Champaign, IL, 1991, pp 524–554
17. Baxter GP, Wallace CC: Changes in volume upon solution in water of the halogen salts of the alkali metals. J Am Chem Soc 38: 70–105, 1916
18. Lowry OH, Rosebrough NJ, Farr AL, Randall RJ: Protein measurement with the Folin phenol reagent. J Biol Chem 193: 265–275, 1951
19. Folch J, Lees M, Sloane-Stanley GH: A simple method for the isolation and purification of total lipids from animal tissues. J Biol Chem 226: 497–509, 1957
20. Seltzer CC, Mayer J: A simple criterion of obesity. Postgraduate Med 38: A101, 1965
21. Eskimo Survey. Nutrition Canada. Information Canada, 1975, p 28
22. Draper HH: A review of recent nutritional research in the arctic. In: R.J. Shephard, S. Itoh (eds). Circumpolar Health. University of Toronto Press, Toronto, 1975, pp 120–129.
23. Miller GJ, Miller NE: Plasma high-density lipoprotein concentration and development of ischemic heart disease. Lancet 1: 16–20, 1975
24. Schaefer O: Are Eskimos more or less obese than other Canadians? A comparison of skin thickness and ponderal index of Canadian Eskimos. Am J Clin Nutr 30: 1623–1628, 1977

Molecular and Cellular Biochemistry **246:** 129–135, 2003.

Effect of DHLA on response of isolated rat urinary bladder to repetitive field stimulation

Robert M. Levin,[1,2] Abby Borow[3], Sheila S. Levin[1] and Niels Haugaard[4]

[1]*Albany College of Pharmacy;* [2]*Stratton VAMC;* [3]*St. Rose College, Albany, NY;* [4]*Division of Urology, University of Pennsylvania, Philadelphia, PA, USA*

Abstract

Lipoic acid is an essential coenzyme in the oxidation of pyruvate and α-ketoglutarate. It is easily converted to its reduced form, dihydrolipoic acid (DHLA), *in vivo* thereby forming a redox pair. DHLA is important in the maintenance and integrity of specific neuronal and subcellular membranes. In the present study we investigated the effect of DHLA on the response of isolated rat bladder strips to repetitive field stimulation (FS), a method used to exhaust synaptic stores of acetylcholine resulting in nerve and synaptic damage.

Isolated strips of rat urinary bladders were separated into 4 groups. Group 1 strips were incubated with choline + acetyl-CoA; Group 2 strips with choline, acetyl-CoA + DHLA; and Group 3 with DHLA. Group 4 strips were controls. All strips in Groups 1–3 were subjected to 2 h of repetitive FS followed by 2 h of recovery.

DHLA had no effect on the progressive decrease in contractile response observed during repetitive stimulation. However, strips incubated in the presence of DHLA showed a significantly greater degree of recovery than strips incubated in the absence of DHLA. We believe that the protection of the contractile response is related to DHLA's ability to protect nerve and/or muscle membranes from oxidative damage. (Mol Cell Biochem **246**: 129–135, 2003)

Key words: rat, bladder, lipoic acid

Introduction

Thioctic or lipoic acid (LA) occurs naturally as an oxidized disulfide (LA) and a reduced sulfhydryl compound, dihydrolipoic acid (DHLA). A critical coenzyme in the oxidation of a-keto acid dehydrogenase complexes, it is an essential, self-generating redox pair; exogenous LA is converted to its reduced form *in vivo* [1–3]. Both forms are potent antioxidants that exhibit reactive oxygen species (ROS) scavenging and metal chelating properties; DHLA, in addition, can regenerate endogenous antioxidants and repair oxidative damage [4]. LA and, especially, DHLA appear to be important in the maintenance and integrity of neuronal and other cell and subcellular membranes [5–13]. A recent study showed that DHLA increased choline acetyltransferase (ChAT) activity *in vitro* by approximately 6-fold, whereas LA was a potent inhibitor of the enzyme [14], indicating that DHLA is beneficial or, perhaps, necessary for generating adequate acetylcholine levels as well as maintaining healthy neuronal membranes *in vivo*.

Clinically, LA administration has been shown to be of benefit in the treatment of diabetes mellitus [15, 16] and to be particularly effective against the development and/or progression of oxidant-induced diabetic neuropathy [17–20]. The consequences of diabetic neuropathy relate directly to urinary bladder function: Diabetic cystopathy involves sensory and motor nerve dysfunctions resulting in increased urine volumes (bladder distention) and decreased ability to void [21–24].

Results of previous studies showed that repetitive field stimulation (FS) of smooth muscle strips isolated from urinary bladder can be used to exhaust synaptic stores of acetylcholine (Ach) and to produce direct neuronal damage that does not reverse upon cessation of repetitive FS [26, 29, 31]. These studies demonstrated that repetitive stimulation results in a significant increase in the accumulation of MDA (prod-

Address for offprints: R.M. Levin, Albany College of Pharmacy, 106 New Scotland Avenue, Albany, NY 12208, USA (E-mail: levinr@acp.edu) uct of peroxidation) within the smooth muscle membrane components of the bladder strips, and that both the MDA concentration and level of contractile dysfunction are directly related to the calcium concentration, i.e. high calcium enhances both the content of MDA in the tissue and the level of contractile dysfunction, whereas low calcium reduced both parameters [26, 29, 31].

In the present study we investigated the effect of DHLA on the contractile responses of isolated rat bladder strips measured during a period of repetitive FS and following a recovery period. If DHLA stimulates Ach synthesis, it should reduce the progressive loss of contractility observed during repetitive stimulation. And/or if it protects synaptic membranes from exhaustion-induced damage, strips incubated in the presence of DHLA should recover to a greater extent than those incubated in its absence.

Materials and methods

Twenty-four male Wistar rats weighing approximately 250 g each were separated into 4 groups of 6 rats each. Each rat was anesthetized with pentobarbital (25 mg/kg), i.p. The urinary bladder was rapidly exposed and removed through a lower midline incision; the bladder body was separated from the base at the level of the ureteral orifices. Three longitudinal muscle strips were cut from each bladder body. Each strip (2 mm wide, 6 mm long) was mounted in a separate 15 ml-bath containing Tyrode's solution with glucose (1 mg/ml) and maintained at 37°C.

Preliminary studies demonstrated that maximal responses were obtained at a basal tension of 2 g, and thus an initial tension of 2.0 g was placed on each strip. All strips were equilibrated for at least 60 min with a gas mixture of 95% O_2 and 5% CO_2, then field stimulated at 80 v, 1 msec duration, at 2, 8 and 32 Hz. Contractions were recorded on a Grass model D polygraph and digitally monitored using a Polyview A-D conversion system. The Polyview data acquisition system provides both maximal contractile response and the maximal rate of tension generation for each contractile response. Prior studies have demonstrated that the rate of tension generation is more sensitive to free-radical damage than is the maximal response to FS.

After initial equilibration, the incubation medium was changed to Tyrode's solution containing glucose and the following additions for each of 4 experimental groups: All tissues in Group 1 were incubated with choline (5 mM) + acetyl-CoA (0.1 mM).; those in Group 2 with choline, acetyl-CoA + DHLA (1 mM); and those in Group 3 with DHLA (1 mM). Group 4 strips were maintained in Tyrode's solution with glucose, only. All strips were equilibrated in these buffers for 30 min then stimulated at 2, 8 and 32 Hz.

Repetitive stimulation

Twelve strips from each group were stimulated for 2 h at 32 Hz for 15 sec every 5 min. At the end of the period of repetitive stimulation, all strips were washed 4 times with fresh buffer (containing the group-specific additions described above). After 2 h of incubation (recovery period), responses to FS (non-repetitive) at 2, 8 and 32 Hz were re-assessed. Note that the 6 strips from each group that were not repetitively stimulated were incubated for a total of 4 h; their responses to FS were determined after 2 h to coincide with the end of the period of repetitive stimulation and after 4 h to coincide with the end of the recovery period.

The effect of repetitive stimulation, and the responses following the recovery period are presented as the mean ± S.E.M. of the percent of the control response (response prior to repetitive stimulation).

Statistical analysis utilized analysis of variance followed by the Neuman Keuls test for individual differences among groups.

Results

Figure 1 shows the contractile responses to 2, 8 and 32 Hz of isolated rat bladder strips incubated in Tyrode's solution containing glucose. Incubating these tissues in this medium with the addition of choline + acetyl-CoA, choline + acetyl-CoA + DHLA or DHLA alone did not significantly affect the contractile responses to FS at any frequency; in the absence of repetitive stimulation the contractile responses to FS was unaffected during 4 h of incubation (data not shown).

Two hours of repetitive stimulation, caused a progressive decrease in maximal contractile response to FS (32 Hz) (Fig. 2) and an even greater rate of decline in maximal rate of tension generation (Fig. 3) than maximal response; no incubation medium tested affected these rates of contractile decline in response to repetitive FS (Figs 2 and 3).

Two hours of repetitive stimulation resulted in approximately a 40% decrease in the maximal contractile response to all frequencies of stimulation (Figs 4A–4C), and approximately a 60% decrease in the rate of tension generation (Figs 5A–5C). The magnitude of the decrease in response following repetitive stimulation was not affected by any of the incubation media (Figs 4 and 5).

Following the 2 h period of recover, the contractile responses to FS in the control group and in the group incubated with choline + acetyl CoA showed no increase. However, contractility of isolated strips incubated in the presence of choline + acetyl-CoA + DHLA or DHLA alone improved significantly, to greater than 85% of control maximal response (Figs 4A–4C) and to greater than 65% of control

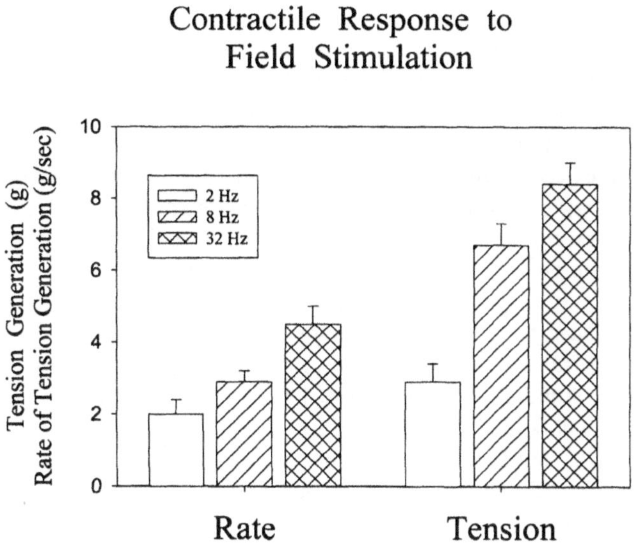

Fig. 1. Maximal tension generation and maximal rate of tension generation of isolated bladder strips in response to FS (2, 8, and 32 Hz). Each value is the mean ± S.E.M.; n = 6 rats.

maximal rate of tension generation in response to 8 and 32 Hz (Figs 5A–5C). Maximal contractile responses following the recovery period were not statistically different from control responses (Figs 4A–4C) for the strips incubated with DHLA.

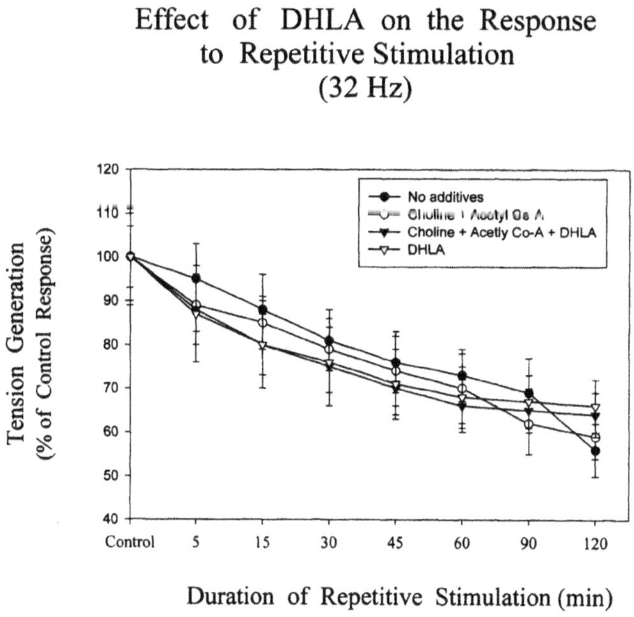

Fig. 2. Maximal tension generation of isolated bladder strips in response to repetitive FS (32 Hz) in the presence of no additives, choline (5 mM) + acetyl-CoA (0.1 mM), choline (5 mM) + acetyl-CoA (0.1 mM) + DHLA (1 mM) or DHLA (1 mM) only. Strips were stimulated for 15 sec at 5 min intervals for 2 h. Each value is the mean ± S.E.M.; n = 6 rats.

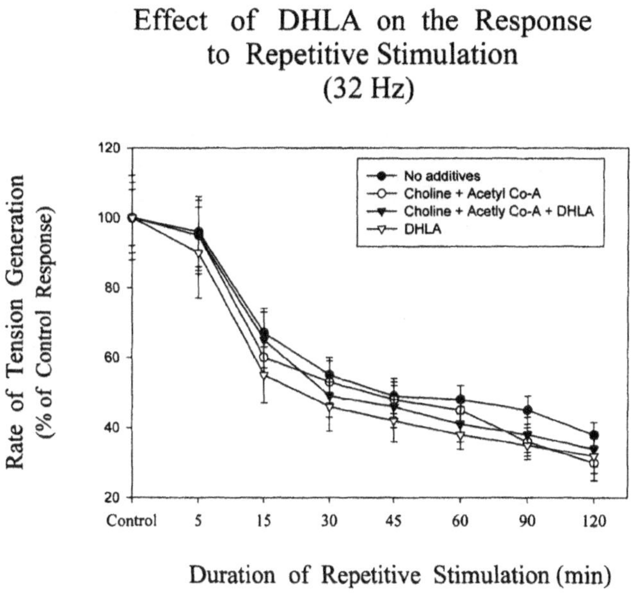

Fig. 3. Maximal rates of tension generation of isolated bladder strips in response to repetitive FS (32 Hz) in the presence of no additives, choline (5 mM) + acetyl-CoA (0.1 mM), choline (5 mM) + acetyl-CoA (0.1 mM) + DHLA (1 mM) or DHLA (1 mM) only. Strips were stimulated for 15 sec at 5 min intervals for 2 h. Each value is the mean ± S.E.M.; n = 6 rats.

The rates of tension generation for all incubation conditions were significantly reduced (Figs 5A–5C), however, the strips incubated with DHLA were significantly greater than the rates of tension generation for the strips incubated without DHLA.

Discussion

There is increasing evidence that LA/DHLA has beneficial effects in the treatment and prevention of specific diseases related to ROS damage and reperfusion injury [1, 2, 15–20, 27, 28]. Although many of these effects have been linked to its potent antioxidant activity, other mechanisms of action may contribute [1, 2, 4]. In addition, DHLA may maintain and protect neural function through its recently shown ability to stimulate ChAT activity [14].

Using isolated rat bladder strips the present study was designed to determine if DHLA could: (1) protect efferent innervation and synaptic transmission from progressive loss due to repetitive stimulation; and (2) improve recovery following a 2 h period of *in vitro* repetitive FS. Results of previous experiments showed that *in vitro* repetitive FS produced a progressive reduction in contractile response to FS that was greater than the reduction in response to bethanechol indicating more pronounced damage to nerve than to muscle cells (rabbit bladder) [25, 26, 29–31]. Glucose deprivation signifi-

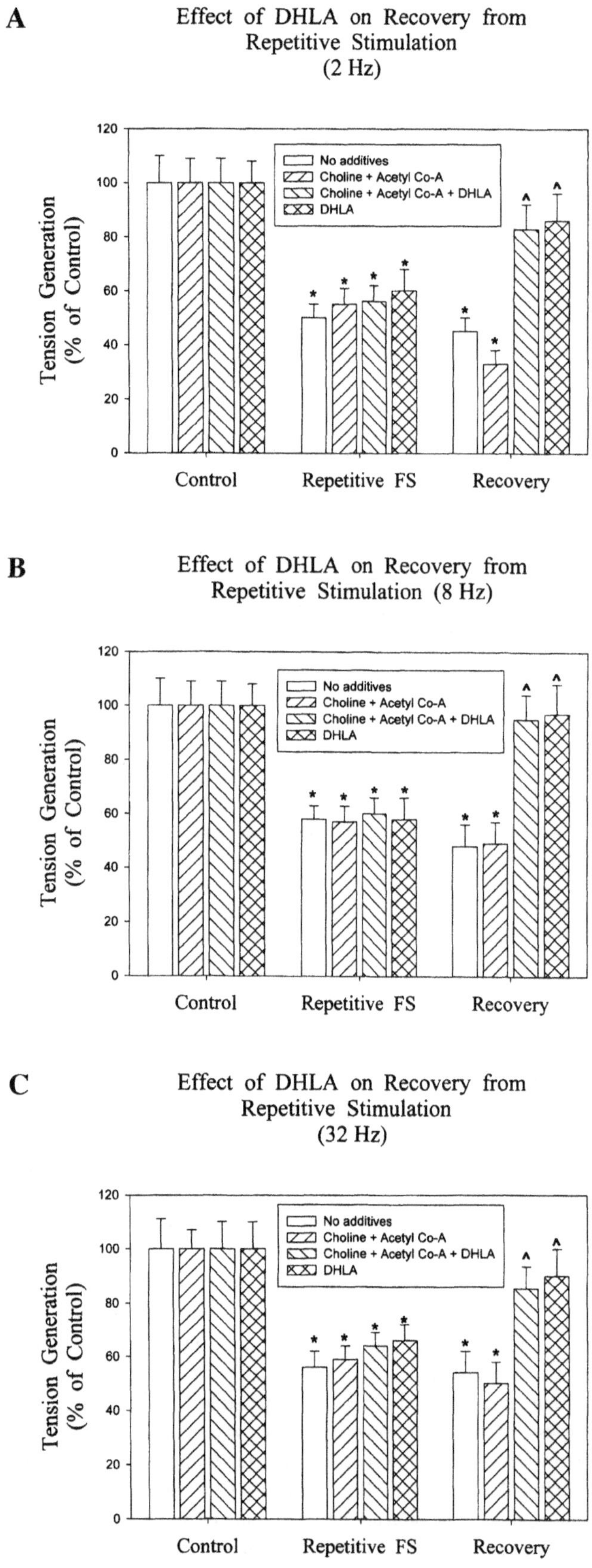

cantly increased the rate of contractile failure during repetitive stimulation [25] and anoxia further exaggerated the rate of failure as did increasing the extracellular calcium concentration [26, 29]. These studies also showed that repetitive stimulation increased lipid peroxidation, which was further enhanced by anoxia followed by reperfusion. The magnitude of lipid peroxidation was further increased by increasing the extracellular calcium concentration [29]. EGTA, diltiazem, pinacidil and low calcium concentration all reduced the rate of contractile dysfunction and amount of lipid peroxidation induced by repetitive stimulation [29, 31].

Results of similar work by Lin *et al.* showed that repetitive stimulation [50 v, 32 Hz, 1 msec, every 2.5 min) of the rat whole bladder *in vitro* reduced pressure generation, rate of pressure generation and emptying ability and, also, reduced tissue ATP and creatine phosphate concentrations [32]. No recovery from fatigue experiments were performed. Results of an *in vivo* study (dogs) showed that spinal nerve electrostimulations of the bladder repeated more frequently than 5 min apart caused decreased maximal pressure generation with each subsequent stimulation that varied directly with a decrease in the time interval between stimulations. Stimulated maximal pressure generation recovered to control values after these nerves/bladders were allowed to rest for 15 min [33].

Based on the recent demonstration that DHLA can markedly increase the activity of ChAT, we incubated individual strips in choline + acetyl-CoA (substrates for the enzyme) ± DHLA [14]. The results, however, showed that none of the incubation media tested prevented the progressive decline in contractile response to repetitive stimulation, suggesting that the synaptic content of Ach was not enhanced in these experiments. On the other hand, both media containing DHLA (choline + acetyl-CoA + DHLA and DHLA alone) significantly improved contractility measured after the 2 h period of recovery.

Several possible mechanisms for the positive effect of DHLA on rat bladder strip recovery from fatigue in this experiment are proposed: DHLA may have (1) maintained the integrity of synaptic membranes (and postsynaptic nerves) during repetitive stimulation, (2) prevented the putative generation and buildup of ROS induced by repetitive stimulation, (3) regenerated the endogenous antioxidant pool (4) and/

←

Fig. 4. (A–C) Maximal tension generation of isolated bladder strips in response to repetitive FS and after 2 h of recovery from repetitive FS in the presence of no additives, choline (5 mM) + acetyl-CoA (0.1 mM), choline (5 mM) + acetyl-CoA (0.1 mM) + DHLA or DHLA (1 mM) only. After 2 h of repetitive stimulation (15 sec, 5 min intervals) all strips were washed 4 times with fresh buffer and incubated for 2 more h, then stimulated once. FS was applied at 2 Hz (A), 8Hz (B) and 32 Hz (C). Each bar is the mean ± S.E.M.; n = 6 rats. *Significantly different from control; ^significantly different from repetitive stimulation. $p < 0.05$.

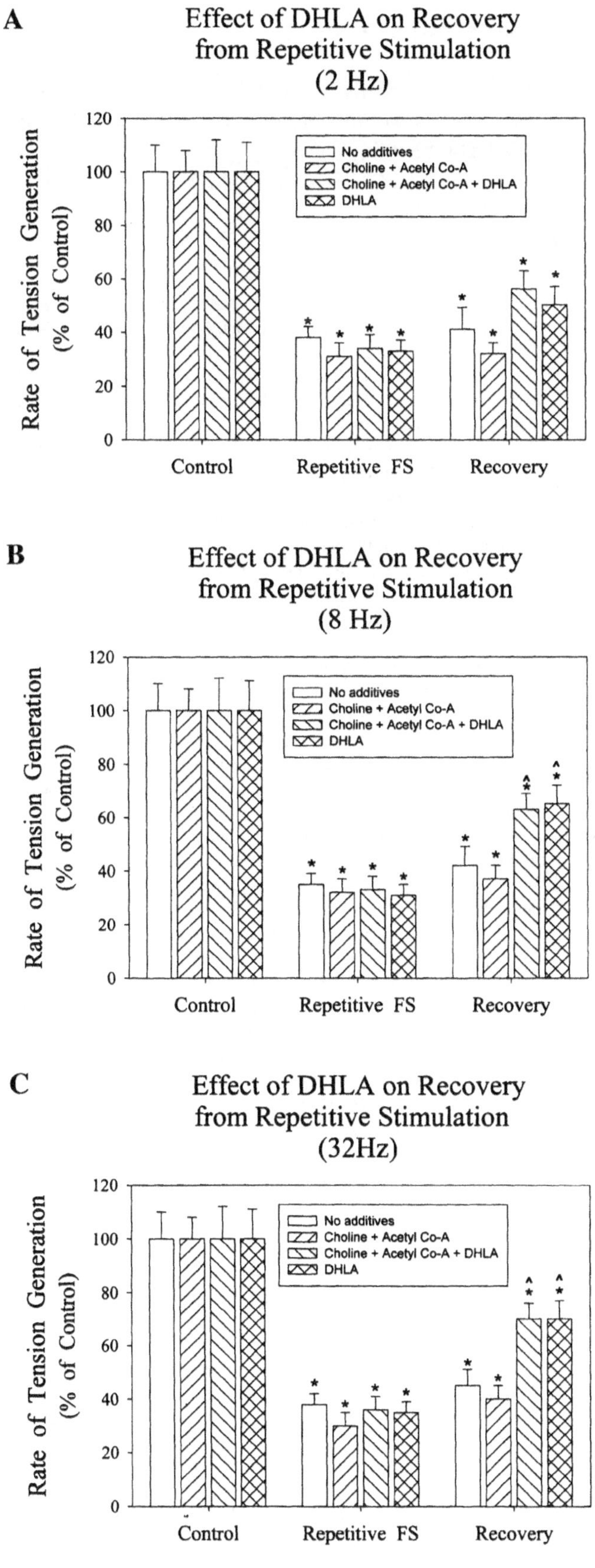

or protected sulfhydryl-containing enzymes critical for energy production from oxidation. Based on evidence that LA is effective in the treatment and prevention of diabetic neuropathy which has been linked to ROS damage to nerves and synapses (neural tissue is particularly vulnerable to peroxidation because of its high lipid content) and that DHLA is a potent reducing agent, all of these may be likely mechanisms for the enzyme's beneficial effects.

ROS are generated from increased electron flux through rapidly respiring, electron 'leaky' mitochondria as well as other sources, e.g. NADPH oxidase activity, xanthine oxidase activity, polymorphoneutrophil infiltration and NO synthase activity during exhaustive exercise (fatigue) of skeletal muscle [34, 35]. Exhaustive exercise induced significant free radical formation (measured in human serum by electron spin resonance spectroscopy), and increased lipid peroxidation [34]. Exhaustive treadmill exercise (rats) increased free radical concentration of skeletal muscle and liver, decreased mitochondrial respiratory control and caused SR and other membrane damage [36]. Functionally, these radicals decreased contractility and increased the rate of fatigue. Results of several other studies showed that strenuous exercise can induce oxidative stress, thereby damaging lipids, proteins and DNA [36].

Protein sulfhydryl groups are also highly reactive with the membrane lipid peroxidation product, 4-hydroxy-2-nonenal (HNE) [38–40]. HNE inactivated mitochondrial alpha-ketoglutarate dehydrogenase (KGDH) *in vitro* [38, 39], and is thought to be responsible for mitochondrial dysfunction following reperfusion in rat hearts [40].

Our prior studies demonstrated that repetitive stimulation of isolated bladder strips results in an increase in membrane peroxidation products, and that the rate of peroxidation is enhanced in the presence of high calcium and reduced in the presence of low calcium is consistent with the above studies on muscle exhaustion [26, 29, 31].

In addition to it's role as an essential coenzyme in oxidative metabolism, the low molecular weight vicinal thiol DHLA is a potent reductant that participates in a series of redox cycles that regenerate the antioxidant network of vitamins C and E, ubiquinol and glutathione, and is then regenerated

Fig. 5. (A–C) Maximal rates of tension generation of isolated bladder strips in response to repetitive FS and after 2 h of recovery from repetitive FS in the presence of no additives, choline (5 mM) + acetyl-CoA (0.1 mM), choline (5 mM) + acetyl-CoA (0.1 mM) + DHLA or DHLA (1 mM) only. After 2 h of repetitive stimulation (15 sec, 5 min intervals) all strips were washed 4 times with fresh buffer and incubated for 2 more h, then stimulated once. FS was applied at 2 Hz (A), 8Hz (B) and 32 Hz (C). Each bar is the mean ± S.E.M.; n = 6 rats. *Significantly different from control; ^significantly different from repetitive stimulation. $p < 0.05$.

from the disulfide LA by NADH/NAD(P)H [36]. And, by providing reducing equivalents, it can protect/regenerate sulfhydryl-containing proteins from oxidant-induced damage. DHLA is also a ROS scavenger and metal chelator.

ROS-induced disruption of intracellular calcium homeostasis results from glutathione depletion and modification of protein thiols: Mammalian cells treated with pro-oxidants exhibit a loss of thiol homeostasis followed sequentially by impaired calcium transport, disrupted intracellular calcium homeostasis and a sustained increase in cytosolic calcium concentration. [41]. DHLA, which is generated from exogenous LA intracellularly, can reduce (regenerate) oxidized glutathione: In a number of cell systems, α-lipoic acid has shown to be protective of thiol-containing proteins [42–44].

Putative mechanism for DHLA prevention of increased intracellular calcium concentration after oxidative stress

Oxidative stress, which oxidizes mitochondrial glutathione and pyridine nucleotides, can activate bovine liver mitochondrial ADP-ribosyl cyclase (NAD^+ glycohydrolase [NADase]) to catalyze the formation of cyclic ADP ribose, a potent intracellular calcium-mobilizing agent, from NAD^+; reducing agents completely inactivate the enzyme. [45]. The authors hypothesize that previously observations of oxidant-induced calcium efflux from mitochondria may result from activation of the NADase and production of cADP ribose after oxidation of intramitochondrial glutathione and pyridine nucleotides. Furthermore, ADP-ribosyl cyclase may be reversibly switched from the active to inactive state depending on its thiol status. DHLA, a powerful reducing agent, may help maintain a high mitochondrial $NADH/NAD^+$ ratio, regenerate glutathione oxidized by ROS and, putatively, inactivate ADP-ribosyl cyclase or maintain the enzyme in its inactive form, thereby preserving intracellular calcium homeostasis.

Conclusion

In the current experiments, DHLA promoted recovery from repetitive stimulation-induced fatigue. Taking the previous discussions into consideration it appears that DHLA exerted its protective effects by several mechanisms, ultimately reducing ROS generated by exhaustively exercised tissue and, also, the deleterious consequences of ROS activity. Exogenously administered LA might improve bladder function in patients with hyperreflexia (for which repetitive stimulation is a model) and, also, in obstructive bladder disorders in which ROS-producing cyclical ischemia/reperfusion probably occurs with each micturition [46, 47], especially when obstruction is accompanied by hyperreflexia, a condition that may, itself, result from oxidant-induced nerve damage.

Acknowledgement

This work was supported in part by a grant from the Department of Veterans Affairs, and NIH grants RO-1-DK 26508 and RO-1-DK 53965

References

1. Packer L, Roy S, Sen CK: α-Lipoic acid: A metabolic antioxidant and potential redox modulator of transcription. Adv Pharmacol 38: 79–101, 1997
2. Bustamante J, Lodge JK, Marcocci L, Tritschler HJ, Packer L, Rihn BH: α-Lipoic acid in liver metabolism and disease. Free Radic Biol Med 24: 1023–1039, 1998
3. Constantinescu A, Pick U, Handelman GJ, Haramaki N, Han D, Podda M, Tritschler HJ, Packer L: Reduction and transport of lipoic acid by human erythrocytes. Biochem Pharmacol 50: 253–261, 1995
4. Biewenga GP, Haenen GR, Bast A: The pharmacology of the antioxidant lipoic acid. Gen Pharmacol 29: 315–331, 1997
5. Hofman M, Mainka P, Tritschler H, Fuchs J, Zimmer G: Decrease of red cell membrane fluidity and -SH groups due to hyperglycemic conditions is counteracted by alpha-lipoic acid. Arch Biochem Biophys 324: 85–92, 1995
6. Scheer B, Zimmer G: Dihydrolipoic acid prevents hypoxic/reoxygenation and peroxidative damage in rat heart mitochondria. Arch Biochem Biophys 302: 385–390, 1993
7. Rudich A, Tirosh A, Potashnik R, Khamaisi M, Bashan N: Lipoic acid protects against oxidative stress induced impairment in insulin stimulation of protein kinase B and glucose transport in 3T3-L1 adipocytes. Diabetologia 42: 949–957, 1999
8. Kilic F, Handelman GJ, Serbinova E, Packer L, Trevithick JR: Modelling cortical cataractogenesis 17: *In vitro* effect of α-lipoic acid on glucose-induced lens membrane damage, a model of diabetic cataractogenesis. Biochem Mol Biol Int 37: 361–370, 1995
9. Muller L, Menzel H: Studies on the efficacy of lipoate and dihydrolipoate in the alteration of $cadmium^{2+}$ toxicity in isolated hepatocytes. Biochim Biophys Acta 1052: 386–391, 1990
10. Packer L, Tritschler HJ, Wessel K: Neuroprotection by the metabolic antioxidant alpha-lipoic acid. Free Radic Biol Med 22: 359–378, 1997
11. Kagan VE, Shvedova A, Serbinova E, Khan S, Swanson C, Powell R, Packer L: Dihydrolipoic acid – a universal antioxidant both in the membrane and in the aqueous phase. Reduction of peroxyl, ascorbyl and chromanoxyl radicals. Biochem Pharmacol 44: 1637–1649, 1992
12. Lockhart B, Jones C, Cuisinier C, Villain N, Peyroulan D, Lestage P: Inhibition of L-homocysteic acid and buthionine sulphoximine-mediated neurotoxicity in rat embryonic neuronal cultures with alpha-lipoic acid enantiomers. Brain Res 855: 292–297, 2000
13. Khamaisi M, Potashnik R, Tirosh A, Demshchak E, Rudich A, Tritschler H, Wessel K, Bashan N: Lipoic acid reduces glycemia and increases muscle GLUT4 content in streptozotocin-diabetic rats. Metabolism 46: 763–768, 1997
14. Haugaard N, Levin RM: Regulation of the activity of choline acetyl transferase by lipoic acid. Mol Cell Biochem 213: 61–63, 2000
15. Konrad T, Vicini P, Kusterer K, Hoflich A, Assadkhani A, Bohles HJ, Sewell A, Tritschler HJ, Cobelli C, Usadell KH: alpha-Lipoic acid treatment decreases serum lactate and pyruvate concentrations and improves glucose effectiveness in lean and obese patients with type 2 diabetes. Diabetes Care 22: 280–287, 1999
16. Jacob S, Ruus P, Hermann R, Tritschler HJ, Maerker E, Renn W, Augustin HJ, Dietze GJ, Rett K: Oral administration of RAC-alpha-

lipoic acid modulates insulin sensitivity in patients with type-2 diabetes mellitus: A placebo-controlled pilot trial. Free Radic Biol Med 27: 309–314, 1999
17. Ruhnau KJ, Meissner HP, Finn JR, Reljanovic M, Lobisch M, Schutte K, Nehrdich D, Tritschler HJ, Mehnert H, Ziegler D: Effects of 3-week oral treatment with the antioxidant thioctic acid (alpha-lipoic acid) in symptomatic diabetic polyneuropathy. Diabetes Med 16:1040–1043, 1999
18. Ziegler D, Reljanovic M, Mehnert H, Gries FA: Alpha-lipoic acid in the treatment of diabetic polyneuropathy in Germany: Current evidence from clinical trials. Exp Clin Endocrinol Diabetes 107: 421–430, 1999
19. Haak ES, Usadel KH, Kohleisen M, Yilmaz A, Kusterer K, Haak T: The effect of alpha-lipoic acid on the neurovascular reflex arc in patients with diabetic neuropathy assessed by capillary microscopy. Microvasc Res 58: 28–34, 1999
20. Androne L, Gavan NA, Veresiu IA, Orasan R: *In vivo* effect of lipoic acid on lipid peroxidation in patients with diabetic neuropathy. In Vivo 14: 327–330, 2000
21. Watanabe T, Miyagawa I: Characteristics of detrusor contractility during micturition in diabetics. Neurourol Urodyn 18: 163–171, 1999
22. Turner WH, Brading AF: Smooth muscle of the bladder in the normal and the diseased state: Pathophysiology, diagnosis and treatment. Pharmacol Ther 75: 77–110, 1997
23. Kaplan SA, Te AE, Blaivas JG: Urodynamic findings in patients with diabetic cystopathy. J Urol 153: 342–344, 1995
24. Kaplan SA, Blaivas JG: Diabetic cystopathy. HNO 2: 133–139, 1988
25. Kwon H-Y, Longhurst PA, Parsons K, Wein AJ, Levin RM: Effects of glucose deprivation on the contractile response of the rabbit bladder to repetitive stimulation. Neurourol Urodyn 15: 71–78, 1996
26. Ohnishi N, Liu S-P, Horan P, Levin RM: Effect of repetitive stimulation on the contractile response of rabbit urinary bladder subjected to *in vitro* hypoxia or *in vitro* ischemia followed by reoxygenation. Pharmacology 57: 139–147, 1998
27. Gutierrez-Cabano CA: Thioctic acid protection against ethanol-induced gastric mucosal lesions involves sulfhydryl and prostaglandin participation. Acta Gastroenterol Latinoam 27: 31–37, 1997
28. Serbinova E, Khwaja S, Reznick AZ, Packer L: Thioctic acid protects against ischemia-reperfusion injury in the isolated perfused Langendorff heart. Free Radic Res Commun 17: 49–58, 1992
29. Levin RM, Leggett R, Whitbeck C, Horan P: Effect of calcium and calcium chelators on the response of the bladder to *in vitro* ischemia. Br J Urol 82: 882–887, 1998
30. Levin RM, Leggett RE, Whitbeck C, Horan P: Effect of pinacidil on the response of the rabbit urinary bladder to repetitive stimulation and *in vitro* ischemia. Neurourol Urodyn 18: 129–137, 1999
31. Levin RM, Leggett R, Whitbeck C, Horan P: Correlation of EGTA and calcium-blocking agents on the response of the bladder to *in vitro* ischemia. Pharmacology 58: 113–119, 1999
32. Lin AT-L, Yang C-H, Chang LS: Impact of aging on rat urinary bladder fatigue. J Urol 157: 1900–1904, 1997
33. Bross S, Schumacher S, Scheepe JR, Seif C, Junemann KP, Alken P: Smooth muscle fatigue due to repeated urinary bladder neurostimulation: An *in vivo* study. Neurourol Urodyn 18: 41–53, 1999
34. Ashton T, Rowlands CC, Jones E, Young IS, Jackson SK, Davies B, Peters JR: Electron spin resonance spectroscopic detection of oxygen-centred radicals in human serum following exhaustive exercise. Eur J Appl Physiol 77: 498–502, 1998
35. Bejma J, Ji LL: Aging and acute exercise enhance free radical generation in rat skeletal muscle. J Appl Physiol 87: 465–470, 1999
36. Sen CK, Packer L: Thiol homeostasis and supplements in physical exercise. Am J Clin Nutr 72: 653S–669S, 2000
37. Uchida K, Stadtman ER: Covalent attachment of 4-hydroxynonenal to glyceraldehyde-3-phosphate dehydrogenase. A possible involvement of intra- and intermolecular cross-linking reaction. J Biol Chem 268: 6388–6393, 1993
38. Humphries KM, Szweda LI: Selective inactivation of alpha-ketoglutarate dehydrogenase and pyruvate dehydrogenase: Reaction of lipoic acid with 4-hydroxy-2-nonenal. Biochemistry 37: 15835–15841, 1998
39. Lucas DT, Szweda LI: Decline in mitochondrial respiration during cardiac reperfusion: Age-dependent inactivation of α-ketoglutarate dehydrogenase. Proc Natl Acad Sci USA 96: 6689–6693, 1999
40. Sadek H, Humphries KM, Szweda PA, Szweda LI: *In vivo* cardiac reperfusion: Free radical mediated loss in mitochondrial function. FASEB J 14: (abstr 1452), A1565, 2000
41. Orrenius S, Burkitt MJ, Dass GE, Dypbukt JM, Nicotera P: Calcium ions and oxidative cell injury. Ann Neurol 32: S33–S42, 1992
42. Sen CK, Packer L: Antioxidant and redox regulation of gene transcription. FASEB J 10: 709–720, 1996
43. Dimpfel W: Effect of thioctic acid on pyramidal cell responses in the rat hippocampus *in vitro*. Eur J Med Res 1: 523–527, 1996
44. Vasdev S, Ford CA, Parai S, Longerich L, Gadag V: Dietary alpha-lipoic acid supplementation lowers blood pressure in spontaneously hypertensive rats. J Hypertens 18: 567–573, 2000
45. Ziegler M, Jorcke D, Schweiger M: Identification of bovine liver mitochondrial NAD^+ glycohydrolase as ADP-ribosyl cyclase. Biochem J 326: 401–405, 1997
46. Greenland JE, Hvistendahl JJ, Andersen H, Jorgensen TM, McMurray G, Cortina-Borja M, Brading AF, Frokiaer J: The effect of bladder outlet obstruction on tissue oxygen tension and blood flow in the pig bladder. BJU Int 85:1109–1114, 2000
47. Greenland JE, Brading AF: The effect of bladder outflow obstruction on detrusor blood flow changes during the voiding cycle in conscious pigs. J Urol 165: 245–248, 2001

Molecular and Cellular Biochemistry **246**: 137–141, 2003.

Decreased level of 2,3-diphosphoglycerate and alteration of structural integrity in erythrocytes infected with *Plasmodium falciparum in vitro*

M.L. Dubey,[1] Ramakrishna Hegde,[1] N.K. Ganguly[2] and R.C. Mahajan[1]

Departments of [1]Parasitology; [2]Experimental Medicine and Biotechnology, Post Graduate Institute of Medical Education and Research, Chandigarh, India

Abstract

2,3-Diphosphoglycerate (2,3-DPG), an intracellular metabolite of glycolytic pathway is known to affect the oxygen binding capacity of haemoglobin and mechanical properties of the red blood cells. 2,3-DPG levels have been reported to be elevated during anaemic conditions including visceral leishmaniasis. 2,3-DPG activity in *P. falciparum* infected red blood cells, particularly in cells infected with different stages of the parasite and its relationship with structural integrity of the cells is not known. Chloroquine sensitive and resistant strains of *P. falciparum* were cultured *in vitro* and synchronized cultures of ring, trophozoite and schizont stage rich cells along with the uninfected control erythrocytes were assayed for 2,3-DPG activity and osmotic fragility. It was observed that in both the strains, in infected erythrocytes the 2,3-DPG activity gradually decreased and osmotic fragility gradually increased as the parasite matured from ring to schizont stage. The decrease in 2,3-DPG may probably be due to increased pyruvate kinase activity of parasite origin, which has been shown in erythrocytes infected with several species of Plasmodium. The absence of compensatory increase in 2,3-DPG in *P. falciparum* infected erythrocytes may aggravate hypoxia due to anaemia in malaria and probably may contribute to hypoxia in cerebral malaria. As 2,3-DPG was not found to be increased in erythrocytes parasitized with *P. falciparum*, the increased osmotic fragility observed in these cells is not due to increased 2,3-DPG as has been suggested in visceral leishmaniasis. (Mol Cell Biochem **246**: 137–141, 2003)

Key words: chloroquine resistant, chloroquine sensitive, 2,3-diphosphoglycerate (2,3-DPG), osmotic fragility, *P. falciparum*

Introduction

Malaria still remains a major public health problem in majority of tropical and subtropical countries including India. Present global incidence of malaria is estimated to be 300–500 million cases every year with 1.5–2.7 million deaths annually which includes more than a million of children below 5 years of age in Africa [1]. Almost all the deaths are due to *P. falciparum* infections.

2,3-DPG is an important intermediate metabolite in the glycolytic pathway of erythrocytes. 2,3-DPG is formed from 1,3-DPG by phosphoglycerate mutase [2].

The cytoskeletal complex of erythrocyte membrane is largely responsible for maintaining the structural integrity and proper functioning of the red cells [3, 4]. 2,3- DPG, may have a role in the physiological regulation of mechanical properties of the red cell membrane [5].

2-3-DPG exerts an effect on oxyhaemoglobin dissociation curve similar to that of CO_2 (the Bohr effect) [2, 6] 2,3-DPG binds to a specific site in the β-chain of haemoglobin and by changing the conformation of haemoglobin molecule, it decreases its oxygen affinity and causes the shift of the dissociation curve to right resulting in increased delivery of oxygen to tissues [2, 7]. In anaemic patients the intra erythrocytic concentration of 2,3-DPG is generally elevated, facilitating oxygen delivery to tissues [8–10]. The destabilization of erythrocyte membrane in visceral leishmaniasis patients has been attributed to the increase in 2,3-DPG concentrations in red cells in patients of visceral leishmaniasis [11]. As the visceral leishmaniasis is a disease of reticuloendothelial system and

Present address: N.K. Ganguly, Indian Council of Medical Research, Ansari Nagar, P.O. Box 4911, New Delhi-110029, India
Address for offprints: M.L. Dubey, Department of Parasitology, PGIMER, Chandigarh-160012, India (E-mail: mdubey@glide.net.in)

Leishmania donovani amastigotes do not affect the erythrocytes directly, in the present study we investigated 2,3-DPG activity and status of erythrocyte membranes by osmotic fragility test in red blood cells infected with *P. falciparum* at different stages during erythrocytic schizogony.

Two strains, one chloroquine sensitive and one chloroquine resistant of *P. falciparum* were used to investigate if there was any biological advantage to chloroquine resistant parasites in terms of erythrocytic 2,3-DPG concentration or it's structural integrity during the course of infection.

Materials and methods

Parasite strains and cell preparations

Two strains of *P. falciparum* (MRC-20 chloroquine sensitive; MRC-76 chloroquine resistant) originally obtained from Malaria Research Centre, Delhi, were maintained *in vitro* in continuous culture according to conventional candle jar method [12]. Normal erythrocytes from same donors whose cells were used for parasite culture, were used as control cell population and were kept under the same *in vitro* conditions as the parasitized red blood cell. To obtain different growth stages of the parasite the cultures were synchronised by 5% D-sorbitol treatment [13]. Following the second sorbitol treatment the cells were recultured to obtain ring, trophozoite and schizont infected RBCs after further incubation for 2–6, 12–18 and 26–40 h respectively (Fig. 1). Further concentration of stage specific parasite was done using 75, 70 and 65% percoll density gradient for ring, trophozoite and schizont stages respectively [14]. With this method it was possible to obtain an average stage specific density of 65–75%.

From both chloroquine sensitive and resistant strains the ring, trophozoite and schizont rich parasitized cells along with the uninfected control erythrocytes were used to estimate 2,3-DPG and osmotic fragility.

2,3-DPG estimation

For determining 2,3-DPG activity in the erythrocytes, various cell preparations were treated with phosphoglyceromutase in presence of glycolate-2-phosphate and the released inorganic phosphate was assayed as described by Rose and Liebowitz [15].

The results of DPG assays were expressed in terms of haemoglobin content of red cell volume. Diphosphoglycerate content was expressed as µmol/µmol haemoglobin. The haemoglobin content of each sample was measured by spectrophotometer using cyanmethaemoglobin method [16].

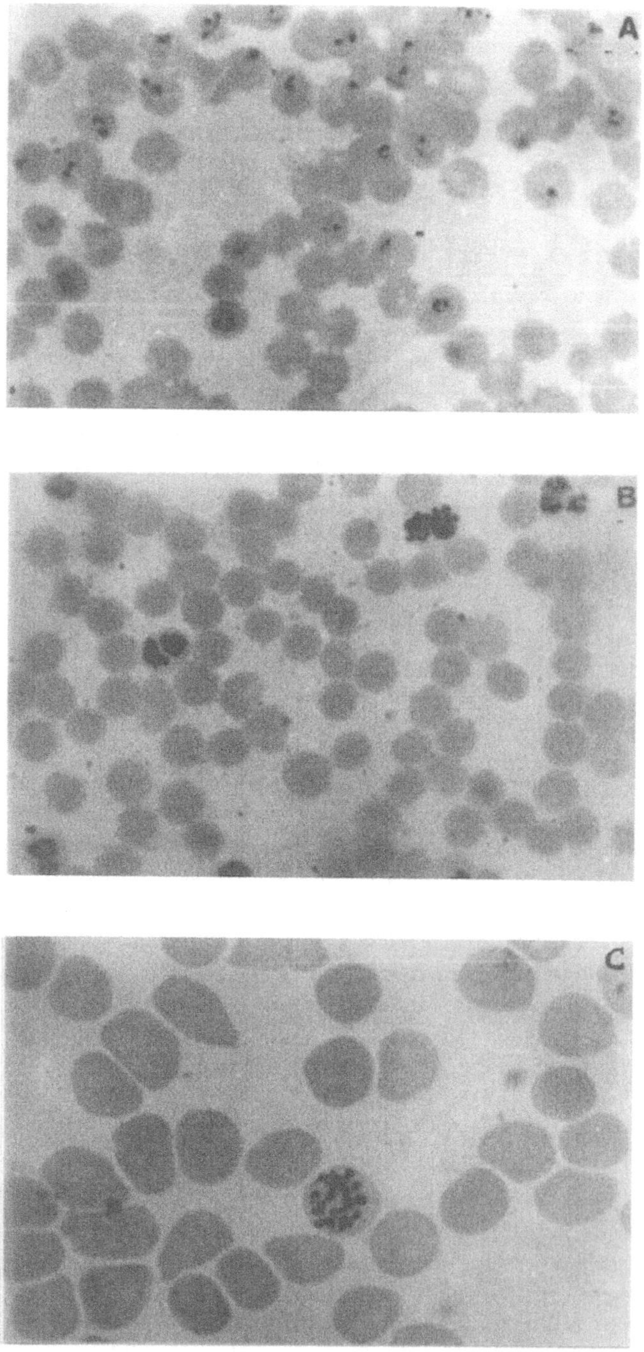

Fig. 1. Photomicrograph of blood smears from synchronized *in vitro* cultures of *P. falciparum* showing infected cells with ring (A), trophozoite (B) and schizonts (C) stages.

Osmotic fragility assay

This was done by observing the cell lysis in various grades of hypotonic saline [17]. Briefly, small volumes (20 µl) of infected and control red blood cells were mixed with excess of buffered saline (5 ml) solutions of varying concentration ranging from 0.1–0.8%. The suspensions were equilibrated for 45 min at room temperature. Further hemolytic process was stopped by addition of 5 ml of corresponding complimentary solutions to each tube and tubes were centrifuged at 1200 g for 5 min. The fraction of red cells lysed at each saline concentration was determined colorimetrically by recording optical density (OD) at a wavelength of 540 nm [17]. The value of 100% lysis was assigned to the reading with the supernatant of tube where distilled water was used for cell lysis and the readings from the other tubes were expressed as the percent of the value of lysis in distilled water [17].

The results were analyzed statistically by ANOVA and student's *t*-test.

Results

2,3-DPG

The 2,3-DPG activity in infected erythrocytes showed a gradual decrease as the parasites attained maturity in both chloroquine sensitive and resistant strains (Table 1). When the 2,3-DPG activity of parasitized RBC was compared with that of the uninfected red cells, the ring infected red cells (RIRBC) did not show statistically significant decrease in both strains. However, the trophozoite infected (TIRBC) and schizont infected red blood cells (SIRBC) showed a statistically significant decrease in both strains. Further, the differences between the 2,3-DPG activity of RIRBC and SIRBC were significant in both the strains while that between TIRBC and SIRBC was statistically significant in chloroquine sensitive strain and not in resistant strain (Table 1). When comparisons were made between respective parasite stages of chloroquine sensitive and resistant strains there were no significant differences in 2,3-DPG activity (Table 1).

Osmotic fragility

The concentration of saline causing 50% haemolysis was used to compare the osmotic fragility of various cell preparations. In uninfected cells 50% haemolysis was observed at 0.43% saline. In infected cells the salt concentration at which 50% haemolysis occurred increased as the parasite matured from ring to schizonts stage (Table 2). On the other hand the amount of haemolysis seen in 0.43% saline also increased gradually with the maturation of parasites from ring to schizonts stage in both Chloroquine sensitive and resistant strains. This ranged from 57–78% depending on the stage of the parasite, maximum being at the schizonts stage (Table 2). In both the strains at this salt concentration the difference in haemolysis between RIRBC and TIRBC and between RIRBC and SIRBC was significant, while between TIRBC and SIRBC it was not statistically significant (Table 2).

The degree of haemolysis is 0.43% saline and the saline concentration causing 50% haemolysis was higher at all the three parasite stages in Chloroquine sensitive strain when compared to that of resistant strain showing that the red cells infected with sensitive strain were comparatively more fragile. But these differences were not statistically significant (Table 2).

Table 1. 2,3-DPG profile of uninfected and infected erythrocytes at different parasite stages in chloroquine sensitive and resistant strains of *P. falciparum*

	Chloroquine sensitive strain			Chloroquine resistant strain		
Uninfected	Ring (R)	Trophozoite (T)	Schizont (S)	Ring (R)	Trophozoite (T)	Schizont (S)
43.02 ± 14.7	33.2 ± 19.1	18.5 ± 10.6^{*+}	12.3 ± 7.1$^{**++\#}$	31.5 ± 21.5	16.2 ± 6.6^{*++}	14.4 ± 10.3^{**++NS}

Values represented in the table are mean ± S.D. of measurements in 6 experiments expressed as µmol of 2,3-DPG/µmol Hb. *p < 0.01, **p < 0.001 when compared uninfected cells with cells infected with parasite stages. +p < 0.01, ++p < 0.001 when compared ring stage with other two stages of maturation (R vs. T, R vs. S). $^{\#}$p < 0.01, NS p > 0.05 when compared T vs. S.

Table 2. Osmotic fragility of uninfected and infected erythrocytes at different parasite stages in chloroquine sensitive and resistant strains of *P. falciparum*

		Chloroquine sensitive strain			Chloroquine resistant strain		
	Uninfected	Ring	Trophozoite	Schizont	Ring	Trophozoite	Schizont
Haemolysis in 0.43% Nacl (%, w/v)	50	64.5	72^{*+}	78^{*+NS}	57	69^{*+}	76.7^{*+NS}
Salt concentration at which 50% haemolysis occurred (%, w/v)	0.43	0.53	0.55	0.56	0.48	0.52	0.54

Values are means of 6 independent observations. *p < 0.05 when compared with uninfected cells. $^{+}$p < 0.5 when compared with ring stage. ^{NS}p > 0.05 when schizont stage compared with trophozoite. Manuscript $^{\#}$ND-01-5542 (M.L. Dubey *et al.*).

Discussion

2,3-DPG is synthesized from glycolytic intermediates by a pathway known as Rapoport-Luebering shunt [18]. The production of 2,3-DPG depends on the relative amounts of its precursor, 1,3-DPG going into Rapoport-Leubering pathway and into the ATP-forming glycolytic pathway. Its concentration also depends on the rate of hydrolysis of 2,3-DPG.

The most important function of 2,3-DPG is its effect on the oxygen affinity of haemoglobin (Hb). Normal haemoglobin (Hb A) in deoxygenated state can bind 2,3-DPG in a molar ratio of 1:1 which results in reduced oxygen affinity of Hb and enhanced oxygen delivery to tissues. In oxygenated state the ability of Hb to bind 2,3-DPG is decreased and most DPG is in free state. As the result the affinity of Hb to oxygen is increased and it binds more of oxygen [18]. Changes in 2,3-DPG levels have important role in adaptation to hypoxia. In anemia and certain hypoximic conditions, 2,3-DPG levels in erythrocytes increase and oxygen affinity of haemoglobin is reduced resulting in improved delivery of oxygen to tissues [2, 18].

Apart from this beneficial effect on oxygen delivery to tissues, 2,3-DPG has adverse effect on the mechanical stability of RBC membrane. This effect is reported to be produced by increased lateral mobility of integral membrane proteins and disruption of links between the cell membrane and cytoskeletal protein network notably the spectrin-actin-band 4.1 complex [19–21].

2,3-DPG levels in erythrocytes is increased in anaemia due to various causes including visceral leishmaniasis (Kala-azar) [8–1]. In Kala-azar the anemia is reported to be due to short life span of red blood cells [22]. The short life span of RBCs in kala-azar is at least in part due to weakened RBC membrane [11].

While studying the enzymes of glycolytic pathway in RBCs infected with *P. falciparum in vitro*, Roth *et al.* [23] reported sharp decline in erythrocytic 2,3-DPG content and have attributed it to an increase in pyruvate kinase (PK) and phosphoglycerate kinase (PGK) activity in absence of concomitant increase in diphosphoglycerate mutase (DPGM) activity. In RBCs with a hereditary pyruvate kinase hyperactivity, Max-Audit *et al.* [24] have also reported associated low levels of 2,3-DPG and high levels of ATP.

In present study, the 2,3-DPG content of erythrocytes infected with *P. falciparum* showed a gradual decrease as the parasite matured from ring to schizont stage, in both chloroquine sensitive and chloroquine resistant strains. The levels were lowest at schizont-stage showing greater decay of 2,3-DPG in later stages of parasite maturation. These observations may suggest that probably the gradually increasing pyruvate kinase activity of mature plasmodium falciparum trophozoites inside the red cell with normal or decreased DPGM activity may have a role in decreasing the 2,3-DPG content of the host cell, as more of 1,3-DPG, the precursor of 2,3-DPG, is diverted to ATP forming glycolytic pathway. Though PK and DPGM activity in parasitised erythrocytes has not been assayed in present study, significantly increased pyruvate kinase and normal or marginally decreased DPGM activity has been reported in *P. berghei* [25] and in *P. falciparum* [23] infected erythrocytes. This probably is favourable to the parasite because it will increase red cell ATP at the expense of red cell, 2,3-DPG [24].

Some reports suggest that osmotic fragility of human red cells remains unchanged during malaria infection [26]. In contrast, some other investigators have indicated that the osmotic fragility of parasitized cells might be increased in human malaria [27]. The results of present study showed that the osmotic fragility of parasitized erythrocytes increases gradually with the maturation of parasite and is significantly high at schizont stage. The procedure for measuring erythrocyte osmotic fragility used in the present study minimizes the potential variables [28]. Thus the present investigations not only confirmed the results of earlier studies but also extended the observations on alteration of erythrocyte fragility at different stages of the parasites maturation. It was evident that the intracellular parasites in some way contributed significantly to the increased osmotic fragility of the erythrocyte as it matured. The present investigation also indicated that the erythrocytes infected with chloroquine sensitive strain were more fragile than those of the chloroquine resistant strain, though the difference was not statistically significant.

Elevated levels of 2,3-DPG have been suggested to have a role in the altered mechanical properties of the red cell membrane resulting in anaemia in leishmaniasis [11]. As there is no increase in 2,3-DPG activity in red blood cells parasitized with *P. falciparum* cells (rather, it is decreased), the increased osmotic fragility observed in these cells is probably due to other factors. The findings of the study also suggest that the compensatory increase in 2,3-DPG levels seen in other anaemic conditions may not be there in anaemia due to falciparum malaria. This decreased level of 2,3-DPG in malaria may affect the oxygen delivery to tissues in malaria and may have a role in hypoxia in cerebral malaria patients.

Acknowledgement

Authors thank Dr. C.R. Pillai, Malaria Research Center, New Delhi for providing chloroquine sensitive and resistant strains of *Plasmodium falciparum.* The financial assistance provided by PGIMER, Chandigarh is gratefully acknowledged.

References

1. World Health Organization, Geneva: World malaria situation in 1994. Weekly Epidemiol Record 72: 269–276, 1997
2. Mayes PA: Glycolysis and the oxidation of pyruvate. In: K. Murray, D.K. Granner, P.A. Mayes, V.W. Rodwell (eds). Harper's Biochemistry, 25th edn. Appleton & Lange, Stamford, Connecticut, USA, 2000, pp 190–198
3. Goodman SR, Schiffe KA: The spectrin membrane skeleton of normal and abnormal human erythrocytes: A review. Am J Physiol 244: C121–C141, 1983
4. Bennett V: The membrane skeleton of human erythrocytes and its implications for more complex cells. Ann Rev Biochem 54: 273–304, 1985
5. Agre P, Orringer FP, Chen DHK, Bennet V: A molecular defect in two families with haemolytic poikilocytic anemia: Reduction of high affinity membrane binding sites for ankyrin. J Clin Invest 68: 1566–1576, 1981
6. Nelson DL, Cox MM: Glycolysis and the catabolism of hexoses. In: Lehninger's Principles of Biochemistry, 3rd edn. Worth Publishers, New York, USA, 2000, pp 527–566.
7. Eaton JW, Brewer GJ, Schultz JS: Variation in 2,3-diphosphoglycerate and ATP levels in human erythrocytes and effects on oxygen transport. Exp Med Biol 6: 21–38, 1970
8. Edwards UJ, Canon B: Oxygen transport during erythropoietic response to moderate blood loss. N Engl J Med 287: 115, 1972
9. Torrance J, Jacobs P, Restrepo A, Eschbach J, Lenfant C, Finch CA: Intraerythrocytic adaptation to anaemia. N Engl J Med 283: 165–169, 1970
10. Bellingham AJ, Huehns ER: Compensation in haemolytic anaemias caused by abnormal haemoglobins. Nature (London) 218: 924, 1968
11. Biswas T, Ghosh DK, Mukherjee N, Ghosal J: Elevated 2,3-diphosphoglycerate concentrations and alteration of structural integrity in erythrocytes of Indian cases of visceral leishmaniasis. Ann Trop Med Parasitol 89: 391–398, 1995
12. Trager W, Jensen JB: Human malaria parasites in continuous culture. Science 193: 674–675, 1976
13. Lambros E, Vanderberg JP: Synchronization of *P. falciparum* erythrocytic stages in culture. J Parasitol 65: 418–420, 1979
14. Rivadeneira EM, Wasserman M, Espinal CT: Separation and concentration of schizonts of *Plasmodium falciparum* by percoll gradients. J Protozool 30: 367–370, 1983
15. Rose ZB, Liebowitz J: Direct determination of 2,3-diphosphoglycerate. Anal Biochem 35: 177–180, 1970
16. Dacie JV, Lewis SM: In: Practical Haematology, 6th edn. Churchill Livingstone, London, 1984, pp 28–29
17. Parpart AK, Lorenz RB, Parpart ER, Gress JR, Chase AM: The osmotic resistance (fragility) of human red cells. J Clin Invest 26: 636–638, 1947
18. Telen MJ: The mature erythrocyte. In: R.G. Lee, T.C. Bithell, J. Foerster, J.W. Athens, J.N. Lukens (eds). Wintrobe's Clinical Haematology, 9th edn. Lea & Febiger, Philadelphia/London, 1993, pp 101–133
19. Sheetz MP, Casaly J: 2,3-diphosphoglycerate and ATP dissociated erythrocyte membrane skeletons. J Biol Chem 200: 9955–9960, 1980
20. Cohen CM, Foley SF: Biochemical characterization of complex formation by human erythrocyte spectrin, protein 4.1, an actin. Biochemistry 23: 6091–6098, 1984
21. Chasis JA, Mohandas N: Erythrocyte membrane deformability and stability. Two distinct membrane properties that are independently regulated by skeletal protein association. J Cell Biol 103: 343–350, 1986
22. Knight R, Woodruff AW, Pettitt LE: The mechanism of anaemia in kala-azar. Trans Roy Soc Trop Med Hyg 61: 701–705, 1967
23. Roth EF Jr, Calvin MC, Max-Audit I, Rosa J, Rosa R: The enzymes of the glycolytic pathway in erythrocytes infected with *Plasmodium falciparum* malaria parasites. Blood 72:1922–1925, 1988
24. Max-Audit I, Rosa R, Marie J: Pyruvate kinase hyperactivity genetically determined: Metabolic consequences and molecular characterization. Blood 56: 902–909, 1980
25. Kruckeberg WC, Sander BJ, Sullivan DC: *Plasmodium berghei*: Glycolytic enzymes of the infected mouse erythrocyte. Exp Parasitol 51: 438–443, 1981
26. Zuckerman A: Autoimmunization and other types indirect damage to host cells as factors in certain protozoan diseases. Exp Parasitol 15: 138–183, 1964
27. Kondo Y: Investigations on the genesis of black water fever. Trop Dis Bull 24: 658–659, 1926
28. Fogel BJ, Shields C, Von Doenhoff A Jr: The osmotic fragility of erythrocytes in experimental malaria. Am J Trop Med Hyg 15: 269–275, 1966

Molecular and Cellular Biochemistry **246:** 143–153, 2003.

Aminosalicylic acid reduces the antiproliferative effect of hyperglycaemia, advanced glycation endproducts and glycated basic fibroblast growth factor in cultured bovine aortic endothelial cells: Comparison with aminoguanidine

Yasotha Duraisamy,[1] John Gaffney,[1] Mark Slevin,[1] Christopher A. Smith,[1] Kenneth Williamson[2] and Nessar Ahmed[1]

Departments of [1]Biological Sciences; [2]Chemistry and Materials, Manchester Metropolitan University, Chester Street, Manchester, UK

Abstract

Hyperglycaemia reduces proliferation of bovine aortic endothelial cells *in vitro*. A similar effect *in vivo* may contribute to long-term complications of diabetes such as impaired wound-healing and retinopathy.

We report the effect of increased glucose concentrations, glycated basic fibroblast growth factor (FGF-2) and bovine serum albumin-derived advanced glycation endproducts (BSA-AGE) on the proliferation of bovine aortic endothelial cells.

Glucose (30 and 50 mmol/l) had an antiproliferative effect on endothelial cells. This effect may be mediated through reduced mitogenic activity of FGF-2. The glycation of FGF-2 with 250 mmol/l glucose-6-phosphate led to reduced mitogenic activity compared to native FGF-2. BSA-AGE at concentrations of 10, 50 and 250 μg/ml had an antiproliferative effect on cultured endothelial cells.

Aminosalicylic acid at a concentration of 200 μmol/l proved to be more effective than equimolar concentrations of aminoguanidine in protecting endothelial cells against the antiproliferative effects of both high (30 mmol/l) glucose and 50 μg/ml BSA-AGE. FGF-2 glycated in the presence of 4 mmol/l aminosalicylic acid or aminoguanidine retained mitogenic activity compared to that glycated in their absence.

Compounds like aminoguanidine and, in particular, aminosalicylic acid protect endothelial cells against glucose-mediated toxicity and may therefore have therapeutic potential. (Mol Cell Biochem **246**: 143–153, 2003)

Key words: glycation, advanced glycation endproducts, aminosalicylic acid, aminoguanidine, basic fibroblast growth factor, free radicals, antioxidants

Abbreviations: AGE – advanced glycation endproducts; FGF-2 – basic fibroblast growth factor; AG – aminoguanidine; ASA – aminosalicylic acid; BSA – bovine serum albumin; 3-DG – 3-deoxyglucosone; RAGE – receptors for advanced glycation endproducts; IR – infrared; DMEM – Dulbecco's Modified Eagle Medium; FCS – foetal calf serum; PBS – phosphate buffered saline; ANOVA – one way analysis of variance; NS – not significant

Address for offprints: N. Ahmed, Department of Biological Sciences, Manchester Metropolitan University, Chester Street, Manchester, M1 5GD, UK (E-mail: N.Ahmed@mmu.ac.uk)

Introduction

Diabetes mellitus is a syndrome characterised by hyperglycaemia. It predisposes individuals to complications such as diabetic retinopathy where there is loss of capillary endothelial cells [1]. The location of endothelial cells and their slow turnover means they are targets for glucose-induced damage during hyperglycaemia. It has been demonstrated that exposure of human umbilical vein [2, 3], human dermal microvascular [4] and bovine retinal [5] endothelial cells to a high glucose concentration decreases their proliferation. This effect is believed to be independent of osmotic effects, although the precise mechanism remains unclear [2, 4, 6]. However, a number of mechanisms have been proposed to account for glucose-induced cellular damage. Glucose can react directly with protein amino groups (glycation) to form labile Schiff bases which subsequently rearrange to a more stable Amadori product. In time, glycated proteins yield reactive dicarbonyl compounds, e.g. 3-deoxyglucosones (3-DG) that can form advanced glycation endproducts (AGE). Increased glycation and tissue AGEs are implicated in long-term complications of diabetes [7]. Free glucose and Amadori products also undergo transition-metal catalysed autoxidation (glycoxidation) to yield free radicals such as the superoxide and hydroxyl radicals [8, 9].

Increased intracellular glycation of cytosolic proteins has been demonstrated in pericytes and retinal endothelial cells cultured in a high glucose medium although the same study could not demonstrate a build-up of intracellular AGE [10]. In contrast, a 13-fold increase in the AGE content of endothelial cells cultured in a high glucose has been shown after only 1 week [1].

Endothelial cell proliferation is promoted by a number of growth factors in particular FGF-2 because of its mitogenic nature [11, 12]. FGF-2 lacks a signal peptide, is stored intracellularly and released following endothelial cell injury making it a target for intracellular glycation and AGE formation [13]. During hyperglycaemia, glycation of FGF-2 stored in endothelial cells may occur due to the presence of high concentrations of reactive sugars, such as, glucose-6-phosphate, fructose-6-phosphate and glyceraldehyde-3-phosphate [14]. One study has shown over a 6-fold increase in cytosolic FGF-2-AGE isolated from bovine endothelial cells cultured in a high glucose medium with a corresponding reduction in mitogenic activity [1].

In diabetes, endothelial cells are exposed to circulating AGEs that are toxic to cultured cells [15, 16] and have been implicated in vascular disease [17]. Circulating AGEs exert their toxicity especially when they interact with their receptors (RAGE) which have been detected on a number of cell types including endothelial cells [18, 19]. The toxicity of AGEs is believed to be mediated via free radicals which can damage cellular lipids, proteins and nucleic acids.

There is interest in compounds that can delay or prevent the onset of diabetic complications. Aminoguanidine has been shown to reduce the progression of retinopathy [20], cataract [21], nephropathy [22, 23] and neuropathy [24] in diabetic animals. Aminoguanidine reduces both glucose-induced toxicity in cultured pericytes [25] and inhibition of angiogenesis in the rat sponge granuloma model [26]. Aminoguanidine may protect by decreasing the active acyclic form of glucose [27] thereby inhibiting glycation or by blocking carbonyl groups on Amadori products and 3-DG thus preventing formation of AGE [28]. Recent studies report that aminoguanidine acts as a free radical scavenger and may reduce glucose or AGE-derived free radicals [29]. It has also been suggested that aminoguanidine may inhibit the sorbitol pathway [30] or the production of nitric oxide a highly toxic free radical implicated in diabetic vascular disease [31]. However, other studies have shown high concentrations of aminoguanidine impair pancreatic β-cell responses to insulin in cultured rat pancreatic islets [32]. Aminoguanidine does not protect against the antiproliferative effect of high glucose on cultured mesangial cells [33] and has been reported to inhibit catalase activity and produce hydrogen peroxide both of which would increase oxidative stress [34].

Aminosalicylic acid is an established anti-inflammatory drug and a known antioxidant [35] but to the authors knowledge has not been investigated in relation to glucose- or AGE-induced cellular toxicity. Aminosalicylic acid possesses a single free amino group and could inhibit glycation by blocking free carbonyl groups on reducing sugars, Amadori products as well as dicarbonyl intermediates involved in AGE formation. In this study, we compare the protective effects of aminosalicylic acid and aminoguanidine on the proliferation of endothelial cells cultured in a high glucose medium. The abilities of these compounds to protect against *in vitro* glycation of FGF-2 and the antiproliferative effects of AGEs on cultured endothelial cells were also investigated. Infrared (IR) spectroscopy was used to gain some insight into the possible mechanism of action of aminosalicylic acid.

Materials and methods

Materials

Bovine FGF-2 was obtained from R&D (Minneapolis, MN, USA). Dulbecco's Modified Eagle Medium (DMEM) was obtained from Gibco (Paisley, Scotland). [U-^{14}C] Glucose-6-phosphate and foetal calf serum (FCS) were obtained from ICN Biochemicals Ltd. (Basingstoke, UK). Bio-Gel P60 was obtained from BioRad Laboratories Ltd. (Hemel Hempstead, UK). Aminoguanidine, 5-aminosalicylic acid, glucose-6-phosphate, glucose and all other reagents were obtained from Sigma (Poole, UK).

Culture of bovine aortic endothelial cells

Bovine aortic endothelial cells were characterised as described previously [36] and cultured in Dulbecco's Modified Eagle Medium (DMEM) supplemented with 15% foetal calf serum (FCS), 2 mmol/l glutamine, 1 U/l penicillin and 100 µg/ml streptomycin. Cells were cultured at 37°C in a humidified atmosphere of 5% CO_2 and 95% air. All cultures were used between passages 5–10 and within 3 weeks of establishing the culture. Cell viability was 100% as determined by trypan blue exclusion.

Effect of high glucose on cell proliferation

Cells were trypsinised with 1% w/v trypsin, centrifuged at 1000 g for 5 min and resuspended in DMEM containing 15% FCS. Cells were drawn through a 1 ml syringe fitted with a 0.5 × 25 mm sterile needle to prepare a single cell suspension. Approximately 2 × 10^4 cells/ml (2 ml/well) were placed into six well plates and seeded at 37°C for 6 h. The medium was removed and cells were rinsed briefly with phosphate buffered saline (PBS), pH 7.4. Fresh DMEM containing 5.5, 30 or 50 mmol/l glucose was added (2 ml/well). Aminoguanidine and aminosalicylic acid were dissolved in 10 mmol/l sodium phosphate buffer, pH 7.4 and filtered through a 0.2 µm sterile filter before addition to cell cultures. Aminoguanidine and aminosalicylic acid were added to final concentrations of 200 µmol/l to endothelial cells grown in 5.5 and 30 mmol/l glucose. Controls containing 5.5 mmol/l glucose to which mannitol was added to a final concentration of 30 and 50 mmol/l were included to exclude the effects of hyperosmolarity. The final FCS concentration was 2.5%. Proliferation of cultured cells was assessed by determining the mean of 3 experiments each with six wells as described below.

Preparation of glycated FGF-2

FGF-2 (10 µg/ml) was incubated with 250 mmol/l glucose-6-phosphate ± 4 mmol/l aminoguanidine or aminosalicylic acid in 50 mmol/l sodium phosphate buffer, pH 7.4 at 37°C for 24 h under sterile conditions.

Measurement of FGF-2 glycation

FGF-2 (25 µg/ml) was mixed with [U-^{14}C] glucose-6-phosphate (specific activity 336 µCi/ml) in 50 mmol/l sodium phosphate buffer, pH 7.4 under sterile conditions and incubated at 37°C for 0–72 h. Aliquots (200 µl) were removed at 24 h intervals for analysis.

Bio-Gel P-60 (15 g) was swollen in 50 mmol/l sodium phosphate buffer, pH 7.4 and poured to give a column (0.8 cm diameter, 12 cm length). The column was equilibrated with two bed volumes of 50 mmol/l sodium phosphate buffer, pH 7.4 at a flow rate of 1 ml/min. Samples (200 µl) containing 5 µg of glycated FGF-2 were loaded onto the column. The column was eluted with 50 mmol/l sodium phosphate buffer, pH 7.4 and 1 ml fractions were collected, mixed with 4 ml scintillant (Ecoscint A) and the radioactivity counted using an LKB Wallac scintillation counter. The number of glucose-6-phosphate residues bound per FGF-2 molecule was determined from the total radioactivity and the proportion eluted with the FGF-2 fraction.

Dose response and time course to determine proliferative effect of native and glycated FGF-2

To determine the effect of native and glycated FGF-2 on endothelial cell proliferation, cells (4 × 10^4 cells/well) were seeded in triplicate in six well plates. Fresh DMEM supplemented with 2.5% FCS was added (2 ml/well). Native or glycated FGF-2 was added (10 pg/ml–20 ng/ml) to cells in triplicate. Control wells were treated with neither native nor glycated FGF-2. Cells were counted after 72 h.

For time course studies, cells were prepared as above except 5 ng/ml of native or glycated FGF-2 (the optimal concentration; see later) was added and cells cultured for a further 24–120 h. Controls containing no FGF-2 were included and cells counted as below.

Effect of FGF-2 glycated in the presence of aminoguanidine or aminosalicylic acid on cell proliferation

Six well plates seeded with 4 × 10^4 cells/well were prepared as above and fresh medium containing 2.5% FCS was added (2 ml/well). Native, glycated FGF-2 and FGF-2 glycated in the presence of aminoguanidine or aminosalicylic acid (at a final concentration of 5 ng/ml) were added to the cultured cells for 72 h and cell numbers determined (see below).

Preparation of BSA-AGE

Bovine serum albumin (10 mg/ml) was incubated with 250 mmol/l glucose in 100 mmol/l sodium phosphate buffer containing 3 mmol/l sodium azide, pH 7.4 at 37°C for 5 weeks. The glycated albumin was dialysed exhaustively against distilled water at 4°C, lyophilised and stored at –20°C. The formation of AGE was assessed by their characteristic fluorescence emission spectra at 440 nm after excitation at 370 nm using a Luminescence Spectrometer (Model LS30). Before addition to cell cultures, BSA-AGE was dissolved in distilled water and filtered through a sterile 0.2 μm filter.

Effect of BSA-AGE on cell proliferation

DMEM containing 2×10^4 cells/ml (2 ml/well) were transferred to six well plates and cells were seeded for 6 h. The medium was replaced with fresh DMEM containing 5.5 mmol/l glucose. BSA-AGE was added to the culture medium to final concentrations of 0–250 μg/ml and tested negative for endotoxin. Aminoguanidine and aminosalicylic acid were added (final concentration of 200 μmol/l) to the endothelial cells cultured in the presence of 50 μg/ml of BSA-AGE. The final FCS concentration was 2.5%. Proliferation of cells was assessed by determining the mean of 3 experiments each with six wells as described below.

Proliferation of bovine aortic endothelial cells

After 72 h the medium was removed and the wells were rinsed with 2 ml PBS, pH 7.4. Cells were trypsinised with 1% w/v trypsin (500 μl/well) for 5 min. DMEM (500 μl/well) was added to neutralise the trypsin. Single cell suspensions were prepared using a 1 ml syringe and sterile needle. The cell suspensions (1 ml) were mixed with 15 ml of isoton and counted using a Coulter counter (model F2 3AG).

Infrared spectroscopy

Lysozyme (10 mg/ml) was incubated ± 250 mmol/l of glucose-6-phosphate ± 5 mmol/l aminosalicylic acid in 50 mmol/l sodium phosphate buffer containing 3 mmol/l sodium azide, pH 7.4 at 37°C for 72 h. Any unbound sugar or aminosalicylic acid was removed by dialysis against distilled water and the protein samples lyophilised. Unreacted dried mixtures of the above were used as controls to allow changes in chemical groups after reaction to be detected and compared. Approximately 5 mg of the samples were incorporated into potassium bromide discs. The discs were positioned in a sample holder and suitable infrared spectra obtained using an ATI Mattson FT-IR spectrometer.

Statistical analysis

The statistical significance between groups and controls was determined using one way analysis of variance (ANOVA) on cell numbers. Data are expressed as percentages for ease of batch to batch comparisons although the statistical analysis were performed on cell numbers.

Results

Effect of high glucose on cell proliferation

Exposure of bovine aortic endothelial cells to increasing concentrations of glucose for 72 h reduced their proliferation in a dose-dependent manner (Fig. 1). Proliferation of endothelial cells incubated in 30 and 50 mmol/l glucose were reduced to 67 ± 3.5 and 55 ± 5% respectively of those incubated in 5.5 mmol/l glucose ($p < 0.01$). This decreased proliferation was not due to hyperosmolarity because mannitol added to 5.5 mmol/l glucose to final concentrations of 30 and 50 mmol/l did not significantly reduce proliferation. Both 200 μmol/l aminoguanidine and aminosalicylic acid had no significant effect on the proliferation of cells exposed to 5.5 mmol/l glucose (Fig. 2). However, both did protect endothelial cells against the antiproliferative effect of 30 mmol/l glucose. Proliferation in the presence of aminoguanidine and aminosalicylic acid were 79 ± 0.8 and 89 ± 2.9% respectively of those at 5.5 mmol/l glucose (compared to a proliferation of 67 ± 3.5% for cells in 30 mmol/l glucose alone, $p < 0.01$).

Glycation of FGF-2

FGF-2 was glycated *in vitro* using radiolabelled glucose-6-phosphate and the extent of glycation was determined and expressed as residues of glucose-6-phosphate per FGF-2 molecule. Glycation increased from 3–11 residues of glucose-6-phosphate after 24 and 72 h of incubation respectively (Fig. 3).

Dose response and time course for effect of native and glycated FGF-2 on cell proliferation

Proliferation of cells in the presence of native FGF-2 increased in a dose-dependent manner with a maximum at a concentration of 7.5 ng/ml (Fig. 4). Proliferation of cells also increased in a dose-dependent manner in the presence of glycated FGF-2 (Fig. 4) but was significantly reduced compared to native FGF-2 at all concentrations ($p < 0.01$ in each case). The largest difference between the treatments occurred at a concentration of 5 ng/ml and this was used in subsequent experiments.

The proliferation of endothelial cells untreated (control) or treated with native or glycated FGF-2 (5 ng/ml) was measured over a period of 24–120 h (Fig. 5). The untreated cells showed increased proliferation, which was less than cells treated with FGF-2 ($p < 0.05$) at time points 48–120 h. Proliferation in the presence of glycated FGF-2 was significantly greater than the control but reduced compared to those cells treated with native FGF-2.

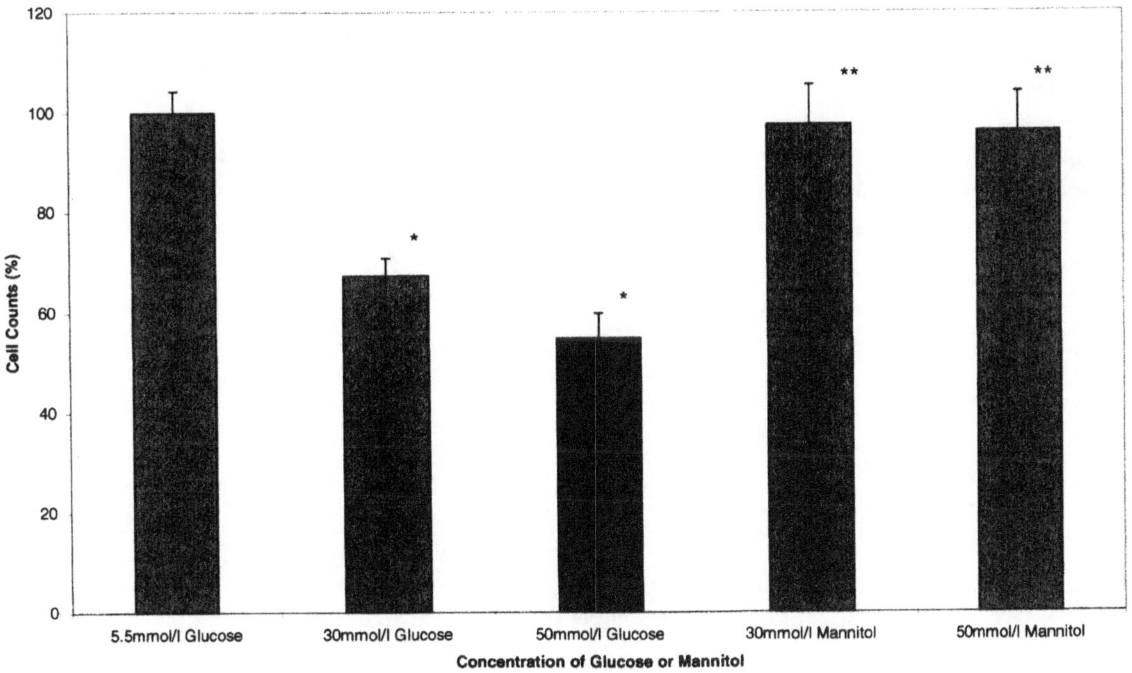

Fig. 1. Effect of high glucose concentration on the proliferation of endothelial cells. Cells were cultured for 72 h in either 5.5, 30 or 50 mmol/l glucose. To discount the effects of hyperosmolarity, cells were also cultured in 5.5 mmol/l glucose made up to 30 or 50 mmol/l using mannitol. Proliferation was expressed as a percentage of that for endothelial cells cultured in 5.5 mmol/l glucose. Results are expressed as mean ± S.D. of 3 experiments. $^{*}p < 0.01$ vs. 5.5 mmol/l glucose, **NS vs. 5.5 mmol/l glucose.

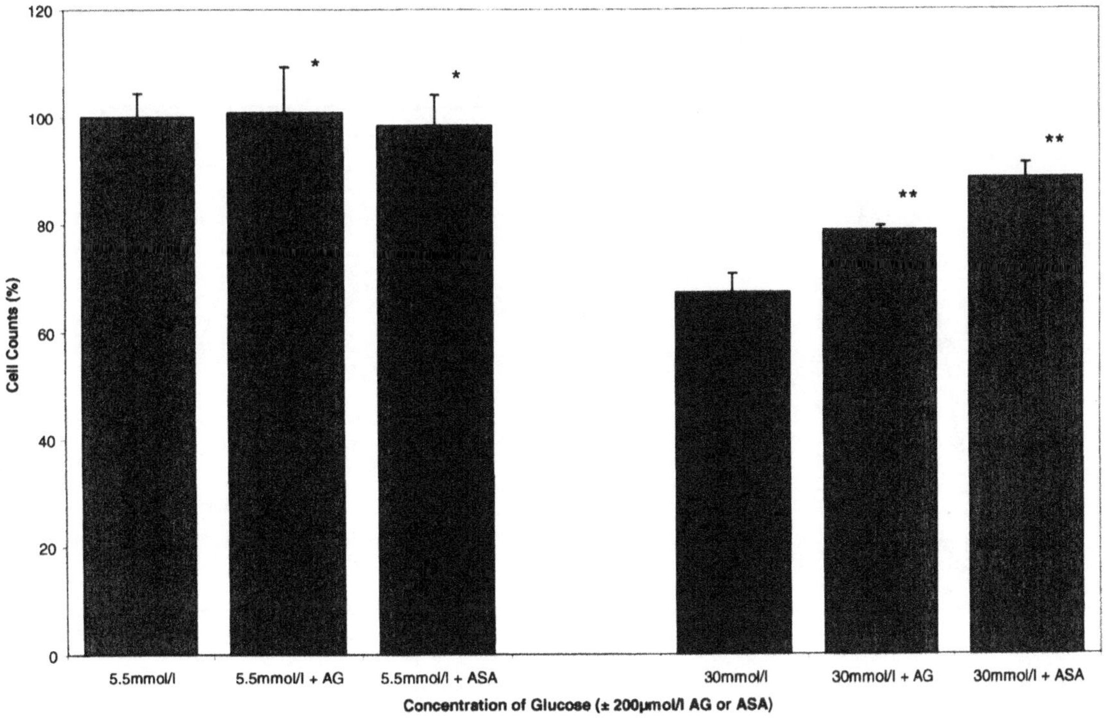

Fig. 2. Effect of 200 µmol/l aminoguanidine (AG) or aminosalicylic acid (ASA) on proliferation of endothelial cells cultured for 72 h in physiological (5.5 mmol/l) or high (30 mmol/l) glucose. Proliferation was expressed as a percentage of that for endothelial cells cultured in 5.5 mmol/l glucose. Results are expressed as mean ± S.D. of 3 experiments. *NS vs. 5.5 mmol/l glucose, $^{**}p < 0.01$ vs. 30 mmol/l glucose.

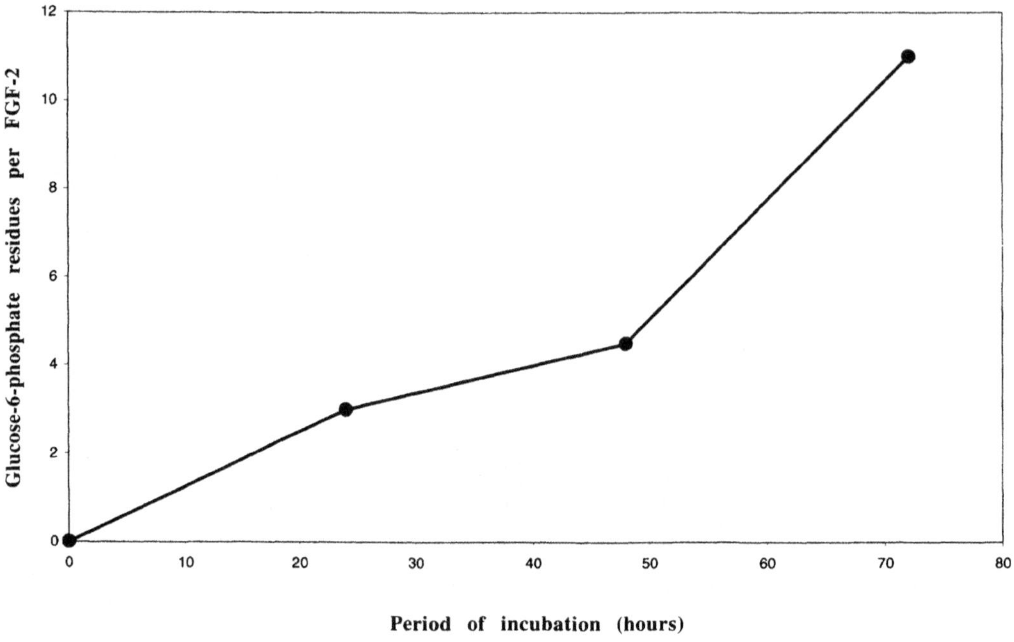

Fig. 3. Extent of glycation expressed as number of residues of glucose-6-phosphate per FGF-2 molecule. FGF-2 (25 μg/ml) was glycated *in vitro* using 250 mmol/l glucose-6-phosphate in 100 mmol/l sodium phosphate buffer, pH 7.4 at 37°C and incubated for 0–72 h.

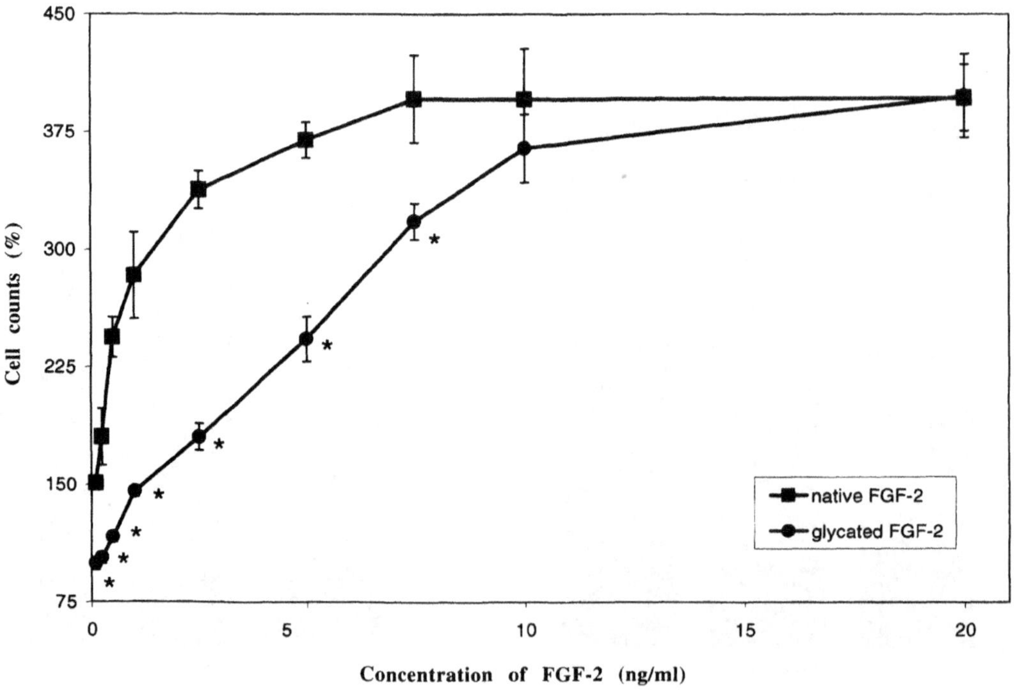

Fig. 4. Proliferation of endothelial cells after 72 h in the presence of 100 pg/ml – 20 ng/ml of native or glycated FGF-2. Proliferation was expressed as a percentage of that for untreated (control) endothelial cells. Results are expressed as mean ± S.D. of 3 experiments. *p < 0.01 vs. native FGF-2.

Effect of glycated FGF-2 on cell proliferation

Endothelial cells were cultured for 72 h in 5.5 mmol/l glucose. Addition of 5 ng/ml of native FGF-2 to the culture medium more than doubled the proliferation of endothelial cells to 268 ± 2.6% compared with untreated controls ($p < 0.01$, Fig. 6). Cells exposed to the same concentration of glycated FGF-2, still proliferated but to a lesser extent (198 ± 5.8% of untreated controls, $p < 0.01$). FGF-2 glycated in the presence of 4 mmol/l aminoguanidine and aminosalicylic acid had increased activities and increased proliferation of cells to 245 ± 4.5 and 239 ± 1.8% respectively of those in 5.5 mmol/l glucose alone ($p < 0.05$ for both).

Effect of BSA-AGE on cell proliferation

The inclusion of 10–250 µg/ml BSA-AGE in the culture medium reduced proliferation of endothelial cells growing in 5.5 mmol/l glucose in a dose-dependent manner over a 72 h period ($p < 0.01$, Fig. 7). BSA-AGE at a concentration of 10 and 50 µg/ml reduced proliferation of endothelial cells to 79 ± 6.3 and 54 ± 3.6% of those grown in the absence of BSA-AGE respectively. However, inhibition showed saturation effects because 250 µg/ml BSA-AGE similarly reduced proliferation of endothelial cells to 54 ± 5.9% of controls.

Both 200 µmol/l aminoguanidine and aminosalicylic acid protect endothelial cells against the toxic effects of 50 µg/ml BSA-AGE (Fig. 7). Proliferation of cells in the presence of aminosalicylic acid was 77 ± 2.7% of those in the absence of BSA-AGE ($p < 0.01$) and this compares to that in the presence of aminoguanidine which was 74 ± 4.2% ($p < 0.05$).

Infrared spectroscopy

The results of infrared spectroscopy are not shown in detail. Incubation of aminosalicylic acid with glucose-6-phosphate produced changes in the spectra at about 1650 and 3000 cm^{-1} compared to controls. Glycated lysozyme when reacted with aminosalicylic acid produced marked changes in the spectra at 1650–1680 cm^{-1}.

Discussion

Bovine aortic endothelial cells were cultured in physiological (5.5 mmol/l) and hyperglycaemic concentrations (30 and 50 mmol/l) of glucose. Hyperglycaemic concentrations of 30 and 50 mmol/l glucose are physiologically relevant in that they may be encountered in uncontrolled type I diabetes. Endothelial cells grown in media containing 30 and 50 mmol/l glucose

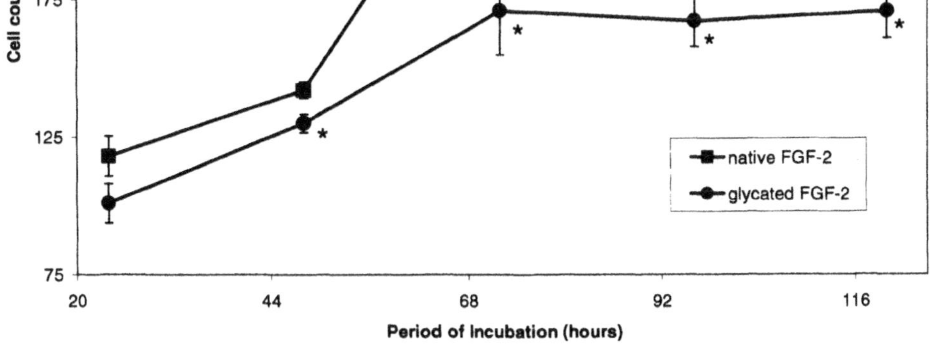

Fig. 5. Proliferation of endothelial cells in the absence or presence of 5 ng/ml of native or glycated FGF-2 for 24–120 h. Proliferation was expressed as a percentage of that for untreated (control) endothelial cells for each time point. Results are expressed as mean ± S.D. of 3 experiments. *$p < 0.01$ vs. native FGF-2.

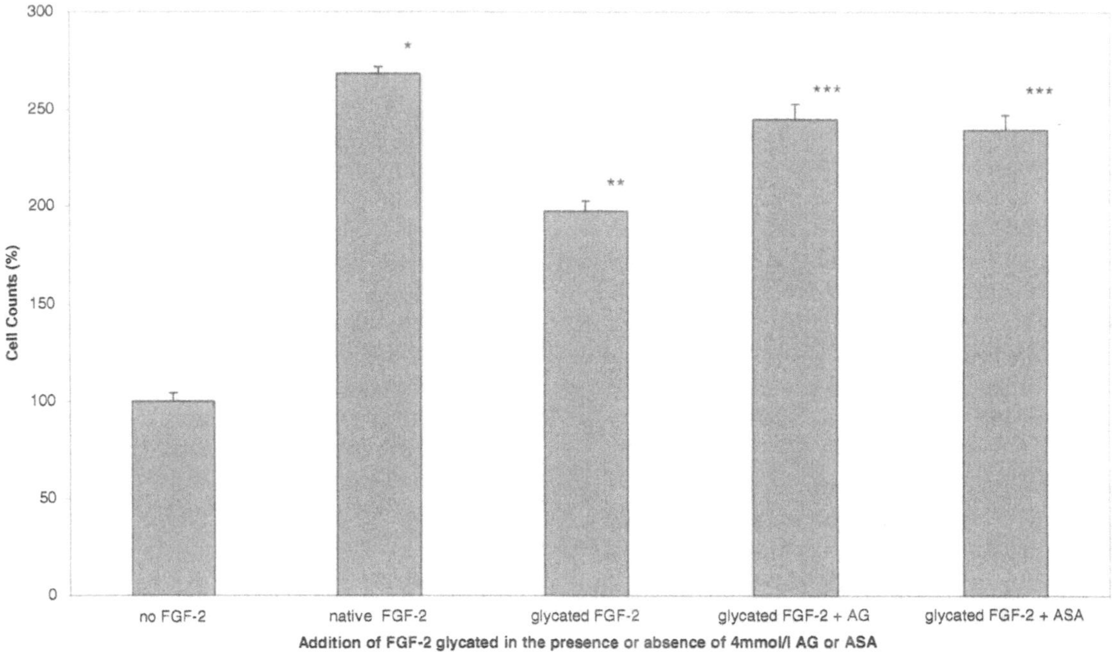

Fig. 6. Effect of FGF-2 on proliferation of endothelial cells after glycation in the presence of 4 mmol/l aminoguanidine (AG) or aminosalicylic acid (ASA). The cells were cultured for 72 h in 5.5 mmol/l glucose ± 5 ng/ml of native or glycated FGF-2. Proliferation was expressed as percentage of that for endothelial cells cultured in 5.5 mmol/l glucose in the absence of FGF-2. Results are expressed as mean ± S.D. of 3 experiments. *$p < 0.01$ vs. no FGF-2, **$p < 0.01$ vs. native FGF-2, ***$p < 0.05$ vs. glycated FGF-2.

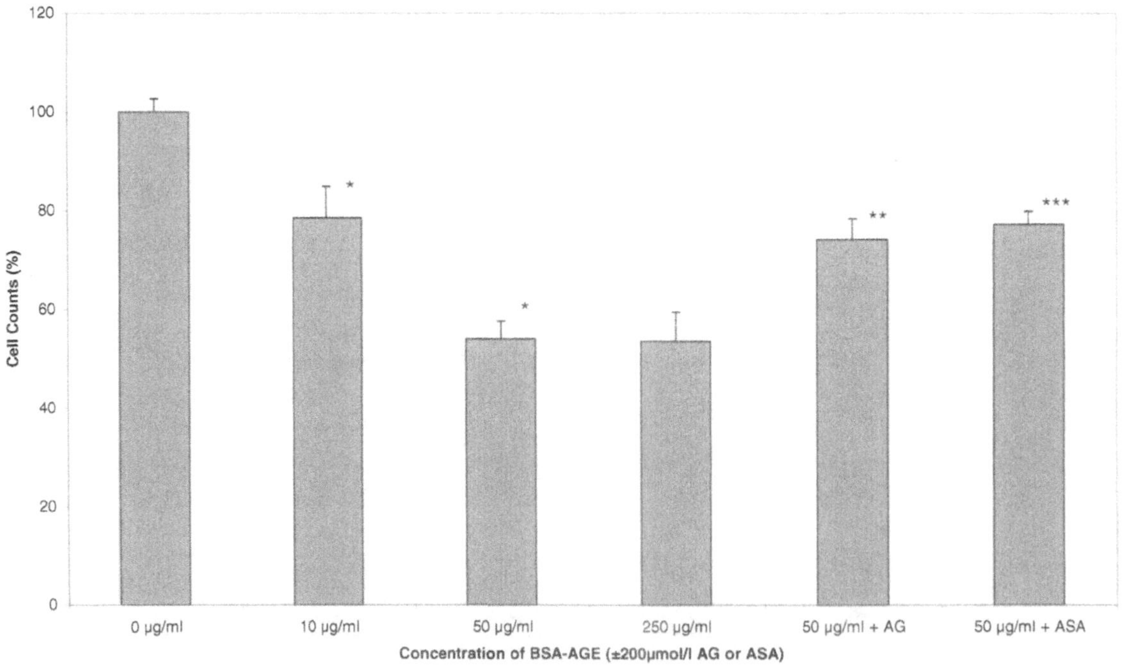

Fig. 7. Effect of 200 µmol/l aminoguanidine (AG) and aminosalicylic acid (ASA) on proliferation of endothelial cells cultured for 72 h in 5.5 mmol/l glucose and 0–250 µg/ml BSA-AGE. Proliferation was expressed as a percentage of that for endothelial cells cultured in 5.5 mmol/l glucose in the absence of BSA-AGE. Results are expressed as mean ± S.D. of 3 experiments. *$p < 0.01$ vs. 0 µg/ml BSA-AGE, **$p < 0.05$ vs. 50 µg/ml BSA-AGE, ***$p < 0.01$ vs. 50 µg/ml BSA-AGE.

showed reduced proliferation compared to those grown in physiological glucose concentrations. Other workers have reported decreased proliferation of endothelial cells [3, 5, 37], human corneal epithelial cells [38] and rat mesangial cells [15] during exposure to a high glucose medium. Precisely how the high glucose mediates cellular toxicity and the sites affected are unclear.

The toxic effects of glucose cannot be accounted for by increased extracellular osmotic pressure because similar concentrations of mannitol fail to inhibit proliferation. This is in accordance with findings by most [15, 38] but not all groups [2] for reasons that are still unclear.

The results of this study suggest that the observed reduction in cell proliferation is due to a direct effect of high glucose. Previous workers have shown that compounds with antioxidant or free radical scavenging properties such as deferoxamine, α-tocopherol and dimethylsulfoxide can protect endothelial cells against glucose-induced damage [39]. Previously, aminoguanidine has been shown to reduce glucose-induced toxicity to pericytes [25]. In this study, both aminoguanidine and aminosalicylic acid protected against glucose toxicity and showed no cellular toxicity at the concentrations used (200 μmol/l). This contrasts with a study using mesangial cells where aminoguanidine did not protect against the antiproliferative effect of high glucose [33].

When endothelial cells are cultured in a high glucose medium, glucose-induced damage may occur to serum proteins and growth factors in the media as well as cell membrane proteins via glycation and glycoxidation reactions. Given the length of exposure and low reactivity of glucose, it is unlikely that the observed cell toxicity in this study was due to modifications of extracellular proteins. Glucose-mediated damage to intracellular biomolecules may be more important under these conditions. During hyperglycaemia, an increase in concentration of a number of reactive intracellular sugars occurs in endothelial cells. For example, the concentrations of glucose-6-phosphate increase 3-fold inside cells grown in a high glucose medium [6], although the involvement of other sugars cannot be discounted. Glucose-6-phosphate is known to glycate proteins and form free radicals via autoxidation much more readily than glucose [40, 41]. Although a number of biomolecules might be damaged via intracellular glycation, FGF-2 is of particular importance because of its mitogenic activity. Indeed, the presence of FGF-2-AGE has been demonstrated in endothelial cells cultured in a high glucose medium after only 1 week [11]. For this reason, FGF-2 was glycated *in vitro* using glucose-6-phosphate. After 24 h there were approximately 3 residues of sugar per FGF-2 molecule, rising to 11 after 72 h.

This study confirms a previous finding that exposure of FGF-2 to high concentrations of reactive sugars reduces its mitogenic activity [1, 42]. However, this study is the first to demonstrate that aminoguanidine and aminosalicylic acid protect the mitogenic activity of FGF-2 during exposure to high concentrations of glucose-6-phosphate. These protective effects might be mediated by a reduction in the glycation of FGF-2 by intracellular sugars such as glucose-6-phosphate.

Bovine aortic endothelial cells showed reduced proliferation when exposed to low (10 μg/ml) or high (250 μg/ml) concentrations of BSA-AGE in a dose-dependent manner. This accords with previous work where addition of BSA-AGE reduced the proliferation of mesangial [15], osteoblast-like [43] and retinal pericyte [16] cells grown in culture.

Antioxidants, such as vitamin E and taurine, have been shown to protect against the antiproliferative effect of both high glucose and BSA-AGE in cultured mesangial cells suggesting the involvement of reactive oxygen species [15]. This study showed that both aminoguanidine and aminosalicylic acid too can protect against AGE-mediated toxicity of cultured endothelial cells in addition to high concentrations of glucose.

The precise mechanisms by which aminoguanidine acts has not been determined, but it may be due to its ability to reduce AGE [28], act as a free radical scavenger [29], or inhibit the sorbitol pathway [30], although the latter mechanism has been discounted by some workers [44]. The protective effect of aminoguanidine may explain why it inhibited the development of retinopathy [20] and prevented impaired wound-healing in diabetic rats [45]. Although aminoguanidine has received the most attention and undergone clinical trials, other compounds have been shown to be at least as effective as aminoguanidine [46].

In this study, aminosalicylic acid proved more effective than aminoguanidine in reducing the antiproliferative effects of both high glucose and BSA-AGE. This could be because aminosalicylic acid is a more potent antioxidant than aminoguanidine, although the ability of aminoguanidine to act as an antioxidant has been questioned [47].

Aminosalicylic acid could react with carbonyl groups on free sugars or on glycated proteins, reducing glycation and AGE formation respectively. Indeed, we interpret our infrared spectroscopy studies to indicate glucose-6-phosphate reacts with aminosalicyclic acid. The decrease in concentration of free sugar could partially explain the protective effects of aminosalicylic acid. The changes in the infrared spectra following incubation of glycated lysozyme with aminosalicylic acid imply a change from ketone to imino groups. It is therefore possible that aminosalicylic acid could prevent glycated proteins forming AGEs. Again, this would afford a protective effect against AGE-induced cellular toxicity, since less AGEs would be available to recognise and bind to their RAGEs.

It is not known whether aminosalicylic acid reacts directly with cells and whether it interferes with second messengers. However previous studies from our laboratory have shown reduced proliferation of endothelial cells in the presence of glycated FGF-2 due to altered expression of second messengers [48] whereas this study has shown that aminosalicylic acid can protect FGF-2 against sugar-induced damage.

Increasing evidence suggests that the glucose-induced cellular damage in diabetes is mediated by a combination of AGE and free radical species. Thus compounds like aminosalicylic acid with both antiglycation and antioxidant properties are likely to have greater therapeutic potential. Indeed, other compounds such as carnosine have both antioxidant and antiglycation properties and also protect against AGE mediated toxicity [49].

Acknowledgements

This research was supported in part by a grant from Diabetes UK. Miss Yasotha Duraisamy was supported by a research studentship from the Faculty of Science and Engineering, Manchester Metropolitan University, UK. We are grateful for the assistance provided by Mr. Mick Hoult in preparing this paper.

References

1. Giardino I, Edelstein D, Brownlee M: Nonenzymatic glycosylation *in vitro* and in bovine endothelial cells alters basic fibroblast growth factor activity. J Clin Invest 94: 110–117, 1994
2. Lorenzi M, Cagliero E, Toledo S: Glucose toxicity for human endothelial cells in culture: Delayed replication, disturbed cell cycle and accelerated death. Diabetes 34: 621–627, 1985
3. Curcio F, Ceriello A: Decreased cultured endothelial cell proliferation in high glucose medium is reversed by antioxidants: New insights on the pathophysiological mechanisms of diabetic vascular complications. In Vitro Cell Dev Biol 28A: 787–790, 1992
4. Kamal K, Du W, Mills I, Sumpio BE: Antiproliferative effect of elevated glucose in human microvascular endothelial cells. J Cell Biochem 71: 491–501, 1998
5. La Selva M, Beltramo E, Pagnozzi F, Bena E, Molinatti PA, Molinatti GM, Porta M: Thiamine corrects delayed replication and decreases production of lactate and advanced glycation endproducts in bovine retinal and human umbilical vein endothelial cells cultured under high glucose conditions. Diabetologia 39: 1263–1268, 1996
6. Lorenzi M, Nordberg JA, Toledo S: High glucose prolongs cell cycle traversal of cultured human endothelial cells. Diabetes 36: 1261–1267, 1987
7. Brownlee M: Glycation and diabetic complications. Diabetes 43: 836–841, 1994
8. Hunt JV, Dean RT, Wolf SP: Hydroxyl radical production and autoxidative glycosylation: Glucose autoxidation as the cause of protein damage in the experimental glycation model of diabetes mellitus and ageing. Biochem J 256: 205–212, 1988
9. Hunt JV, Bottoms MA, Mitchinson MJ: Oxidative alterations in the experimental glycation model of diabetes mellitus are due to protein-glucose adduct oxidation. Biochem J 291: 529–535, 1993
10. Chibber R, Molinatti PA, Kohner EM: Intracellular protein glycation in cultured retinal capillary pericytes and endothelial cells exposed to high glucose concentration. Cell Mol Biol 45: 47–57, 1999
11. Maciag T, Hoover GA, Stemerman MB, Weinstein R: Serial propagation of human endothelial cells *in vitro*. J Cell Biol 91: 420–426, 1981
12. Schweigerer L, Neufeld G, Friedman J, Abraham JA, Fiddes JC, Gospodarowicz D: Capillary endothelial cells express basic fibroblast growth factor, a mitogen that promotes their own growth. Nature 325: 257–259, 1987
13. Muthukrishnan L, Warder E, McNeil PL: Basic fibroblast growth factor is efficiently released from a cytosolic storage site through plasma membrane disruptions of endothelial cells. J Cell Physiol 148: 1–16, 1991
14. Beutler E: Glycosylation of haemoglobin in diabetics. Dev Biochem 9: 489–492, 1980
15. Trachtman H, Futterweit S, Prenner J, Hanon S: Antioxidants reverse the antiproliferative effect of high glucose and advanced glycosylation endproducts in cultured rat mesangial cells. Biochem Biophys Res Commun 199: 346–352, 1994
16. Chibber R, Molinatti PA, Rosatto N, Lambourne B, Kohner EM: Toxic action of advanced glycation endproducts on cultured retinal capillary pericytes and endothelial cells: relevance to diabetic retinopathy. Diabetologia 43: 156–164, 1997
17. Bucala R, Vlassara H: Advanced glycosylation endproducts in diabetic renal and vascular disease. Am J Kidney Dis 26: 875–888, 1995
18. Schmidt A-M, Vianna M, Gerlach M, Brett J, Ryan J, Kao J, Esposito C, Hegarty H, Hurley W, Clauss M, Wang F, Pan Y-C, Tsang T, Stern D: Isolation and characterisation of binding proteins for advanced glycation endproducts from lung tissue which are present on the endothelial cell surface. J Biol Chem 267: 14987–14997, 1992
19. Yan S-D, Schmidt A-M, Anderson GM, Zhang J, Brett J, Zou Y-S, Pinsky D, Stern D: Enhanced cellular oxidant stress by the interaction of advanced glycation endproducts with their receptors/binding proteins. J Biol Chem 269: 9889–9897, 1994
20. Hammes HP, Martin S, Federlin K, Geisen K, Brownlee M: Aminoguanidine treatment inhibits the development of experimental diabetic retinopathy. Proc Natl Acad Sci USA 88: 11555–11558, 1991
21. Swamy-Mruthinti S, Green K, Abraham EC: Inhibition of cataracts in moderately diabetic rats by aminoguanidine. Exp Eye Res 62: 505–510, 1996
22. Ellis EN, Good BH: Prevention of glomerular basement membrane thickening by aminoguanidine in experimental diabetes mellitus. Metabolism 40: 1016–1019, 1991
23. Soulis T, Cooper ME, Vranes D, Bucala R, Jerums G: Effects of aminoguanidine in preventing experimental diabetic nephropathy are related to the duration of treatment. Kidney Int 50: 627–634, 1996
24. Yagihashi S, Kamijo M, Baba M, Yagihashi N, Nagai K: Effect of aminoguanidine on functional and structural abnormalities in peripheral nerves of STZ-induced diabetic rats. Diabetes 41: 47–52, 1992
25. Chibber R, Molinatti PA, Wong JSK, Mirlees D, Kohner EM: The effect of aminoguanidine and tolrestat on glucose toxicity in bovine retinal capillary pericytes. Diabetes 43: 758–763, 1994
26. Teixeira AS, Andrade SP: Glucose-induced inhibition of angiogenesis in the rat sponge granuloma is prevented by aminoguanidine. Life Sci 64: 655–662, 1999
27. Lewis BS, Harding JJ: The effects of aminoguanidine on the glycation (non-enzymic glycosylation) of lens protein. Exp Eye Res 50: 463–467, 1990
28. Bucala R, Vlassara H: Advanced glycosylation endproducts in diabetic renal disease: Clinical measurement, pathophysiological significance

and prospects for pharmacological inhibition. Blood Purif 13: 160–170, 1995

29. Courderot-Masuyer C, Dalloz F, Maupoil V, Rochette L: Antioxidant properties of aminoguanidine. Fund Clin Pharmacol 13: 535–540, 1999
30. Kumari K, Umar S, Bansal V, Sahib MK: Inhibition of diabetes-associated complications by nucleophilic compounds. Diabetes 40: 1079–1084, 1991
31. Corbett JA, Tilton RG, Cheng K, Hasan KS, Iodo Y, Wang JC, Sweetland MA, Lancaster JR, Williamson JR, McDaniel ML: Aminoguanidine a novel inhibitor of nitric oxide formation, prevents diabetic vascular dysfunction. Diabetes 41: 552–556, 1992
32. Tasaka Y, Nakaya F, Matsumoto H: Effects of aminoguanidine on insulin release from pancreatic islets. Endocrine J 41: 309–313, 1994
33. Cosio FG: Effects of high glucose concentrations on human mesangial cell proliferation. J Am Soc Nephrol 5: 1600–1609, 1995
34. Ou P, Wolff SP: Aminoguanidine: A drug proposed for prophylaxis in diabetes inhibits catalase and generates hydrogen peroxide *in vitro*. Biochem Pharmacol 46: 1139–1144, 1992
35. Simmonds NJ, Millar AD, Blake DR, Rampton DS: Antioxidant effects of aminosalicylates and potential new drugs for inflammatory bowel disease: Assessment in cell-free systems and inflamed human colorectal biopsies. Aliment Pharmacol Ther 13: 363–372, 1999
36. Sattar A, Kumar S, West DC: Does hyaluronan have a role in endothelial cell proliferation of the synovium? Semin Arthritis Rheum 22: 37–43, 1992
37. Porta M, La Selva M, Bertagna A, Molinatti GM: High glucose concentrations inhibit DNA synthesis and replication without causing death or impairing injury repair in cultured human endothelial cells. Diabetes Res 7: 59–63, 1988
38. McDermott AM, Kern TS, Murphy CJ: The effect of elevated extracellular glucose on migration, adhesion and proliferation of SV40 transformed human corneal epithelial cells. Curr Eye Res 17: 924–932, 1998
39. Giardino I, Edelstein D, Brownlee M: BCL-2 expression or antioxidants prevent hyperglycaemia-induced formation of intracellular advanced glycation endproducts in bovine endothelial cells. J Clin Invest 97: 1422–1428, 1996
40. Brownlee M, Vlassara H, Cerami A: Nonenzymatic glycosylation reduces the susceptibility of fibrin to degradation by plasmin. Diabetes 32: 680–684, 1983
41. Thornally P, Wolff S, Crabbe J, Stern A: The autoxidation of glyceraldehyde and other simple monosaccharides under physiological conditions catalysed by buffer ions. Biochim Biophys Acta 797: 276–287, 1984
42. Nissen NN, Shankar R, Gamelli RL, Singh A, DiPietro LA: Heparin and heparin sulphate protect basic fibroblast growth factor from nonenzymic glycosylation. Biochem J 338: 637–642, 1999
43. McCarthy A, Etcheverry SB, Bruzzone L, Cortizo AM: Effects of advanced glycation endproducts on the proliferation and differentiation of osteoblast-like cells. Mol Cell Biochem 170: 43–51, 1997
44. Cameron NE, Cotter MA, Dines K, Love A: Effects of aminoguanidine on peripheral nerve function and polyol pathway metabolites in streptozotocin-diabetic rats. Diabetologia 35: 946–950, 1992
45. Teixeira AS, Caliari MV, Rocha OA, Machado RD, Andrade SP: Aminoguanidine prevents impaired healing and deficient angiogenesis in diabetic rats. Inflammation 23: 569–581, 1999
46. Booth AA, Khalifah RG, Hudson BG: Thiamine pyrophosphate and pyridoxamine inhibit the formation of antigenic advanced glycation endproducts: Comparison with aminoguanidine. Biochem Biophys Res Commun 220: 113–119, 1996
47. Baynes JW: Role of oxidative stress in development of complications in diabetes. Diabetes 40: 405–412, 1991
48. Duraisamy Y, Slevin M, Smith N, Bailey J, Zweit J, Smith C, Ahmed N, Gaffney J: Effect of glycation on basic fibroblast growth factor induced angiogenesis and activation of associated signal transduction pathways in vascular endothelial cells: Possible relevance to wound healing in diabetes. Angiogenesis 4: 277–288, 2001
49. Hipkiss AR, Carnosine, a protective, anti-ageing peptide? Int J Biochem Cell Biol 30: 863–868, 1998

Molecular and Cellular Biochemistry **246:** 155–162, 2003.

High levels of palmitic acid lead to insulin resistance due to changes in the level of phosphorylation of the insulin receptor and insulin receptor substrate-1

Rosalía Reynoso,[1,2] Luis M. Salgado[1] and Víctor Calderón[1]

[1]*Departamento de Bioquímica, Centro de Investigación y de Estudios Avanzados del IPN, México;* [2]*Departmento de Investigación y Posgrado en Alimentos PROPAC, Facultad de Química, Universidad Autónoma de Querétaro, Querétaro, Qro*

Abstract

Insulin resistance is defined as the decrease in the glucose disposal in response to insulin by the target tissues. High concentrations of nonesterified fatty acids (NEFA) in plasma have been implicated with many insulin resistance states. We evaluated several aspects of the insulin resistance induced by palmitic acid in rats and found that after treatment with 0.09 g/kg of palmitic acid there is a delay in the curve of tolerance to glucose. We measured the changes in protein phosphorylation in samples from *abdominus rectus* muscle and there was a decrease of 64 and 75% in the levels of phosphorylation in tyrosine of the insulin receptor and insulin receptor substrate-1, respectively. This diminution in the tyrosine phosphorylation is consistent with a decrease in the main pathway known to be activated after insulin treatment, the mitogen activated protein kinases (MAPKs). If the animals were treated with inhibitors of PKC, like sphingosine, there was a prevention of the effect of palmitic acid determined at the level of tyrosine phosphorylation. According with this result, we found an increase in the phosphorylations in serine of the insulin receptor after the treatment with palmitate. These results suggest that PKC has a role as negative regulator (by phosphorylation in serine) of the insulin receptors activation in the insulin resistance induced by palmitic acid. (Mol Cell Biochem **246**: 155–162, 2003)

Key words: insulin resistance, palmitic acid, free fatty acids, insulin receptor, insulin receptor substrate-1

Introduction

Type II diabetes, obesity, hypertension and cardiovascular diseases are frequently characterized by target tissue resistance to insulin reflected as a reduction in the glucose uptake by the cells in response to the insulin [1]. These pathologies are often associated with a high concentration in serum of nonesterified fatty acids (NEFA) [2]. Due to this correlation, much of the effort in the field had been concentrated on the effect of the NEFA in the insulin action and glucose metabolism. Randle *et al.* [3] showed that high levels of NEFA reduce the glucose oxidation and uptake in perfused muscle from rat heart and diaphragm. However, some authors did not find a reduction in glucose metabolism due to high concentration of plasma lipids [4, 5] but others had found that acute elevation of NEFA in plasma decrease insulin sensitivity in muscle [6, 7]. The mechanism responsible for this alteration is unknown but it includes changes on the insulin-signaling pathway.

In the insulin-signaling cascade, the hormone activates the tyrosine kinase activity of its receptor that lead to tyrosine phosphorylation of cellular substrates such as the insulin receptor substrate-1 (IRS-1). Once phosphorylated, IRS-1 is able to bind other proteins which activate different signal

Address for offprints: L.M. Salgado, CINVESTAV–IPN, Department of Biochemistry, Apartado Postal 14-740, 07000 México, D.F. (E-mail: Lsalgado@mail.cinvestav.mx)

transduction pathways. Phosphatidyl inositol 3 kinase (PI3K) is activated through the binding of the regulatory subunit (p85) to IRS-1 and it catalyses the formation of PtdIns (3, 4, 5) P_3, necessary for the glucose transport [8, 9].

In humans, a 5-h infusion of lipid/heparin produces an inhibition of glucose transport. This may be a consequence of the decrease of PI 3-kinase activity due to a direct effect of intracellular NEFA on this kinase activity and/or due to secondary alterations upstream in the insulin signaling pathway [10], possibly at the level of IR and IRS-1 phosphorylation. The phosphorylation of some serine positions in the cytoplasmic domain of the insulin receptor seems to be the main point of regulation and there is a strong correlation between the level of phosphorylation and the inhibition of the pathway [11]. The PKC isoenzymes had been implicated in the pathogenesis of insulin resistance [12]. The inhibitory effect of PKC over the insulin receptor kinase was tested, both *in vitro* and *in vivo* [13, 14]. The activity of PKC in the membrane fraction of fat cells is increased under resistance conditions and such resistance is prevented with the use of some inhibitors of PKC [15]. The membrane associated PKC activity was increased in humans with NIDDM and humans and animals with obesity [16, 17]. There is a direct relationship between insulin resistance and the increase in the availability of lipids, some studies indicate that PKC may be activated by cis-unsaturated and their CoA esters, causing a perturbation of different metabolic pathways [18].

However, the majority of the data reported came from *in vitro* or from cell culture studies and, therefore, it is not always possible to extrapolate to the *in vivo* situation. One of the main NEFA found in the plasma of the humans is palmitic acid and it was used in studies *in vitro* which demonstrated that this acid is able to induce insulin resistance [19, 20, 21]. Nevertheless, the molecular mechanism by which the NEFA lead to insulin resistance *in vivo* is not known. In the present work, we evaluated several aspects of the insulin resistance induced by palmitic acid in rats after intra peritoneal administration of this acid.

Materials and methods

Animals

Male Wistar rats weighing 200–250 g were maintained in rooms with constant light-dark cycles and fed *ad libitum*. The rats were fasted overnight (12 h) prior to the experiment.

Plasma NEFA concentration

To test the effects of free fatty acids, different groups of animals fasted overnight were administered intraperitoneally (i.p.) with palmitic acid at 0.045, 0.09 and 0.137 g/kg and blood samples were obtained at 0, 30, 60, 90 and 120 min, two animals received only DMSO and were used as controls. Plasma NEFA concentrations were determined by the method of Itaya and UI [22].

Glucose tolerance

To test the effects of free fatty acids on glucose tolerance, the animals were treated with palmitic acid (i.p.) at 0.09 g/kg and 30 min later the glucose (0.7 g/kg) was given orally. The rats were anesthetized with sodium pentobarbital and after 10 min the insulin (Humulin Regular; Eli Lilly and Co., IN, USA) was injected via cardiac puncture and blood samples were drawn at 0, 15, 30, 45, 60 and 120 min, two animals received only DMSO and were used as controls. Plasma glucose was measured in a glucose analyzer (Boehringer).

Muscle processing

The *rectus abdominus* muscle was rapidly dissected and immediately frozen in liquid nitrogen. The frozen tissues were powdered and homogenized in a polytron (Brinkman Instruments, Westbury, NY, USA) for 30 sec in ice-cold buffer (50 mM Hepes pH 7.5, 150 mM NaCl, 10 mM NaPP, 2 mM Na_3VO_4, 1 mM $MgCl_2$, 1 mM $CaCl_2$, 10 mM NaF, 2 mM EDTA, 2 mM PMSF, 5 μg/ml leupeptin, 1% NP-40, 10% glycerol). The homogenate was incubated for 45 min at 4°C with gentle mixing and then centrifuged at 35,000 g for 60 min [23]. The supernatant was collected and assayed for protein concentration using the Lowry method.

Insulin signaling

The rats were anesthetized and insulin was injected via cardiac puncture at different concentrations (0.1, 1.0 and 10 U/100 g), after 2 min the muscle was removed and frozen. Others animals were treated with insulin (1.0 U/100 g) and the muscle was extracted at 1, 3, 5 and 15 min and frozen.

For palmitic acid treatment, the rats were administered i.p. with palmitic acid (0.09 g/kg), the animals were anesthetized and 30, 60 and 90 min later the insulin was injected via cardiac puncture (1 U/100 g) and 3 min later the muscle was removed and frozen. Other animals were treated with palmitic acid (0.09 g/kg) and 30 min later were administered with sphingosine i.p. (0.6 mg/kg) then after 30 min were injected with insulin and the muscle was removed after 3 min, two animals received only DMSO and were used as controls.

Immunoblotting

Equal amounts of solubilized protein (200 μg) were subjected to SDS-PAGE (7% tris acrylamide) in a Bio-Rad miniature slab gel apparatus. Electro transfer of the proteins from the gel to a nylon membrane (Immunoblot PVDF membrane, Bio-Rad) was performed for 1 h at 15 V (constant) using a semidry transfer apparatus (Bio-Rad) with transfer buffer (25 mM Tris and 192 mM glycine). To reduce nonspecific protein binding, the nylon membrane was blocked at 4°C overnight with 1% blocking solution Kit (Boehringer) in Tris-NaCl buffer (pH 7.5). The membrane was incubated for 3 h at room temperature with an anti-phosphotyrosine antibody (αpY) (SC-7020, Santa Cruz, USA) or with an anti-actin antibody (SC-1616, Santa Cruz, USA) diluted in 0.5% blocking buffer. Then, it was washed with TBS-T (TBS with 0.1% Tween 20) 4 times of 5 min each and 2 times with 0.5% blocking buffer for 10 min. The membrane was then incubated for 1 h at room temperature with anti-mouse immunoglobulin G conjugated with horseradish peroxidase (62-6520, Zymed) and washed 6 times of 5 min with TTBS and 1 time of 10 min. The horseradish peroxidase was detected with a chemiluminescence blotting substrate (POD) (Boehringer). Other antibodies utilized: mouse anti-IR (Mg-636-P, Neo Markers), mouse anti-IRS1 (Mg-630-P, Neo Markers), mouse anti-ERK1-2 (9106S, Cell Signaling Technology), anti-phosphoserine (61-8100, Zymed), rabbit anti-goat (SC-2033, Santa Cruz). Quantitation of specific protein bands was done after digitalization in an Eagle Eye II using the One Scan software (Stratagene).

Immunoprecipitation

To identify and quantify tyrosine phosphoproteins, equal amount of protein (200 μg) was used. It was immunoprecipitated with anti-IR (Mg-636-P, Neo Markers) or –IRS1 (Mg-630-P, Neo Markers) for 1 h at 4°C. The immuno-complexes were adsorbed with protein A-agarose beads overnight at 4°C. The resulting immunocomplexes were washed 5 times with homogenization buffer with 0.1% NP-40. The protein was then solubilized in Laemmli buffer. Immunoblotting procedures for the detection of these proteins were as previously described for immunoblotting with αpY or αpS.

Statistical analyses

Three samples of muscle from two rats (either control or treated) were used in each experiment and the results showed came from at least 3 independent experiments (n = 6). The results were analyzed statistically by one-way and the differences were determined using Student's *t*-test.

Results

High levels of palmitic acid induce a delay in the curve of glucose utilization in vivo

The most common saturated fatty acid found in the human plasma is palmitic acid. Therefore, we used the administration of this acid to recreate the conditions of high concentrations of NEFA and to evaluated its effect on the induction of insulin resistance. Palmitate was dissolved in DMSO and administrated by intraperitoneal injection. Two doses were tested 0.045 g/kg and 0.09 g/kg and the average levels of NEFA after 120 min of administration were 730 ± 20 and 970 ± 45 μM, respectively (Fig. 1A), the administration of only DMSO had no effect and the animals behave as the non treated controls (data not shown). Higher concentrations of palmitate were toxic for the animals and we decided to use

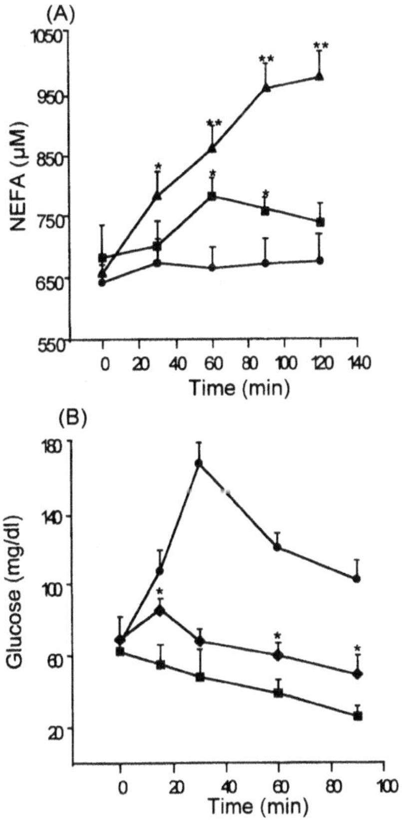

Fig. 1. Effect of palmitic acid on plasma concentration of non-esterified fatty acids and glucose in Wistar rats. (A) Each group received: vehicle (●), palmitic acid 0.045 g/kg (■) and 0.09 g/kg (▲). The data are mean ± SE. $*p < 0.01$, $**p < 0.05$ comparing palmitic acid with vehicle. (B). Animals treated with glucose (●), glucose plus insulin (■), glucose, insulin plus palmitic acid (0.09 g/kg) (◆). *Indicated significantly difference with respect to animals treated with glucose plus insulin ($p < 0.05$).

0.09 g/kg to measure their effect on the plasma glucose concentrations.

High concentrations of palmitate change the behavior of the glucose utilization curve (Fig. 1B). After 90 min of the glucose intake, the animals treated with insulin showed glucose concentrations lower than basal. At the same time, the animals treated with palmitate (0.09 g/kg) and insulin had glucose concentration two times higher, even when both curves showed a decrease at the same rate. The effect of palmitic acid was not caused by substrate competition, since the administration of other metabolic precursors of citrate or acetyl-CoA such as pyruvate and glutamine, did not affect the glucose plasma concentration (data not shown). The results are in agreement with those of Yki-Jarvinen [20], who showed that lipid/heparin infusion in healthy humans raises plasma NEFA levels and this produces an inhibition of the insulin dependent glucose uptake by human forearm tissues *in vivo*. It has been reported also that NEFA induces insulin resistance in humans [10].

Palmitic acid treatment induces changes in protein phosphorylation in vivo

The insulin resistance had been associated with alterations in the glucose uptake by the cells and most of the last reports point to changes on the signal transduction pathways activated after insulin treatment. The PI3K-PKB signaling pathway that regulates the movement of the Glut4 transporters was the most studied and the last reports agree that alterations on it correlate with the insulin resistance. Nevertheless, there is not much information about the alterations before that point. We were interested in detect and measure the changes in tyrosine phosphorylation of the insulin receptor and the insulin receptor substrate-1 under our experimental conditions.

Different concentrations of insulin had been used, both *in vitro* and *in vivo* in insulin resistance studies. We determined first, which insulin concentration was able to give us the best response measured as changes on the pattern of tyrosine phosphorylation. We found that 1 U/100 g gave the strongest response (Fig. 2A), higher concentrations did not increase the phosphorylation levels over this limit and some were toxic for the animals. As expected, the response is very fast and we could detect the maximal stimulation after 3 min of administration (Fig. 2B). The decay in the response is very quick, and after 20 min of stimulation the tyrosine phosphorylation is almost the same as the one found under non-stimulated conditions. The decay on tyrosine phosphorylation could reflect the activity of specific tyrosine phosphatases, as had been proposed [21].

High concentrations of palmitic acid (0.09 g/kg) in plasma were very effective to block the signaling pathway activated

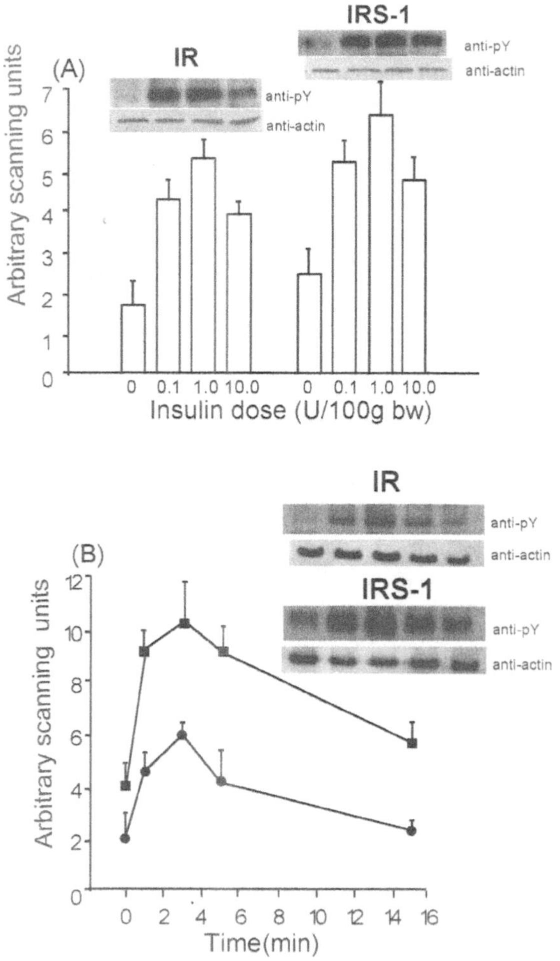

Fig. 2. Effect of insulin on tyrosine phosphorylation of IR and IRS-1 in skeletal muscle of Wistar rats. (A) Dose response effect of insulin. (B) Time Course of stimulation with insulin (1 U/100 g body wt). In both cases, a representative experiment from antiphosphotyrosine and actin is shown in superior panel and the bar graph was constructed after the quantification, by scanning densitometry, of autoradiograms from 8 experiments. The data are mean ± S.E.

by insulin (Fig. 3). Its effect, measured as the level of tyrosine phosphorylation of IR and IRS-1, was evident after 15 min of administration. The peak of inhibition was after 60 min, in such conditions, the levels of tyrosine phosphorylation were similar to the ones found in control non-treated animals, and the inhibitory effect was similar for both proteins studied. The treatment with palmitic acid did not change significantly the basal insulin levels or the levels induced by glucose intake (data not shown).

To evaluate the functionality of our system and to search for changes downstream of the receptors, we measured the activity of other members of the signaling pathway. We decided to measure the activity of ERK1 and ERK2 because one of the main pathways activated by insulin is the MAPKs. We

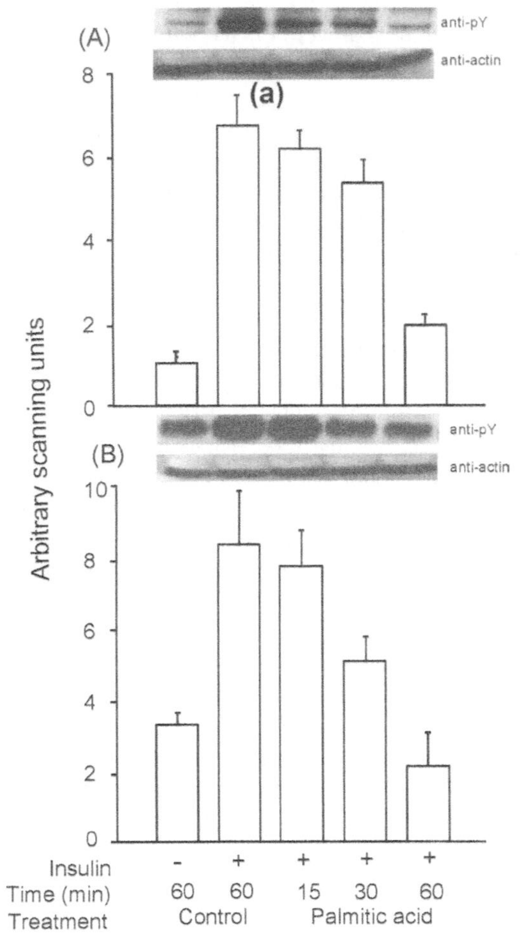

Fig. 3. Effect of palmitic acid (0.9 g/kg) on tyrosine phosphorylation in skeletal muscle of Wistar rats stimulated with insulin. (A) Insulin receptor (IR). (B) Insulin receptor substrate-1 (IRS-1). A representative experiment from antiphosphotyrosine and actin is shown in superior panel and the bar graph was constructed after the quantification, by scanning densitometry, of autoradiograms from 6 experiments. The data are mean ± S.E. *Indicated significantly difference with respect to the animals treated with insulin but without palmitic acid ($p < 0.05$).

used an antibody that recognizes the active (pThr202/pTyr204) form of both ERK1 and ERK2. As expected, there was an increase in the phosphorylation of ERK1 and ERK2 after insulin stimulation (Fig. 4), more evident in ERK2. Palmitate alone did not show an increase on the amount of active enzymes (the discrete increase was not significantly different with the non-treated tissue). Palmitic acid administrated with insulin was able to inhibit, 60–70%, the appearance of active enzymes. The inhibitory effect was more pronounced for ERK2 than for ERK1 (Fig. 4).

The level of phosphorylation in serine residues, after immunoprecipitation of the insulin receptor, was measured with specific antibodies (Fig. 5). We found an increase in the amount of serine phosphorylation in the animals treated

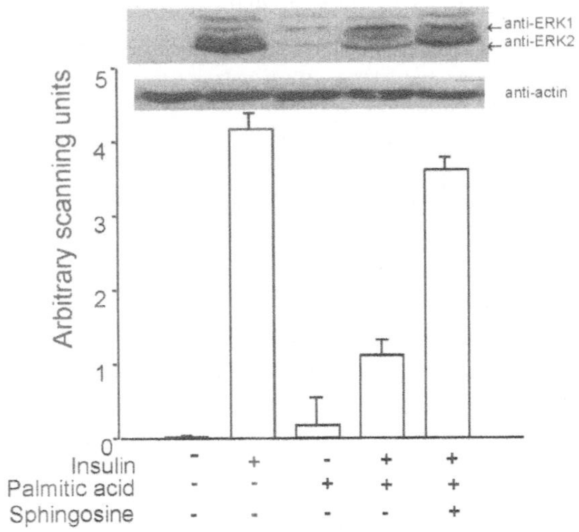

Fig. 4. Effect of palmitic acid on ERK phosphorylation in skeletal muscle of Wistar rats stimulated with insulin. A representative experiment of ERK phosphorylation and actin is shown in superior panel and the bar graph was constructed after the quantification, by scanning densitometry, of autoradiograms from 6 experiments. The data are mean ± S.E.

with insulin. Interestingly, animals treated with palmitic acid only shown the same level of serine phosphorylation than the rats treated with insulin. The animals that received the two treatments, insulin and palmitic acid, at the same time showed

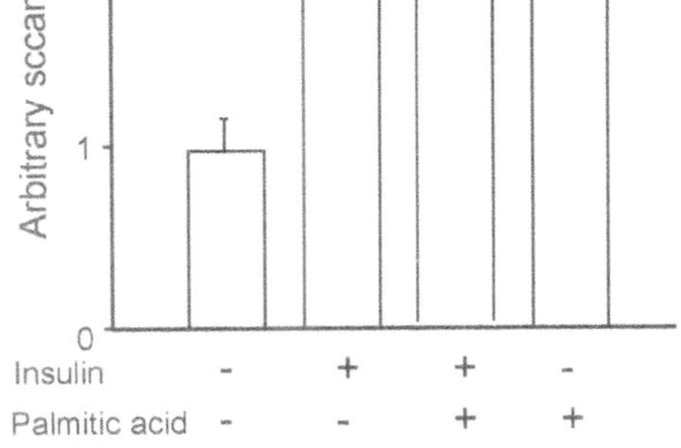

Fig. 5. Effect of palmitic acid on phosphorylation in serine residues of IR in skeletal muscle of Wistar rats stimulated with insulin. A representative experiment of antiphosphoserine and actin is shown in superior panel and the bar graph was constructed after the quantification, by scanning densitometry, of autoradiograms from 6 experiments. The data are mean ± S.E.

phosphorylation levels higher than the individual treatments.

Sphingosine is able to revert the effects of the palmitic acid treatment in vivo

There is a direct correlation between the level of phosphorylation on serine residues of the insulin receptors and the diminution of its activity *in vitro*. This fact suggested that one way to regulate the insulin pathway could be the phosphorylation of such residues [11, 24, 25]. Another piece of evidence is the inverse correlation found between the activities of PKC and IR [26]. And one interesting conclusion from those facts is that PKC could act as negative regulator of the activity of the insulin receptor. We tested this hypothesis in our system. First, we used sphingosine a natural inhibitor of PKC and determine its effect on the tyrosine phosphorylation of IR and IRS-1. The administration of sphingosine (0.6 mg/kg) induced a significant increase (16%) on the tyrosine phosphorylation of IR after 15 min (data not shown). When sphingosine (0.6 mg/kg) was used together with palmitate, it was able to revert, at least partially, the inhibition caused by palmitate (Fig. 6). The recover was clearly evident for both receptor proteins, IR and IRS-1.

ERK activity was modified also after the treatment with sphingosine (Fig. 4). Animals treated with insulin, palmitic acid and sphingosine showed a phosphorylation pattern similar to the animals that were treated with insulin only.

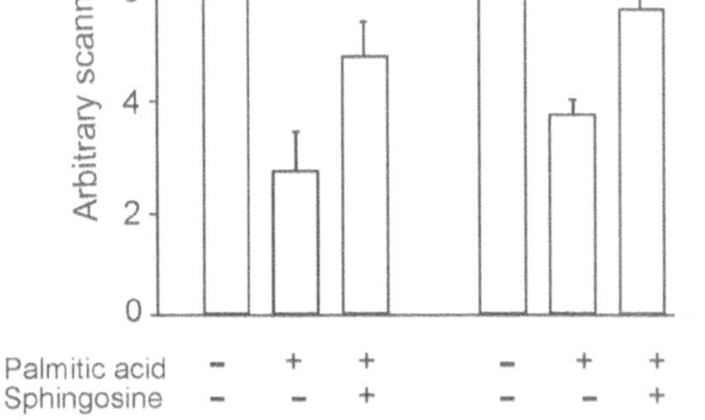

Fig. 6. Effect of palmitic acid and sphingosine on tyrosine phosphorylation of IR and IRS-1 of Wistar rats stimulated with insulin. A representative experiment of antiphosphotyrosine and actin is shown in superior panel and the bar graph was constructed after the quantification, by scanning densitometry, of autoradiograms from 7 experiments. The data are mean ± S.E.

Discussion

The effect of lipids on the glucose metabolism is well documented and much of the effort in the field has been concentrated on the effect of the NEFA on it. The principal approach used to afford the problem was the treatment of cells or tissues with NEFA to increase their levels in the cell. The interpretation of such results has been some time contradictory. For example, some of them suggested that an increase of NEFA in the cell induce changes in the affinity of insulin by its receptor [26, 27], but the consensus now is that there is not effect on the binding or in the kinetic parameters of it [28]. The next step where the effects were investigated was those on the signal transduction pathways activated by insulin. Among them, these that lead to the activation of PKB and the movement of the glucose transporter-4 (Glut-4) have been studied [9]. The changes in the receptor phosphorylation and its activity were also investigated. Many researchers have attempted to demonstrate that high plasma NEFA concentrations inhibit whole body glucose disposal, most have used lipid/heparin infusions to maintain those levels of NEFA. However, the heparin may inhibit both insulin binding and insulin-stimulated autophosphorylation of insulin receptor [28].

In the current study the experimental conditions were design to reduce other parameters that contribute to insulin resistance such as heparin and triglycerides. Between one and two hours after the administration of palmitic acid high levels of NEFA were detected in serum. These levels were comparable to the ones found in diabetic and obese animals. The acute increase of NEFA induces some metabolic changes in the cell, some of them that affect directly the metabolism of the glucose. For example those described for Randle et al. [3], who postulated that elevated NEFA oxidation produced an increase of the intramitochondrial acetyl-CoA/CoA and NADH/NAD$^+$ ratios and citrate and glucose 6-phosphate concentration, that could inhibit hexokinase II resulting in decreased glucose uptake. However, in the current study pyruvate and glutamine did not change the glucose blood levels, suggesting that not metabolic effects were involved.

There are several points at which we could search for changes induced by high levels of NEFA. One of the most important could be the autophosphorylation of the receptor (IR) or the activation of the insulin receptor substrate-1 (IRS-1). In animals treated with palmitic acid we detected a strong diminution on the tyrosine phosphorylation levels, both on IR and IRS-1. The diminution observed could be enough to explain the resistance to insulin showed for some tissues when there is an increase in plasma of NEFA. They were not the only alterations that we detected, a decrease in phosphorylation of ERK2 was also observed. And, as the changes on tyrosine phosphorylation, they were evident after the increase of NEFA but strongly influenced by insulin. It could be possible that these two results were related in some way.

As the changes in the signaling seem to correlate with the metabolic changes, one interesting possibility could be that after the increase of palmitic acid in serum it could activate a system which could induce alterations on the insulin activated pathway. From the known effects of NEFA in the cell, the activation of PKC could be an important step. It has been proposed that PKC could regulate negatively systems activated by insulin [30, 31]. The regulation was proposed to include the phosphorylation of serine residues of IR and IRS-1 [13, 29, 31]. According with that, the animals treated with palmitic acid showed an increase on the phosphorylation in serine residues of IR after treatment. It suggested that this point could be an important negative regulator as it had been proposed. We found that sphingosine, the natural inhibitor of PKC, could prevent some of the effects induced by palmitic acid. Even that the effects of sphingosine *in vivo* are difficult to analyze due to its fast metabolization, taking together these two results strongly suggest that PKC could be one of the key regulating points. It is not clear now if there is a direct regulation by PKC or there are other components activated by it (like ERK), which lead to the phosphorylation of the serine residues.

The MAPK pathway has been proposed as negative regulator of systems activated by insulin [29]. They could be doing it by at least two ways, one coming from PKC other through Shc and Ras [32]. *In vitro*, ERK phosphorylates IRS-1 on serine/threonine residues and it was associated with diminished tyrosine phosphorylation by the insulin receptor. The increase of active ERK found in animals treated with palmitic acid, with and without insulin, supports such model of regulation. Also the reversion of the effect of palmitic acid by sphingosine (the PKC inhibitor) strongly suggested that PKC and ERK could mediate the regulation of the pathway.

The pathology of insulin resistance is complex and it must be multifactorial. Our work is a direct approximation to understand, at the molecular level, the disease *in vivo*. From our results we can suggest that after an acute increase of NEFA, the glucose metabolism changes, probably due to a direct effect of one of the nonesterified fatty acids present in serum (palmitic acid?). One of the first changes than can be detected is at the level of the activation of the receptor itself. The change provokes alterations at several points of the signal transduction pathways activated by insulin, mostly IRS-1 and MAPK. And, these changes are due to down-regulation mediated by PKC. The next step will be to determine if PKC directly phosphorylates the receptor or it does it by an indirect way (through other kinases of serine residues?). It could be also interesting to determine if the effect of NEFA are mostly due to palmitic acid or there is a synergetic effect from more than one fatty acid.

Acknowledgements

This work was supported by a fellowship from CONACYT to RR. We thank Margarita Rosas Flores and J. Luis Chávez for their excellent experimental technical assistance, David Heath for his help with the manuscript and two anonymous referees for their comments.

References

1. Reaven GM: Role of insulin resistance in human disease. Diabetes 37: 1595–1607, 1988
2. Boden G: Role of fatty acids in the pathogenesis of insulin resistance and NIDDM. Diabetes 46: 3–10, 1997
3. Randle PJ, Garland PB, Hales CN, Newsholme EA: The glucose fatty acid cycle its role in insulin sensitivity and the metabolic disturbance of diabetes mellitus. Lancet 1: 785–789, 1963
4. Bevilacqua S, Bonadonna R, Buzzigoli G, Boni C, Ciociaro D, Maccari F, Giorico MA, Ferrannini E: Acute elevation of free fatty acid levels leads to hepatic insulin resistance in obese subjects. Metabolism 36: 502–506, 1987
5. Saloranta C, Koivisto V, Widen E, Falholt K, DeFronzo RA, Harkonen M, Groop L: Contribution of muscle and liver to glucose-fatty acid cycle in humans. Am J Physiol 264: E599–E605, 1993
6. Roden M, Price TB, Perseghin G, Petersen KF, Rothman DL, Cline GW, Shulman GA: Mechanism of free fatty acid-induced insulin resistance in humans. J Clin Invest 97: 2859–2865, 1996
7. Boden G, Chen X, Ruiz J, White JV, Rossetti L: Mechanisms of fatty acid-induced inhibition of glucose uptake. J Clin Invest 93: 2438–2446, 1994
8. Whitehead JP, Clark SF, Urso B, James DE: Signaling trough the insulin receptor. Curr Opin Cell Biol 12: 222–228, 2000
9. Czech MP, Corvera S: Signaling mechanisms that regulate glucose transport. J Biol Chem 274. 1865–1868, 1999
10. Dresner A, Laurent D, Marcucci M, Griffin ME, Dufour S, Cline GW, Slezak LA, Andersen DK, Hundal RS, Rothman DL, Petersen KF, Shulman GI: Effects of free fatty acids on glucose transport and IRS-1 associated phosphatidylinositol 3-kinase activity. J Clin Invest 103: 253–259, 1999
11. Schmitz-Peiffer C: Signaling aspects of insulin resistance in skeletal muscle: Mechanisms induced by lipid oversupply. Cell Signal 12: 583–594, 2000
12. Considine RV, Caro JF: Protein kinase C: Mediator or inhibitor of insulin action? J Cell Biochem 52: 8–13, 1993
13. Takayama S, White MF, Kahn MF: Phorbol ester-induced serine phosphorylation of the insulin receptor decreases its tyrosine kinase activity. J Biol Chem 263: 3440–3457, 1988
14. Chin JE, Dickens M, Tavare JM, Roth RA: Over expression of protein kinase C isoenzymes alpha, beta I, gamma and epsilon in cells over expressing the insulin receptor. J Biol Chem 268: 6338–6347, 1993
15. Muller HK, Kellerer M, Ermel B, Muhlhofer A, Obermaier-Kusser B, Vogt B, Haring HU: Prevention by protein kinase C inhibitors of glucose-induced insulin-receptor tyrosine kinase resistance in rat fat cells. Diabetes 40: 1440–1448, 1991

16. Considine RV, Nyce MR, Allen LE, Morales LM, Triester S, Serrano J, Colberg J, Lanzacoby S, Caro JF: Protein kinase C is increased in the liver of humans and rats with non-insulin-dependent diabetes mellitus: Alterations not due to hyperglycemia. J Clin Invest 95: 2938–2944, 1995
17. Avignon A, Yamada K, Zhou X, Spencer B, Cardona O, Saba-Siddique S, Galloway L, Standaert ML, Farese RV: Chronic activation of protein kinase C in soleus muscle and tissues of insulin resistance type II diabetic Goto-Kakizaki (GK) obese/aged, and obese/zucker rats. A mechanism for inhibiting glycogen synthesis. Diabetes 45: 1396–1404, 1996
18. Nesher M, Boneh A: Effect of fatty acids and their acyl-CoA esters on protein kinase C activity in fibroblasts: Possible implications in fatty acid oxidation defects. Biochim Biophys Acta 1221: 66–72, 1994
19. Storz P, Döppler H, Werning A, Pfizenmaier K, Müller G: Cross-talk mechanisms in the development of insulin resistance of skeletal muscle cells, palmitate rather than tumor necrosis factor inhibits insulin-dependent protein kinase B (PKB)/Akt stimulation and glucose uptake. Eur J Biochem 266: 17–25, 1999
20. Yki-Jarvinen H, Puhakainen I, Koivisto VA: Effect of free fatty acids on glucose uptake and monoxidative glycolysis across human forearm tissues in the basal state and during insulin stimulation. J Clin Endocrinol Metab 72: 1268–1277, 1992
21. Rothenberg PL, Lane WS, Karasik A, Backer J, White M, Kahn CR: Protein purification and partial sequence analysis of pp185, the major substrate of the insulin receptor tyrosine kinase. J Biol Chem 266: 8302–8311, 1991
22. Itaya K, Ui M: Colorimetric determination of free fatty acids in biological fluids. J Lipid Res 6: 16–20, 1965
23. Goodyear LJ, Giorgino F, Sherman LA, Carey J, Smith RJ, Dohm GL: Insulin receptor phosphorylation, insulin receptor-substrate-1 phosphorylation and phosphatidylinositol 3-kinase activity are decreased in intact skeletal muscle strips from obese subjects. J Clin Invest 95: 2195–2204, 1995
24. Liu F, Roth RA: Identification of serines-1035/1037 in the kinase domain of the insulin receptor as protein kinase C alpha mediated phosphorylation sites. FEBS Lett 352: 389–392, 1994
25. Lingamanaidu V, Ravichandran DL, Esposito JC, Quon MJ: PKC-ζ phosphorylates insulin receptor substrate-1 and impairs its ability to activate phosphatidylinositol 3-kinase in response to insulin. J Biol Chem 276: 3543–3549, 2001
26. Svedberg J, Björntorp P, Smith U, Lönnroth P: Free-fatty acid inhibition of insulin binding, degradation and action in isolated rat hepatocytes. Diabetes 39: 570–574, 1990
27. Grunfeld C, Baird KL, Kahn CR: Maintenance of 3T3-L1 cells in culture media containing saturated fatty acids decreases insulin binding and insulin action. Biochem Biophys Res Commun 103: 219–226, 1981
28. Kriauciunas KM, Grigorescu MF, Kahn CR: Effects of heparin on insulin binding and biological activity. Diabetes 36: 163–168, 1987
29. Defea K, Roth RA: Modulation of insulin receptor substrate-1 tyrosine phosphorylation and function by mitogen-activated protein kinase. J Biol Chem 272: 31400–31406, 1997
30. Berti L, Mosthaf L, Kroder G, Kellerer M, Tippmer S, Mushack J, Seffer E, Seedorf K, Haring HU: Glucose-induced translocation of protein kinase C isoforms in rat-1 fibroblasts is paralleled by inhibition of the insulin receptor tyrosine kinase. J Biol Chem 269: 3381–3386, 1994
31. Strack V, Hennige AM, Krützfeldt J, Bossenmaier B, Klein HH, Keller M, Lammers R, Häring HU: Serine residues 994 and 1023/25 are important for insulin receptor kinase inhibition by protein kinase C isoforms beta2 and theta. Diabetologia 43: 443–449, 2000
32. Nystrom FH, Quon MJ: Insulin signaling: Metabolic pathways and mechanisms for specificity. Cell Signal 11: 563–574, 1999

Molecular and Cellular Biochemistry **246:** 163–170, 2003.

Diabetes and mitochondrial oxidative stress: A study using heart mitochondria from the diabetic Goto-Kakizaki rat

Dario Loureiro Santos,[1] Carlos Marques Palmeira,[2,3] Raquel Seiça,[2,4] José Dias,[5] José Mesquita,[5] António Joaquim Moreno[2,3] and Maria Sancha Santos[2,3]

[1]*University of Trás-os-Montes e Alto Douro, Vila Real;* [2]*Center for Neurosciences and Cell Biology of Coimbra,* [3]*Department of Zoology,* [4]*Faculty of Medicine,* [5]*Department of Botany, University of Coimbra, Coimbra, Portugal*

Abstract

Increasing evidence shows that the overproduction of reactive oxygen species, induced by diabetic hyperglycemia, contributes to the development of several cardiopathologies. The susceptibility of diabetic hearts to oxidative stress, induced *in vitro* by ADP-Fe^{2+} in mitochondria, was studied in 12-month-old Goto-Kakizaki rats, a model of non-insulin dependent diabetes mellitus, and normal (non-diabetic) Wistar rats. In terms of lipid peroxidation the oxidative damage was evaluated on heart mitochondria by measuring both the O_2 consumption and the concentrations of thiobarbituric acid reactive substances. Diabetic rats display a more intense formation of thiobarbituric acid reactive substances and a higher O_2 consumption than non-diabetic rats. The oxidative damage, assessed by electron microscopy, was followed by an extensive effect on the volume of diabetic heart mitochondria, as compared with control heart mitochondria. An increase in the susceptibility of diabetic heart mitochondria to oxidative stress can be explained by reduced levels of endogenous antioxidants, so we proceeded in determining α-tocopherol, GSH and coenzyme Q content. Although no difference of α-tocopherol levels was found in diabetic rats as compared with control rat mitochondria, a significant reduction in GSH (21.5% reduction in diabetic rats) and coenzyme Q levels of diabetic rats was observed. The data suggest that a significant decrease of coenzyme Q_9, a potent antioxidant involved in the elimination of mitochondria-generated reactive oxygen species, may be responsible for an increased susceptibility of diabetic heart mitochondria to oxidative damage. (Mol Cell Biochem **246**: 163–170, 2003)

Key words: heart mitochondria, diabetes, reactive oxygen species, coenzyme Q, α-tocopherol, Goto-Kakizaki rat

Introduction

Diabetes mellitus Type 2 or non-insulin-dependent diabetes mellitus (NIDDM) is a metabolic disorder characterised by hyperglycemia due to a decrease in the response of peripheral tissues (e.g. liver, skeletal muscle, adipose tissue, pancreatic β-cells) to insulin [1]. It has been well established that patients with diabetes mellitus, exhibit a high incidence of cardiac dysfunction and mortality [2]. Clinical studies suggested that diabetic patients have a significantly greater incidence and severity of several cardiopathies (e.g. angina, acute myocardial failure, atherosclerosis) [3, 4], and approximately 80% of all patients with diabetes die of cardiovascular disease [3].

Mitochondria play an important role in the development of NIDDM, which agrees with an age-related decline in the capacity for oxidative phosphorylation, and consequently contributes to its pathophysiology [5, 6]. In several animal models, a relationship has been established between diabetes and some dysfunctions in mitochondrial oxidative events [7, 8], suggesting the occurrence of bioenergetic alterations at the mitochondrial level [7, 9]. Mitochondria which play

Address for offprints: M.S. Santos, Center for Neurosciences and Cell Biology of Coimbra, Department of Zoology, University of Coimbra, 3004-517 Coimbra, Portugal (E-mail: mssantos@ci.uc.pt)

an important role in cell function, appear to constitute one of the main sources of reactive oxygen species (ROS) [10, 11]. It has been estimated that up to 2% of the total mitochondrial oxygen consumption results in ROS generation [9, 11]. Mitochondria are also the cellular component most extensively affected by increased concentrations of ROS. The inner mitochondrial membrane is believed to be particularly susceptible to oxidative damage even under physiological conditions because it is a major site of ROS production and has a high content in polyunsaturated fatty acids (PUFA) [9, 10]. ROS generated in the electron transport chain of mitochondria may react with PUFA which cause lipid peroxidation and alterations in the structure integrity of mitochondrial membranes, causing irreversible swelling and disruption, which are a cause of perturbation of mitochondrial functions [12–14]. Oxidative damage is suspected to be involved in the pathogenesis of several diseases (for review see [9]) particularly in myocardial injury and several reports indicate that oxidative stress could be responsible for the susceptibility of diabetic heart to cardiopathies [6, 7, 15].

Vitamin E (α-tocopherol), an unsaturated lipid found in the heart mitochondrial membrane, was reported to display an higher antioxidant activity, by inhibiting lipid peroxidation [16]. In the presence of lipoperoxides α-tocopherol (TOH) can donate one electron, and create a tocopheroxyl radical ($TO^{\bullet}$) which can be reduced back to TOH by ascorbate or coenzyme Q (CoQ) [17]. The abundant occurrence of CoQ in mitochondria, due to is role in mitochondrial electron transport, indicates CoQ as a preferential regenerator of α-tocopherol [17]. However, research from several groups reported that the antioxidant effects of CoQ are independent of α-tocopherol recycling [17, 18]. The emergence of CoQ as an antioxidant has a special meaning since it is the only lipid-soluble antioxidant known, so far, that is manufactured by the human body [18, 19].

The Goto-Kakizaki (GK) rat has been raised as a non-obese animal model of NIDDM. This animal is spontaneously diabetic, is obtained by selective inbreeding of normal Wistar rats with the highest glucose values during oral glucose tolerance tests [20]. The GK rat exhibits a moderate but stable fasting hyperglycemia, as early as 2–4 weeks after birth, which does not progress to the ketotic state. Our work evaluates the susceptibility of heart mitochondria from Goto-Kakizaki rats and normal Wistar rats to oxidative stress injury induced by the oxidant pair ADP/Fe^{2+}, as well as the contribution of the antioxidant defenses.

Materials and methods

This study conforms with the *Guide for the Care and Use of Laboratory Animals* published by the US National Institutes of Health (NIH Publication No. 85–23, revised 1996).

Chemicals

Ubiquinone 10 (CoQ_{10}) was obtained from Fluka (Germany), ubiquinone 9 (CoQ_9) and α-tocopherol were obtained from Sigma Chemicals (St. Louis, MO, USA). All other chemicals used were of analytical grade.

Animals and blood glucose determination

Male spontaneously diabetic GK rats (46–54 weeks of age; body wt. 384.7 ± 2.4 g; non-fasting blood glucose levels 198.6 ± 13.5 mg/dl) were obtained from a local breeding colony of Coimbra, established in 1995 with breeding couples from the colony at the Tohoku University School of Medicine (Sendai, Japan; courtesy of Dr. K.I. Susuki). Control animals were non-diabetic male Wistar rats of similar age (44–53 weeks of age; body wt. 672.5 ± 60.5 g; non-fasting blood glucose levels 93.4 ± 2.9 mg/dl) obtained from our local colony. Animals were kept under controlled light and humidity conditions and with free access to powdered rodent chow (diet CRF 20; Charles Rivers, France) and water. Glucose tolerance test was used as a selection index. Blood glucose levels were determined, immediately after animal's sacrifice, through the glucose oxidase reaction, using a glucometer (Glucometer Elite, Bayer, Portugal).

Isolation of rat heart mitochondria

Mitochondria were isolated by the conventional differential centrifugation method from the hearts of Wistar and GK rats [21], with some modifications. The animals were killed by decapitation and the heart was immediately excised and finely minced with a pair of sharp scissors in an ice-cold isolation medium containing 250 mM sucrose, 1 mM EGTA, and 5 mM HEPES–KOH, pH 7.4. The minced blood-free tissue was then resuspended in 5 ml of isolation medium and transferred to a 50-ml glass homogenizer vessel and homogenized with a tightly fitted Potter–Elvehjen homogenizer and a Teflon pestle driven by a motor, during 30 sec. Protease (1 mg subtilisine prepared in 1 ml of isolation medium) was added and the suspension was re-homogenized for an additional 30 sec. Finally 30 ml medium was added, followed by a last 30-sec period homogenization. The homogenate was equally divided into two 50 ml centrifuge tubes and each aliquot was made up to 40 ml with isolation medium and centrifuged at 11,000 × g for 10 min. The supernatant fluid was completely decanted and the pellet, devoid of protease, was gently homogenized to its original volume with isolation medium using a loose-fitting homogenizer. The suspension was centrifuged at 900 × g during 10 min and the obtained low-speed supernatant was transferred through two layers of cheesecloth to different centrifuge tubes and submitted to centrifugation at 9000 × g for 10 min. The high-speed supernatant was discarded and the pellet was resuspended using a paintbrush and repelleted at 9000 × g for 10 min. EGTA was omitted from the final washing medium. Mitochondria were resuspended in isolation medium (without EGTA) in a final concentration of 20–40 mg/ml.

All the isolation procedure was performed at 0–4°C and the mitochondrial suspension was maintained on ice (0–4°C) throughout the assays. The concentration of mitochondrial membrane proteins was estimated according to the biuret method [22], using bovine serum albumin as standard.

Induction and assay of lipid peroxidation

Lipid peroxidation in rat heart mitochondria was measured polarographically by monitoring the rates of oxygen consumption with an YSI 5300 oxygen monitor (Yellow Spring Instruments) equipped with a Clark oxygen electrode, as previously described [23]. Reaction was run at 30°C in a 1-mL reaction chamber equipped with magnetic stirring and containing 1 mL medium (175 mM KCl, 10 mM Tris-HCl, pH 7.4), supplemented with 3 μM rotenone and 1 mg mitochondrial protein. After 1-min pre-incubation, peroxidation was started by adding 1 mM ADP and 0.1 mM $FeSO_4$. The oxygen solubility at 30°C was considered 232 μM when the medium was air-equilibrated at 760 Torr.

Lipid peroxidation was also measured by determining the amount of lipid peroxides formed during incubation as the amount of thiobarbituric acid reactive species (TBARS) formed, according to a modified procedure described by Ernster and Nordenbrand [24]. Mitochondrial protein (1 mg) was incubated for 45 min, at 30°C, in 1 ml of a medium consisting of 175 mM KCl, 10 mM Tris-HCl, pH 7.4, supplemented with 3 μM rotenone. Membrane lipid peroxidation was started by adding simultaneously ADP/$FeSO_4$ (1 mM/0.1 mM). Lipoperoxidation was interrupted by placing the tubes in ice, which lowered the temperature to 0–4°C. To measure lipid peroxidation, 0.5 mL of cold 40% trichloroacetic acid and 2 mL of 0.67% TBA containing 6.8 mM 2,6-diterbutyl-4-methylphenol (BHT) were added to 0.5 mL of the test material. The mixture was heated at 100°C during 10 min, and was allowed to cool in ice before centrifugation in a bench centrifuge (1500 × g, 5 min). The supernatant was collected and lipid peroxidation was estimated by the appearance of TBARS spectrophotometrically quantified at 535 nm. The amount of TBARS formed was calculated using a molar extinction coefficient of 1.56×10^5 M^{-1} cm^{-1} and expressed as nmol TBARS/mg protein [25].

Extraction and quantification of coenzyme Q and α-tocopherol

CoQ were extracted from aliquots of mitochondria containing 1 mg protein/ml according to the previously described method [26]. The extract was evaporated to dryness under a stream of N_2, and suspended in ethanol for HPLC analysis. Liquid chromatography was performed using a Gilson high-performance liquid chromatography apparatus with a reverse phase column (RP18; Spherisorb; S5 ODS_2) as earlier described [27]. Samples were eluted from the column with methanol:heptane (10:2, by volume), at a flow of 2 ml/min. Detection was performed with a UV detector, at 269 nm. Identification and quantification were based in pure standards by their retention times and peak areas, respectively. CoQ levels (CoQ_9 and CoQ_{10}) in mitochondrial membranes were expressed in pmol/mg protein.

The extraction and separation of α-tocopherol was performed following the method described by Vatassery *et al.* [28]. The extract, evaporated to dryness under a stream of N_2 and kept at –80°C, was dissolved in n-hexane and the α-tocopherol contents was analyzed by reverse phase HPLC. A 4.6 × 200 mm Spherisob S10w column was eluted with n-hexane modified with 0.9% methanol, at a flow of 1.5 ml/min. Detection was performed with a UV detector, at 287 nm. The levels of α-tocopherol in mitochondrial membranes were expressed in pmol/mg protein.

Assay for GSH contents in mitochondria

Determination of GSH in samples were carried out with fluorescence detection after reaction of the supernatant of the H_3PO_4/EDTA-NaH_2PO_4 deproteinized mitochondria solution with the reagent *o*-phthaldeyde, at pH 8.0, according to Hissin and Hilf [29].

Electron microscopy

Mitochondria were fixed for electron microscopy by adding 3% glutaraldehyde in 0.1 M sodium cacodylate buffer, pH 7.3, and were incubated during 2 h at 4°C. After centrifugation (15,000 × g, 3 min), pellets were washed with 0.1 M sodium cacodylate buffer (pH 7.3). The fixed and washed pellet was resuspended in 1% OsO_4 buffered with sodium cacodylate 0.1 M, pH 7.3. After a 2 h incubation, membranes were washed with cacodylate buffer pH 7.3 (without osmium tetroxide). Following pre-inclusion in 1% agar, the samples were dehydrated in grade ethanol and were embedded in Spurr. The ultrathin sections were obtained with a LKB ultra-microtome Ultrotome III and stained with methanolic uranyl acetate followed by lead citrate and examined with a Jeol Jem-100SX electron microscope operated at 80 kV. For quantitative studies, ultrathin sections from 4 blocks were obtained and 5 micrographs were taken at random on them. So, 20 micrographs with the final magnification of ×20,000 from peroxidized mitochondria and the same number from control mitochondria were analyzed. The relative volumes of mitochondria were calculated according with the Weibel method [30], for which a test area 14 × 13.5 cm was used. This grid contains 112 lines of 1 cm doing 224 points and the total line length of 112 cm.

Data analysis and statistics

Data are presented as mean ± S.E.M. of the indicated number of experiments, each one obtained from a different animal. Statistical evaluation was performed using the paired Student's *t*-test and by using the one way ANOVA Student–Newmann–Keuls post-test for multiple comparisons. Values of $p < 0.05$ were considered significant.

Results

Mitochondrial lipid peroxidation

Heart mitochondria isolated from diabetic GK rats show a higher susceptibility to lipid peroxidation induced by ADP/Fe^{2+}, as assessed by the oxygen consumption (Fig. 1) and the TBARS production (Fig. 2), when compared to Wistar control rats. The time-dependent change in peroxidation related to oxygen concentrations of heart mitochondria was different in preparations isolated from diabetic rats compared to control rats. In both mitochondrial preparations a slow oxygen consumption occurred for several minutes after ADP-Fe^{2+} addition, after which oxygen consumption drastically increased. When mitochondria begun to peroxidize, it occurred at the same rate as control mitochondria. As shown on Fig. 2, there was no variation in the absolute mean of TBARS levels in non-oxidized mitochondria from Wistar and GK rats (0.78 ± 0.69 and 0.97 ± 0.69 nmol/mg protein, respectively). In the presence of ADP/Fe^{2+} both mitochondrial fractions showed a marked increase in TBARS, from 0.78 ± 0.69 to 27.49 ± 2.37 nmol/mg protein in Wistar mitochondria and from 0.97 ± 0.69 to 45.15 ± 2.05 nmol/mg protein in GK mitochondria, indicating the occurrence of oxidative damage to mitochondrial lipids induced by the pro-oxidant pair. The analysis of the TBARS production after treatment of mitochondria with ADP/Fe^{2+} showed a higher value in mitochondria from diabetic rats compared with the controls (45.15 ± 2.05 and 27.49 ± 2.37 nmol/mg protein, respectively). A statistically significant difference ($p < 0.05$) was determined between both preparations, indicating a higher susceptibil-

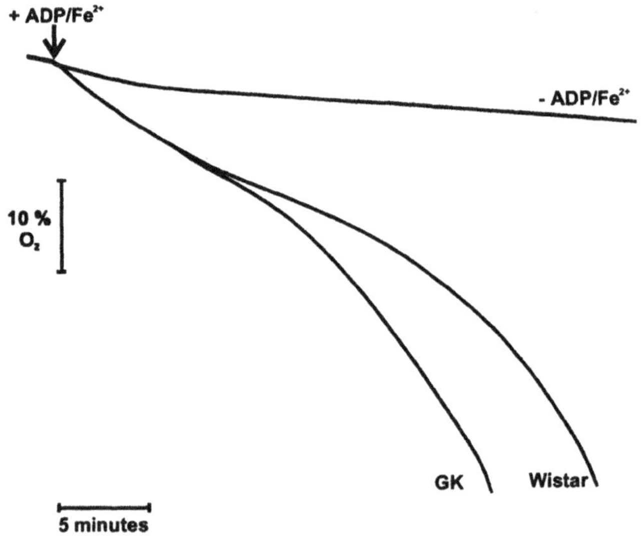

Fig. 1. O_2 consumption due to lipid peroxidation induced by ADP/Fe^{2+} in rat heart mitochondrial membranes isolated from Wistar and GK rats. Rat heart mitochondria (1 mg) were incubated at 30°C, for 30 min, in 1 ml of medium consisting of 175 mM KCl, 10 mM Tris (pH 7.4) supplemented with 3 μM rotenone. Peroxidation was started by adding 1 mM ADP and 0.1 mM $FeSO_4$. The lines represent typical recordings from seven different experiments.

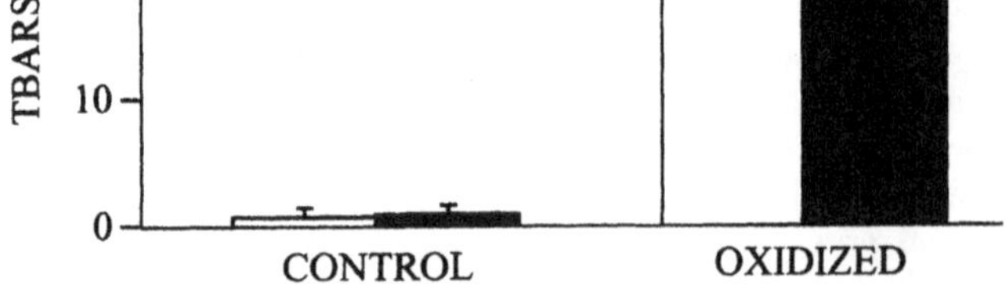

Fig. 2. TBARS formed *in vitro* during lipid peroxidation induced by ADP/Fe^{2+} in rat heart mitochondria of control (Wistar) and diabetic (GK) rats induced by ADP/Fe^{2+} (1 mM/0.1 mM). Rat heart mitochondria (1 mg) were incubated in 1 ml of medium consisting of 175 mM KCl, 10 mM Tris (pH 7.4) supplemented with 3 μM rotenone, at 30°C, for 2 min. Peroxidation was started by adding 1 mM ADP and 0.1 mM $FeSO_4$ and incubated for 45 min. TBARS were determined at the end of each incubation period as described in Materials and methods, and were expressed as nmol TBARS/mg protein. Data are expressed as mean ± S.E.M. of seven independent experiments. Statistical significance: *$p < 0.05$ compared with control group; **$p < 0.05$ compared with Wistar rats.

ity to induced oxidative stress in diabetic heart mitochondria.

Electron microscopy of mitochondria after oxidative stress

Heart mitochondria isolated from Wistar and GK rats were exposed to oxidative attack by ADP-Fe^{2+} and structural-morphological changes were examined by electron microscopy. Mitochondria incubated in the absence of ADP-Fe^{2+} at 30°C maintained their inner envelope membrane intact but, in the presence of ADP-Fe^{2+}, membrane discontinuities and volume fluctuations were observed, as compared to control mitochondria (not shown). In both mitochondrial fractions, swelling was followed by vesiculation of the inner membrane, and crenellation of the outer membrane coupled to its dissociation from the inner membrane. Based on several electron micrographs, the volume of rat heart mitochondria from Wistar and GK rats were measured, before and after induction of oxidative stress (Fig. 3). Control mitochondria from Wistar rats displayed a statistically different average volume from diabetic rat mitochondria (1.73 ± 0.21 and 1.17 ± 0.14 μm^3, respectively). Wistar mitochondria incubated with ADP-Fe^{2+} (oxidized mitochondria) presented a significantly larger volume (2.84 ± 0.31 μm^3), compared to control non-oxidized mitochondria (1.73 ± 0.21 μm^3). A drastic increase in the volume can be seen when comparing GK heart oxidized mitochondria to non-oxidized GK heart mitochondria (3.1 ± 0.59 and 1.17 ± 0.14 μm^3, respectively). Additionally, no statistical difference was found when we compared oxidized Wistar to oxidized GK heart mitochondria (2.84 ± 0.31 and 3.1 ± 0.59 μm^3, respectively).

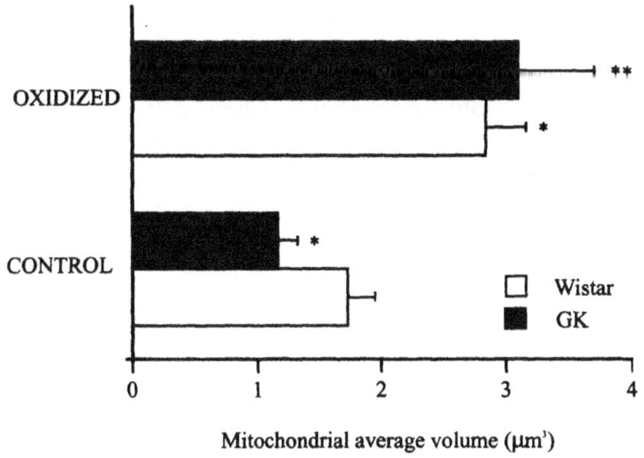

Fig. 3. Average volume of heart mitochondria isolated from Wistar and GK rats, before (control mitochondria) and after (oxidized mitochondria) induction of oxidative damage by ADP/Fe^{2+} (1 mM/0.1 mM). Data are expressed as mean ± S.E.M. of seven independent experiments. Statistical significance: *$p < 0.05$ compared with Wistar-control group; **$p < 0.05$ compared with GK-control rats.

Mitochondrial non-enzymatic antioxidants

Mitochondrial non-enzymatic contents in both preparations were determined: CoQ (Fig. 4), vitamin E (Fig. 5) and GSH (data not shown). Vitamin E (α-tocopherol), GSH and CoQ, as assessed by the two main homologues CoQ_9 and CoQ_{10}, are effective inhibitors of membrane lipid peroxidation. In heart mitochondria of Wistar and GK rats, CoQ_9 represents 96% while CoQ_{10} represents only 4% of the total pool of mitochondrial CoQ (7.21 ± 0.354 and 5.76 ± 0.64 nmol/mg protein, respectively). The content of α-tocopherol in mitochondria from control and diabetic rats display no significant differences (1051.0 ± 193.5 and 1051.7 ± 132.1 pmol/mg protein, respectively), but the GSH contents were significantly ($p < 0.05$) lowered, by 21.5%, in GK rat mitochondria compared to Wistar rat mitochondria (1.42 ± 0.12 and 1.86 ± 0.22 pmol/mg protein, respectively). Additionally, the CoQ_{10} contents have shown no statistical differences between both preparations (0.29 ± 0.039 and 0.23 ± 0.048 nmol/mg, respectively). However, CoQ_9 and consequently, total CoQ ($CoQ_9 + CoQ_{10}$) contents was significantly different ($p < 0.05$) in both mitochondrial preparations (Fig. 4).

Discussion

NIDDM has been studied in several animal models including the GK rat, which was selected for this study because it represents a good model of hereditary non-obese NIDDM that develops into a moderate and stable diabetic state with age [20, 31]. Several reports showed that increased levels of glucose found in NIDDM patients are responsible not only

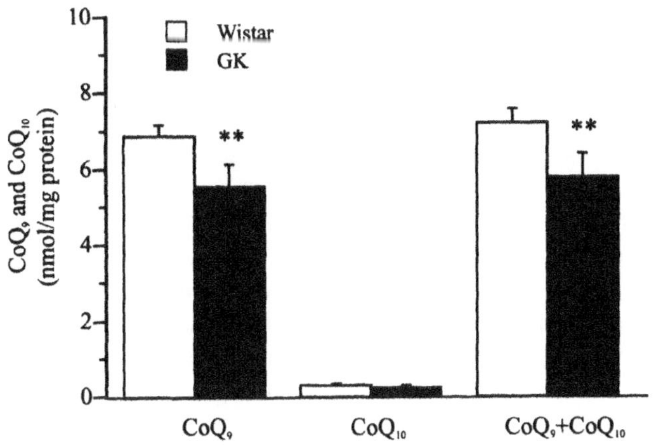

Fig. 4. Coenzyme Q_9 (CoQ_9) and Q_{10} (CoQ_{10}) levels in heart mitochondria isolated from Wistar and GK rats. Coenzyme Q levels were measured by HPLC as described in Materials and methods. Data are the mean ± S.E.M. for seven different samples, each obtained from a different rat. Statistical significance: **$p < 0.05$ as compared to control Wistar rats.

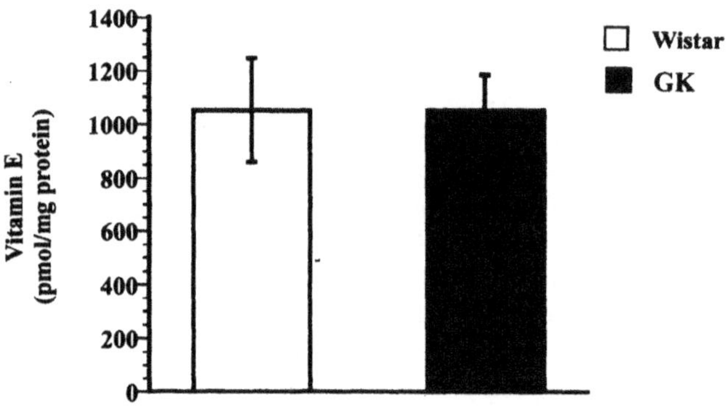

Fig. 5. Vitamin E levels of GK and control heart mitochondria preparations. Vitamin E levels were measured by HPLC as described in Materials and methods. Data are the mean ± S.E.M. for seven different samples, each obtained from a different rat.

for oxygen free radical generation due to non-enzymatic protein glycation, autooxidation of glycation products, ascorbic acid and glucose, but also for changes in tissue content and activity of antioxidant defense systems [32–38]. Elevated levels of glycated products induce an increase of lipid/protein oxidation, that in addition to different levels of antioxidant defences in tissues have been correlated with the extent of tissue damage at the molecular level [33, 36] and with the development of chronic complications of diabetes mellitus [33–35, 38]. Heart tissue has several features that makes it vulnerable to damage from free radicals such as, abundant mitochondria which may leak activated species, and a higher percentage of polyunsaturated fatty acids in mitochondrial and plasma membranes which are susceptible to lipid peroxidation [10, 39–41]. Oxidative stress due to high levels of glucose may disrupt the efficiency of mitochondrial electron transport, reduce ATP availability to cellular function [7, 8, 42] and cause the failure of heart activity as well as myocardial diseases [5, 7].

In this study we found that heart mitochondria isolated from 12-month-old GK rats and Wistar rats display different susceptibilities to oxidative stress induced by ferrous ion in the form of ADP-Fe^{2+}. We observed that oxidative damage induced in heart mitochondria of diabetic rats showed a significant increase of lipid peroxidation (measured by O_2 consumption and TBARS formation). The slow O_2 consumption could be result of mitochondrial non-enzymatic antioxidants consumption or can be ascribed to the formation of an ADP-Fe^{3+}-O_2^- (perferryl) complex [23], and presented a similar rate on both Wistar and GK mitochondria assays. However, the time required for the rapid O_2 uptake (lag time) and the time required to produce an amount of perferryl complex to induce lipid peroxidation [23] is different when diabetic and non-diabetic mitochondrial fractions were tested. A short lag time on GK mitochondrial suspensions indicated a higher susceptibility to lipid peroxidation. The analysis by electron microscopy of morphological changes showed that oxidative stress, induced by ADP-Fe^{2+}, was followed by considerable number of alterations in heart mitochondria with swelling more pronounced in diabetic heart mitochondria. This increase susceptibility is correlated with a decrease in mitochondrial GSH and CoQ_9. The observed decrease in lag phase and the increase in oxygen consumption and TBARS formation, are probably due to the lower contents of CoQ and GSH in GK heart mitochondria. GSH plays a crucial role in the metabolism of hydroperoxides and free radicals [9, 43]. A decrease of GSH content in GK heart mitochondria suggest that diabetic mitochondria could be more susceptible to free radicals damage.

Recently, it became clear that CoQ in the reduced form plays an important role as an antioxidant, and that this might explain its broad distribution in subcellular membranes other than mitochondria [18, 19, 44, 45]. The study of CoQ content is very relevant when studying lipoperoxidation because such a liposoluble antioxidant serves as the first line of defense against lipid peroxidation [18, 46, 47]. The antioxidant function of CoQ, by directly scavenging radicals [46] or by indirectly regenerating vitamin E [45, 47, 48] from its oxidized form, may be correlated with the high susceptibility of diabetic heart mitochondria to oxidative stress. In this study, the concentration of CoQ_9, on a mole base, was estimated to be more than 5–6 times higher than that of heart mitochondria's α-tocopherol; in addition CoQ_9 levels are statistically lower on diabetic mitochondria. A decrease of CoQ levels found in diabetic mitochondria causes a reduction in the antioxidant potential of these membranes and a concomitant increase in the susceptibility to oxidative damage. Vitamin E (α-tocopherol) is suggested to be the most important endogenous antioxidant [47] scavenger of radicals generated in the presence of Fe^{2+} by preventing the propagation of lipid peroxidation chain reactions [42, 43]. The decreased susceptibility to lipid peroxidation in diabetic rats was not correlated to α-tocopherol levels, because diabetic and non-diabetic heart mitochondria did not present statistically different levels.

In conclusion, the present study shows that heart mitochondria from diabetic rats differ significantly from non-diabetic rats in their susceptibility to undergo lipid peroxidation when oxidative damage is induced *in vitro*. A reduced antioxidant potential is responsible for the greater susceptibility of diabetic heart mitochondria to oxidative damage as supported from the lower CoQ and GSH contents of such mitochondria, despite diabetic mitochondria don't have different levels of α-tocopherol compared to non-diabetic mitochondria. Recently, Kucharska *et al.* [49] reported that a deficit of coenzyme Q in heart mitochondria of rats with streptozotocin-induced diabetes mellitus was also associated with an increased lipoperoxidation. Previous reports from our group [50] on the susceptibility of brain and liver mitochondrial preparations to lipid peroxidation *in vitro*, showed that GK mitochondrial preparations from liver had reduced susceptibility to lipoperoxidation while brain had an increase susceptibility to lipid oxidation, in agreement with the heart mitochondria results reported in this study. Considering the high energy requirements of the heart, possible damage to heart mitochondria induced by oxidative stress could cause a collapse on the cardiac pool of energetic compounds and be responsible for the reported impairment of heart function in diabetes.

Acknowledgements

We thank Prof. João Patrício and co-workers (Laboratory of Animal Research Center, University Hospitals, Coimbra) for their help in the maintenance of animals. This research was supported by grants of FCT (Portuguese Research Council; research projects PRAXIS/SAU/1/96 and PRAXIS/SAU/16/96).

References

1. Groop LC: The molecular genetics of non-insulin-dependent diabetes mellitus. J Intern Med 241: 95–101, 1997
2. Ren J, Davidoff AJ: Diabetes rapidly induces contractile dysfunctions in isolated ventricular myocytes. Am J Physiol 272: H148–H158, 1997
3. Feuvray D, Lopaschuk GD: Controversies on the sensitivity of diabetic heart to ischemic injury: The sensitivity of the diabetic heart to ischemic injury is decreased. Cardiovasc Res 34: 113–120, 1997
4. Paulson DJ: The diabetic heart is more sensitive to ischemic injury. Cardiovasc Res 34: 104–112, 1997
5. Luft R: The development of mitochondrial medicine. Biochim Biophys Acta 1271: 1–6, 1995
6. Momiyama Y, Atsumi Y, Ohsuzu F, Ui S, Morinaga S, Matsuoka, Kimura M: Rapid progression of cardiomyopathy in mitochondrial diabetes. Jpn Circ J 63: 130–132, 1999
7. Schmidt AM, Stern D: A radical approach to the pathogenesis of diabetic complications. Trends Pharmacol Sci 21: 367–369, 2000
8. Palmeira CM, Ferreira FM, Santos DL, Ceiça R, Suzuki K, Santos MS: Higher efficiency of the liver phosphorylative system in diabetic Goto-Kakizaki (GK) rats. FEBS Lett 458: 103–106: 1999
9. Halliwell B, Gutteridge JMC: Free Radicals in Biology and Medicine, 3rd Edit. Oxford University Press, Oxford, 1999
10. Kowaltowski AJ, Vercesi AE: Mitochondrial damage induced by conditions of oxidative stress. Free Radic Biol Med 26: 463–471, 1999
11. Chance B, Sies H, Boveris A: Hydroperoxide metabolism in mammalian organs. Physiol Rev 59: 527–605, 1979
12. Reinheckel T, Wiswedel I, Noack H, Augustin W: Electrophoretic evidence for the impairment of complexes of the respiratory chain during iron/ascorbate induced peroxidation in isolated rat liver mitochondria. Biochim Biophys Acta 1239: 45–50, 1995
13. Marshansky VN, Novgorodov SA, Yaguzhinsky LS: The role of lipid peroxidation in the induction of cation transport in rat liver mitochondria. The antioxidant effect of oligomycin and dicyclohexylcarbodiimide. FEBS Lett 158: 27–30, 1983
14. Carbonera D, Azzone GF: Permeability of inner mitochondrial membrane and oxidative stress. Biochim Biophys Acta 943: 245–255, 1988
15. Baynes JW, Thorpe SR: Role of oxidative stress in diabetic complications: A new perspective on an old paradigm. Diabetes 48: 1–9, 1999
16. Scholz RW, Minicucci LA, Reddy CC: Effects of vitamin E and selenium on antioxidant defense in rat heart. Biochem Mol Biol Int 42: 997–1006, 1997
17. Kagan VE, Serbinova E, Packer L: Antioxidant effects of ubiquinone in microsomes and mitochondria are mediated by tocopherol recycling. Biochem Biophys Res Commun 169: 851–857, 1990
18. Ernster L, Dallner G: Biochemical, physiological and medical aspects of ubiquinone function. Biochim Biophys Acta 1271: 195–204, 1995
19. Rauchova H, Lenaz G: Coenzyme Q and its therapeutics used. Cezka Slov Farm 50: 78–82, 2001
20. Goto Y, Kakizaki M: The spontaneous diabetic rat: A model of non-insulin dependent diabetes mellitus. Proc Jpn Acad 57: 381–384, 1981
21. Cain K, Skilleter, DN: Preparation and use of mitochondria in toxicological research. In: K. Snell, B. Mullock (eds). Biochemical Toxicology – A Practical Approach. IRL Press, Washington, DC, 1987, pp 217–254
22. Gornall AG, Bardawill CJ, David MM: Determination of serum proteins by means of the biuret reaction. J Biol Chem 177: 751–766, 1949
23. Sassa H, Takaishi Y, Terada H: The triterpene celastrol as a very potent inhibitor of lipid peroxidation in mitochondria. Biochem Biophys Res Commun 172: 890–897, 1990
24. Ernster L, Nordenbrand K: Microsomal lipid peroxidation. Meth Enzymol 10: 574–580, 1967
25. Buege JA, Aust S: Microsomal lipid peroxidation. Meth Enzymol 52: 302–310, 1978
26. Takada M, Ikenoya S, Yuzuriha T, Katayama K: Simultaneous determination of reduced and oxidized ubiquinones. Meth Enzymol 105: 147–155, 1984
27. Chung AP, Rainey F, Nobre MF, Burghardt J, Costa MS: *Meiothermus cerbereus sp.nov.*, a new slightly thermophilic species with high levels of 3-hydroxy fatty acids. Int J System Bacter 47: 1225–1230, 1997
28. Vatassery GT, Morley JE, Kuskowski MA: Vitamin E in plasma and platelets of human diabetic patients and control subjects. Am J Clin Nutr 37: 641–644, 1983
29. Hissin PJ, Hilf R: A fluorometric method for determination of oxidized and reduced glutathione in tissues. Anal Biochem 74: 214–226, 1976
30. Weibel ER: In: M.A. Hayat (ed). Principles and Techniques of Electron Microscopy: Biological Applications. Van Nostrand Reinhold, New York, 1973
31. Agardh CD, Agardh E, Hultberg B, Qian Y, Östenson CG: The glutathione levels are reduced in Goto-Kakizaki rat retina, but are not influenced by aminoguanidine treatment. Curr Eye Res 17: 251–256, 1998
32. Wolff SP, Dean RT: Glucose autoxidation and protein modification: The potential role of autoxidative glycosylation in diabetes. Biochem J 245: 243–250, 1987
33. Oberley LW: Free radicals and diabetes. Free Radic Biol Med 5: 113–124, 1988
34. Baynes JW: Role of oxidative stress in development of complications in diabetes. Diabetes 40: 405–412, 1991
35. Mullarkey CJ, Edelstein D, Brownlee M: Free radical generation by early glycation products: A mechanism for accelerated atherogenesis in diabetes. Biochem Biophys Res Commun 173: 932–939, 1990
36. Sukalski KA, Pinto KA, Berntson JL: Decreased susceptibility of liver mitochondria from diabetic rats to oxidative damage and associated increase in alpha-tocopherol. Free Radic Biol Med 14: 57–65, 1993
37. Jain SK, Palmer M: The effect of oxygen radicals metabolites and vitamin E on glycosylation of proteins. Free Radic Biol Med 22: 593–596, 1997

38. Aguirre F, Martin I, Grinspon D, Ruiz M, Hager A, De Paoli T, Ihlo J, Farach HA, Poole CP Jr: Oxidative damage, plasma antioxidant capacity, and glucemic control in elderly NIDDM patients. Free Radic Biol Med 24: 580–585, 1998
39. Daum G: Lipids of mitochondria. Biochim Biophys Acta 822: 1–42, 1985
40. Werns SW, Lucchesi BR: Free radicals and ischemic tissue injury. Trends Pharmacol Sci 11: 161–166, 1990
41. Wallace KB: Free radical-mediated chemical cardiomiopathies. In: K. Wallace (ed). Free Radical Toxicology. Taylor and Francis, Washington, DC, 1997, pp 205–219
42. Ferreira FM, Palmeira CM, Matos MJ, Seiça R, Santos MS: Decreased susceptibility to lipid peroxidation of Goto-Kakizaki rats: Relationship to mitochondrial antioxidant capacity. Life Sci 65: 1013–1025, 1999
43. Choliparambil KP, Subramanian M, Devasagayam PA, Singh BB: Study on lipid peroxidation potential in different tissues induced by ascorbate-Fe^{2+}: Possible factors involved in their differential susceptibility. Mol Cell Biochem 178: 197–202, 1998
44. Beyer RE: The participation of coenzyme Q in free radical production and antioxidation. Free Radic Biol Med 8: 545–565, 1990
45. Lass A, Sohal RS: Electron transport-linked ubiquinone-dependent recycling of alpha-tocopherol inhibits autooxidation of mitochondrial membranes. Arch Biochem Biophys 352: 229–236, 1998
46. Forsmark P, Aberg F, Norling B, Nordenbrand K, Dallner G, Ernster L: Inhibition of lipid peroxidation by ubiquinol in submitochondrial particles in the absence of vitamin E. FEBS Lett 285: 39–43, 1991
47. Kagan VE, Packer L: Antioxidants: Function of vitamin E and ubiquinols. In: Methods in Toxicology, Vol. 2. Academic Press, New York, 1993, pp 277–285
48. Stoyanovsky DA, Osipov AN, Quinn PJ, Kagan VE: Ubiquinone-dependent recycling of vitamin E radicals by superoxide. Arch Biochem Biophys 323: 343–351, 1995
49. Kucharska J, Braunova Z, Ulicna O, Zlatos L, Gvozdjakova A: Deficit of coenzyme Q in heart and liver mitochondria of rats with streptozotocin-induced diabetes. Physiol Res 49: 411–418, 2000
50. Santos MS, Santos DL, Palmeira CM, Seiça R, Moreno AJ, Oliveira CR: Brain and liver mitochondria isolated from diabetic Goto-Kakizaki rats show different susceptibility to induced oxidative stress. Diabetes Metab Res Rev 17: 223–230, 2001

Molecular and Cellular Biochemistry **246**: 171–177, 2003.

Interaction of bilirubin with native and protein-depleted human erythrocyte membranes

Huma Rashid and Saad Tayyab
Interdisciplinary Biotechnology Unit, Aligarh Muslim University, Aligarh, India

Abstract

In order to study the role of membrane proteins in bilirubin (BR) binding phenomenon, selective removal of membrane proteins was carried out using various reagents, namely, ethylenediamine tetraacetic acid (EDTA), sodium hydroxide (NaOH), 3,5-diiodosalicylic acid, lithium salt (LIS), dimethylmaleic anhydride (DMMA), sodium iodide (NaI), o-phenanthroline-cupric sulfate (CuP) and phenanthroline-cupric sulfate containing 2-mercaptoethanol (CuP-mercaptoethanol). Effects of these treatments on the conformation and BR binding properties of the membrane were studied using circular dichroism (CD) spectroscopy as well as estimation of membrane-bound BR by diazotised-color reaction. Though a significant amount of protein (ranging from 23–69%) was lost from the membranes upon these treatments, only a small decrease (3–13%) was observed in BR binding, being maximum with NaOH-treated membranes. However, DMMA and NaI treatments produced a little increase in BR binding. Conformation of the membrane was retained to a significant extent as indicated by far-UV CD spectra upon these treatments except in DMMA and NaI treatments which resulted in the perturbation in CD spectra. Taken together, these results suggest that membrane proteins play little role in BR binding, rather act as barriers in BR binding phenomenon. (Mol Cell Biochem **246**: 171–177, 2003)

Key words: human erythrocyte membrane, membrane proteins, cross-linking, bilirubin binding, CD spectra

Introduction

The interaction of bilirubin (BR), a bile pigment produced from the catabolism of hemoproteins, with the erythrocyte membranes has been the subject of study [1–4] towards understanding the mechanism of entry of BR inside the cells which may have important physiological implications in the pathogenesis of kernicterus (BR-induced encephalopathy) in newborns. Since biological membranes have two important constituents, i.e. lipids and proteins, the BR binding sites are supposed to be made up of lipids or proteins or both. It is believed that BR mainly interacts with the lipid bilayer of the erythrocyte membrane [1, 5–7]. However, the direct involvement of erythrocyte membrane proteins in the BR-binding phenomenon has been excluded in an earlier study [1] and it has been suggested that they function as effective barriers to BR binding. Further, affinity labelling of the membrane proteins by bilirubin has failed in identifying any specific BR-binding protein [1]. Contrary to this, membrane proteins seem to have some type of non-specific effect on the properties of the lipids, specially, in the apolar phase of the bilayer, which enhances the capacity of the bilayer to sequester BR as compared to protein-free bilayer [8].

The influence of physico-chemical properties of the membranes such as overall charge on the membrane surface, membrane fluidity, pH and temperature on the binding of BR to membranes are of great concern. Vazquez *et al.* [7] have suggested that interaction of BR with membrane is modulated by the membrane fluidity, being directly correlated. Human erythrocyte membrane is composed of two structural units, viz. membrane skeleton (or cytoskeleton) and membrane bilayer, of which the cytoskeleton is formed from three major (bands 1 and 2, band 5 and band 4.1) and several minor peripheral membrane proteins. The membrane bilayer-skeleton association controls not only the membrane mechanical stability and deformability but also the lateral mobility of the integral membrane proteins [9]. Besides, this association is considered to be a major factor in maintaining the

Address for offprints: S. Tayyab, P.O. Box 255, Department of Plant Sciences, Alemaya University, Ethiopia (E-mail: alemaya.univ@telecom.net.et)

asymmetric phospholipid distribution across the erythrocyte membrane bilayer [10]. Further, proteins like band 3 and Na^+-K^+-ATPase are reported to be strongly associated with phospholipids [11]. Erythrocytes with defective membrane skeleton have also been shown to possess disturbed phospholipid asymmetry [12].

Human erythrocyte membrane is composed of nine major protein components visualized on SDS-polyacrylamide gels [13]. Their removal and interaction of ligands with these protein-depleted membranes can prove to be a simple strategy in the investigation of their role in the functional behavior of membranes. The present communication is an attempt towards identifying the role of membrane proteins in BR binding phenomenon to the membranes. Here, we describe our results on the binding of BR to erythrocyte membranes before and after various treatments, leading to differential release of various membrane components, as studied by CD spectroscopy.

Materials and methods

Materials

Dimethylmaleic anhydride (DMMA), lot 77H2521; 3,5-diiodosalicylic acid, lithium salt (LIS), lot 48H0714; 1,10-phenanthroline, lot 77H0316; 2-mercaptoethanol, lot 53H0741; phenylmethylsulfonylfluoride (PMSF), lot 67H1645 and bovine serum albumin (BSA), fraction V, lot 100F0249 were purchased from Sigma Chemical Company, USA. Sodium hydroxide (NaOH), ethylenediaminetetraacetic acid (EDTA) and copper sulfate ($CuSO_4$) were from Qualigens Fine Chemicals, India. Extrapure, analytical grade bilirubin (BR) (99% pure, 2× crystallized) was the product of Sisco Research Laboratories, India. This preparation was used as such, without any further purification, throughout these studies. Sodium iodide (NaI) was obtained from Loba Chemie, India. All other reagents used were of analytical grade.

Human blood (in 1.32% sodium citrate and 1.47% dextrose) was obtained from the Blood Bank of J.N. Medical College, Aligarh Muslim University, Aligarh, India.

Preparation of erythrocyte membranes and various treatments

Hemoglobin-free human erythrocyte membranes were isolated from recently outdated human blood by the method of Palfrey and Waseem [14] as described elsewhere [15]. Erythrocyte suspension of 50% hematocrit value was hemolysed with 10 volumes of cold 10 mM tris-HCl buffer, pH 7.4 containing 0.1 mM EDTA and 0.01 mM PMSF, followed by gentle swirling and centrifugation at 16,000 × g for 20 min at 4°C. The dark red supernatant was removed carefully by gentle aspiration. Each tube was tilted and rotated to allow the loose ghost pellet to slide off to another tube leaving behind the opaque button, which is very rich in proteolytic activity. This minimized the contamination of ghosts with proteinases.

Various treatments of erythrocyte membranes were carried out by the procedure described by Kahlenberg [16]. Membranes (1.0 ml in 50 mM tris-HCl buffer, pH 7.4) were treated separately with 0.1 mM EDTA (pH 8.0), 20 mM LIS, DMMA (20 mg/ml of incubation mixture), 1.0 M NaI and 1.0 mM NaOH (pH 11.5) in a total volume of 4.0 ml for 1 h at 37°C. The supernatant and pellet fractions were separated by centrifugation at 16,000 × g for 20 min at 4°C. The pellet, so obtained, was washed twice either with 50 mM tris-HCl buffer, pH 7.4 or 5 mM sodium phosphate buffer, pH 7.4 and suspended in 1.0 ml of the same buffer.

Oxidative cross-linking of erythrocyte membrane proteins was carried out according to the method of Kahlenberg [16]. To 1.0 ml of membrane suspension in 50 mM tris-HCl buffer, pH 7.4 was added 1.0 ml of the solution containing 100 μM o-phenanthroline and 20 μM $CuSO_4$. The total volume was made to 4.0 ml with the same buffer and incubated for 1 h at 37°C. Half of the above samples received 100 mM 2-mercaptoethanol after 30 min of incubation and further incubated for additional 30 min. Membrane pellets obtained by centrifugation were washed in the same way as described above.

SDS-polyacrylamide gel electrophoresis

To monitor the changes in the erythrocyte membrane protein profile, membrane samples were prepared as above, centrifuged at 16,000 × g for 20 min at 4°C and washed twice subsequently with 50 mM tris-HCl buffer containing 0.01 mM PMSF, pH 7.4 prior to sample dissolution for SDS-PAGE, and run on 10% gels according to the method of Laemmli [17]. In the cross-linking experiments, unreduced membrane samples were prepared in the same way but in the absence of 2-mercaptoethanol. The gels were stained for proteins with Coomassie blue.

Analytical techniques

Protein concentration was determined by the method of Lowry *et al.* [18] after solubilization of erythrocyte membranes in 1% SDS, using BSA as the standard. BR concentration was determined by the method of Fog [19] whereas modified Fog's method [20] was employed for the estimation of membrane-bound BR as described earlier [15].

Bilirubin (BR) binding experiments

BR solution was prepared just before use, under light protection, by dissolving a few crystals of BR in 38 mM sodium carbonate solution, pH 11.0. This solution was centrifuged at 16,000 × g for 5 min at 4°C and the clear supernatant was aspirated into another tube leaving a bright orange precipitate behind. The concentration of the clear stock solution was determined as described above in 'Analytical techniques'. BR solution was stored in dark and was used within an hour of preparation.

Binding of BR to native and treated membranes was studied in 50 mM tris-HCl buffer, pH 7.4 after incubating these membranes (0.5 ml) with bilirubin (300 μM) in a total volume of 1.5 ml for 30 min at 37°C and determining the membrane-bound BR.

Circular dichroism (CD) spectroscopy

CD spectra of native and treated membranes (in the absence as well as presence of bilirubin) were recorded on a Jasco spectropolarimeter, model J-720, equipped with a microcomputer. All the CD measurements were carried out at 25°C with a thermostatically controlled cell holder attached to a Neslab RTE-110 waterbath with an accuracy of ±0.1°C. Spectra were collected at a scan speed of 20 nm/min and with a response time of 1 sec. Each spectrum was the average of four scans. Far-UV CD spectra were taken in the wavelength range of 200–250 nm using quartz cuvettes of 1 mm path-length. The BR and protein concentrations in these samples were 150 μM and 0.513 mg, respectively in a total volume of 4.0 ml incubation mixture. After incubation of BR with native and treated membranes for 30 min at 37°C, the mixture was centrifuged at 16,000 × g for 20 min at 4°C and the supernatant containing unbound BR was discarded. The membranes were rinsed several times (at least twice) with 5 mM sodium phosphate buffer, pH 7.4 until the last of the supernatant was devoid of yellow color. The BR-bound membranes, resuspended in the same buffer, were used as such for CD measurements. CD values were transformed into molar ellipticity $[\theta]_\lambda$ using the following relationship [21]:

$$[\theta]_\lambda = \theta.MRW/10.l.c \qquad (1)$$

where θ is the observed ellipticity in degrees, c is the protein concentration in g/cm^3, l is the path length in cm and MRW is mean-residue weight taken as 114.3 g/mol [21].

Results and discussion

Incubation of human erythrocyte membranes with alkaline EDTA (0.1 mM, pH 8.0) resulted in the selective release of peripheral membrane proteins, namely, bands 1 and 2 as well as band 5 as shown in lane 2 of Figs 1A and 1B. These results were in line with the previous reports [16, 22]. Measurement of protein in the pellets obtained from native and

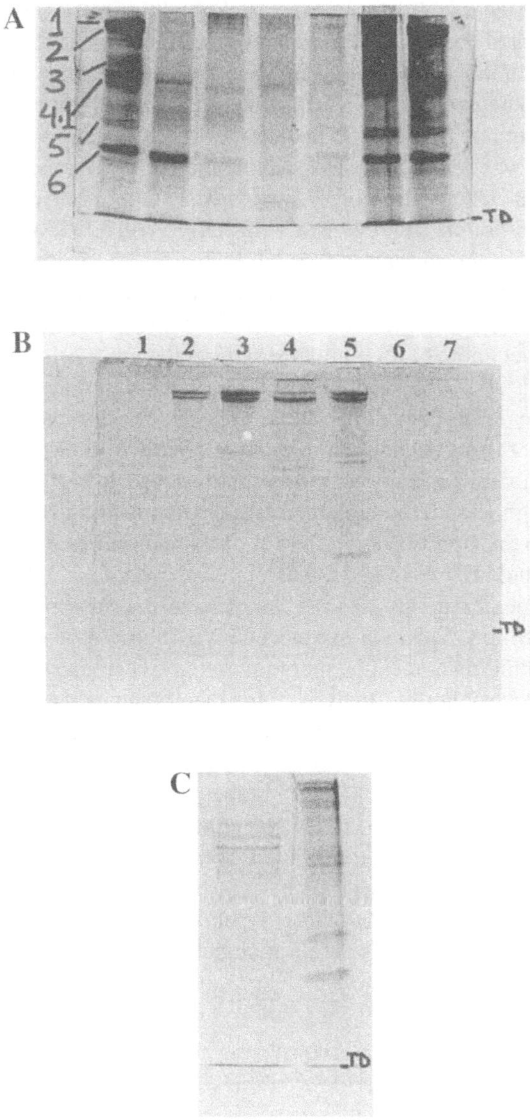

Fig. 1. (A) Protein profiles of human erythrocyte membranes before (lane 1) and after treatment with 0.1 mM EDTA, pH 8.0 (lane 2); 1.0 M NaI (lane 3); 20 mg/ml DMMA (lane 4); 20 mM LIS (lane 5); 100 μM phenanthroline + 20 μM cupric sulfate (lane 6) and phenanthroline + cupric sulfate containing 100 μM 2-mercaptoethanol (lane 7). (B) Protein profiles of released polypeptides as observed in the supernatant of native and treated human erythrocyte membranes. Various lanes represent the same order as shown in (A). (C) Protein profiles of 0.1 mM NaOH-treated erythrocyte membranes (lane 1) and released polypeptides (lane 2). About 25–35 μg of proteins were analysed electrophoretically on 10% SDS-polyacrylamide gels. TD is the position of tracking dye.

EDTA-treated membranes suggested a release of 32% of proteins upon EDTA treatment (see Table 1). Conformational analysis of erythrocyte membranes, both before and after EDTA treatment as studied by CD spectroscopy, suggested significant retention of the CD spectral features in EDTA-treated membranes (see Figs 2A and 2B, Table 2). However, some additional CD bands at 205 and 208 nm appeared in modified membranes. Further, a slight decrease in the MRE values of the CD band minima (Table 2) was noticed in these membranes. Presence of minima around 223 and 208 nm were indicative of α-helical structure of membrane proteins. Binding of BR to these membranes (both native and modified) was studied both by quantification of membrane-bound BR and CD spectroscopy. A comparison of the values of membrane-bound BR obtained with native and EDTA-treated membranes suggested a small decrease (3%) in the membrane-bound BR upon EDTA treatment. Further, CD spectral analysis of the complexes of BR with these membranes showed more or less similar structural features with red shift in one of the band minima and small decrease in ellipticity values (Figs 2A and 2B, Table 2). It can be interpreted from these results that the removal of peripheral proteins from the membranes did not affect the binding of BR significantly though, an alteration in membrane composition is reflected from the significant red shift of the band minima at 214 nm. It should be mentioned that the removal of bands 1 and 2 from the membrane may lead to destabilization of the membrane structure [23] which may account for the small decrease in BR binding and slight alteration in the CD signals.

Electrophoretic analysis of the membrane pellet obtained as a result of alkali treatment of human erythrocyte membranes (see Fig. 1C, lanes 1 and 2) showed the release of about 64% (Table 1) of membrane proteins. It appears that most of the extrinsic proteins are eluted with NaOH treatment since a number of protein bands were visible in the electrophoretic profile of the supernatant (see lane 2 of Fig. 1C). The CD spectrum of NaOH-treated membranes (Fig. 2C) showed a significant decrease in ellipticity at band minima (223 nm) whereas a blue shift of 4 nm in the band minima at 212 nm and appearance of a new band at 203 nm was observed (Table 2). Decrease in ellipticity at 223 nm was more pronounced in NaOH-treated membranes compared to EDTA-treated membranes, suggesting gross conformational changes in NaOH-treated membranes. BR binding studies to these membranes showed a significant decrease (13%) in BR binding (Table 1). This decrease in BR binding can also be seen from the CD spectra of BR-membrane complexes in which the ellipticity at 209 nm was found to be similar to that obtained with treated membranes in the absence of BR. Presence of additional features at 227 and 216 nm and abolishment of minima at 203 nm in the presence of BR indicated conformational alterations in the membranes (Fig. 2C, Table 2). A significant decrease in BR binding observed with NaOH-treated membranes seems to be understandable in view of the depletion of extrinsic polypeptides namely bands 1, 2, 5 and 6 as well as predominant inside-out orientation of these vesicles upon dilute alkali treatment [24]. This is because inside-out vesicles have been shown to bind lesser amount of BR as compared to right side-out vesicles [25].

Treatment of membranes with LIS resulted in the release of a majority of polypeptides (lane 5 of Figs 1A and 1B), which together constituted 69% of proteins (Table 1). These results were in accordance with earlier reports suggesting the release of bands 1, 2, 2.1, 4.2, 5 and 6 from the membranes upon LIS treatment [16]. LIS, in addition to disrupting membranes into small vesicles, releases glycoproteins from the membranes [16, 26]. Conformational analysis of LIS-treated membranes also revealed a marked decrease in ellipticity value at the band minima (223 nm) and appearance of more CD signals in comparison to untreated membranes (Fig. 2D, Table 2). These changes in the CD spectra can be attributed to the noticeable alteration in membrane conformation upon release of various polypeptides. Contrary to this, when LIS-treated membranes were studied for BR binding, only 6% decrease in BR binding was noticed (Table 1). Similarly, CD spectra of BR-membrane complex, though showed the appearance of addi-

Table 1. Effect of treatment of erythrocyte membranes with selective reagents on their protein content and bilirubin binding

Membrane sample	Protein concentration in pellet		Membrane-bound bilirubin
	(mg/ml)	%	(nmoles)
Native	3.08 ± 0.19	100	114.0 ± 0.82
EDTA-treated	2.10 ± 0.27	68	110.6 ± 0.83
NaOH-treated	1.12 ± 0.14	36	99.6 ± 0.50
LIS-treated	0.94 ± 0.17	31	107.9 ± 1.34
[a]CuP-treated	2.37 ± 0.14	77	105.3 ± 0.66
[b]CuP-mercaptoethanol-treated	2.19 ± 0.19	71	110.6 ± 0.93
DMMA-treated	1.29 ± 0.17	42	119.3 ± 1.26
NaI-treated	1.47 ± 0.17	48	115.9 ± 0.80

Each value is the mean of 3 individual experimental values. [a]CuP refers to o-phenanthroline-cupric sulfate cross-linked membranes. [b]CuP-mercaptoethanol refers to o-phenanthroline-cupric sulfate cross-linked membranes reduced by 2-mercaptoethanol.

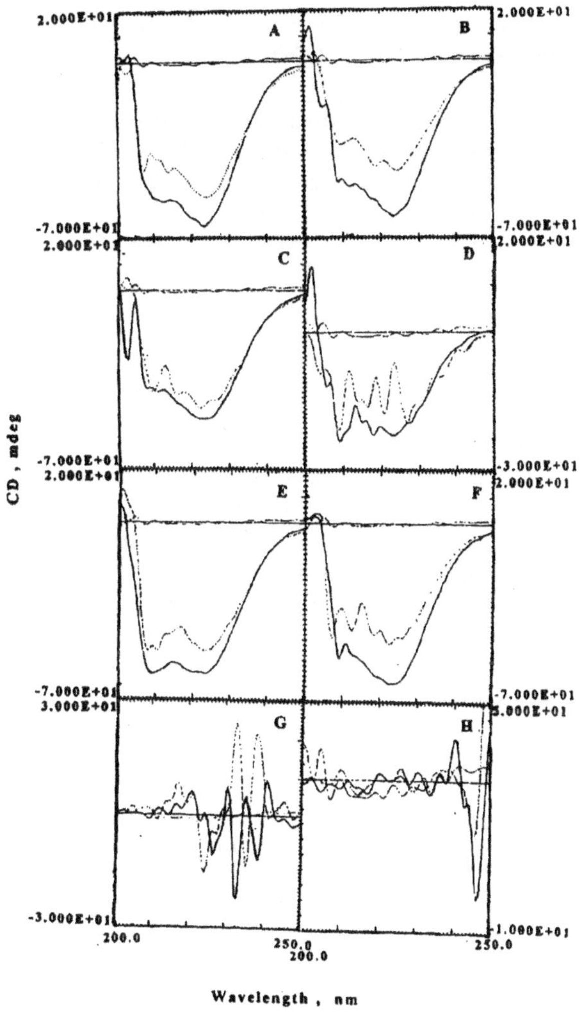

Fig. 2. Far-UV CD spectra of human erythrocyte membranes before (solid line) and after the addition of 150 μM bilirubin (dotted line) in 5 mM sodium phosphate buffer, pH 7.4 at a protein concentration of 0.513 mg/4 ml. Various membrane preparations were: native (A), EDTA-treated (B), NaOH-treated (C), LIS-treated (D), phenanthroline + cupric sulfate-treated (E), phenanthroline + cupric sulfate + 2-mercaptoethanol-treated (F), DMMA-treated (G) and NaI-treated (H), respectively. Treatment conditions were the same as described in the legend to Fig. 1.

tional peaks but the ellipticity values were significantly decreased (Table 2). Greater retention in BR binding by LIS-treated membranes compared to NaOH-treated membranes suggest that some other factors (other than protein loss) were responsible for the decrease in BR binding to the membranes.

In order to study the role of band 3 in the BR binding phenomenon, erythrocyte membranes were treated with o-phenanthroline and cuprie sulfate. This treatment resulted in the cross-linking of band 3 proteins as evident from lane 6 of Fig. 1A in which the band 3 protein was shifted to a higher molecular weight form. Bands 1 and 2 were also found to be cross-

Table 2. Circular dichroism data of native and treated human erythrocyte membranes before and after bilirubin binding

Human erythrocyte membrane	Observed CD band minima/maxima. Position (nm) and corresponding molar ellipticities in parenthesis (deg cm^2 dmol^{-1})	
	Bilirubin (–)	Bilirubin (+)
Native	223 (–14710)	224 (–12030)
	212 (–12480)	214 (–9800)
		208 (–9800)
EDTA-treated	223 (–13820)	224 (–9800)
	212 (–11140)	218 (–9360)
	208 (–10700)	209 (–7580)
	205 (–4010)	
NaOH-treated	223 (–11140)	227 (–9800)
	208 (–8470)	223 (–10250)
	203 (–6240)	216 (–9360)
		209 (–8910)
LIS-treated	223 (–5120)	227 (–4460)
	218 (–5120)	221 (–4010)
	209 (–5350)	216 (–3570)
	205 (–2670)	214 (–3340)
		209 (–4900)
CuP-treated	224 (–13370)	224 (–11140)
	208 (–13370)	208 (–11140)
CuP-mercaptoethanol-treated	223 (–13810)	226 (–10250)
	209 (–11590)	222 (–10700)
		218 (–9800)
		212 (–9360)
		207 (–10250)
DMMA-treated	239 (–2670)	239 (+4900)
	233 (–4900)	233 (+5570)
	227 (–2450)	228 (–1120)
		225 (–3340)
NaI-treated	247 (–1670)	246 (–780)

Presence and absence of bilirubin is represented as (+) and (–), respectively.

linked to a certain extent (see lane 6 of Fig. 1A). These results were in accordance with the earlier observations suggesting the cross-linking of largely band 3, and bands 1 and 2 by an o-phenanthroline-cupric sulfate catalyzed oxidation [16]. About 23% of the membrane proteins were lost in cross-linking experiments (Table 1). Kahlenberg [16] also showed a 25% loss of ghost protein in this cross-linking experiments. CD spectra of cross-linked membranes showed strong α-helical characteristics as evident by the presence of two band minima at 224 and 208 nm with a slight decrease in ellipticity values (Table 2) as compared to native membranes. BR binding results obtained with these membranes indicated a smaller decrease (8%) in BR binding as compared to native membranes.

CD spectra of cross-linked membrane in the presence of BR also showed CD signals and ellipticity values comparable to native membrane (Fig. 2E and Table 2).

The decrease in BR binding upon cross-linking of membrane proteins can be ascribed to either the loss of some membrane proteins or the formation of larger aggregates. To rule out the first possibility, reversal of the cross-linking reaction was carried out by reduction with 2-mercaptoethanol. That phenanthroline-copper induced cross-linking is prevented in the presence of 2-mercaptoethanol is evident from the protein profile of treated membrane with reference to band 3 protein (lane 7 of Fig. 1A). However, 29% loss (Table 1) in protein content was observed upon this treatment, which was 6% higher than the loss observed with cross-linking treatment in the absence of 2-mercaptoethanol. CD spectra of these membranes also showed more or less similar features with a slight decrease in ellipticity at 209 nm to those observed with cross-linking experiments (Fig. 2F, Table 2). Presence of 2-mercaptoethanol resulted in the recovery of BR binding as these membranes showed 97% binding of BR compared to 92% observed with cross-linked membranes (Table 1). This can also be seen from the CD spectra of the complexes of these membranes with BR (Fig. 2F, Table 2). Thus, it appears that the decrease in BR binding, noticed in cross-linked membranes, was due to the formation of larger aggregates rather than the loss of proteins.

Selective removal of the membrane proteins was also carried out by DMMA and NaI treatments. As can be seen from Table 1, about 58 and 52% proteins were released by DMMA and NaI treatments, respectively. This can also be seen from protein profiles of these membranes shown in lanes 3 and 4 of Figs 1A and 1B, respectively. Similar loss of membrane proteins by these treatments has also been reported earlier [16]. Majority of the proteins eluted by these treatments were peripheral proteins as shown in lanes 3 and 4 of Fig. 1B. CD spectral analysis of these membranes showed unusual spectral features as bands were found to be red shifted towards the higher wavelength side (Figs 2G and 2H as well as Table 2). When these membranes were tested for BR binding, slightly higher binding was noticed as compared to native membranes (Table 1). CD spectra of the complexes were of complicated type. From these results, it appears that the loss of proteins from these membranes by DMMA and NaI treatments changed the overall conformation of band 3 as the CD spectra of these membranes were totally different from those of LIS-treated membranes, in which about 69% of the total proteins were released but helical spectral features were retained as shown by native membranes. This unusual perturbation in the membrane may account for the complete retention, rather slightly higher BR binding properties as compared to native membranes. The disappearance of double minima reflects the loss of α-helical structure and the transition towards random coil. It can be assessed that the disruption of self-associated band 3 by DMMA is accompanied with a disruption of helical structure, thereby giving the impression that the CD spectrum of membranes was largely due to the folded structure maintaining a band 3-dimer [27].

Taken together, all these results suggest that removal of membrane proteins does not affect the BR binding properties of erythrocyte membranes to an appreciable extent. These results strengthen previous findings [1, 5–7] suggesting that BR binding sites in the membranes are composed of lipids rather than proteins.

Acknowledgements

This work was financially supported by a research grant (SP/SO/D-52/95) from the Department of Science and Technology, New Delhi, India. Facilities provided by Aligarh Muslim University, Aligarh are gratefully acknowledged. Thanks are due to Yogesh Kumar for his assistance in CD measurements and Mohammed Kutub Ali for helpful discussions.

References

1. Sato H, Kashiwamata S: Interaction of bilirubin with human erythrocyte membranes. Biochem J 210: 489–496, 1983
2. Sato H, Aono S, Semba R, Kashiwamata S: Interaction of bilirubin with human erythrocyte membranes. Bilirubin binding to neuraminidase- and phospholipase-treated membranes. Biochem J 248: 21–26, 1987
3. Gulian J-M, Gonard V, Dalmasso C, Palix C: Bilirubin displacement by ceftriaxone in neonates: Evaluation by determination of 'free' bilirubin and erythrocyte-bound bilirubin. J Antimicrobial Chemother 19: 823–829, 1987
4. Hayer M, Piva MT, Sieso V, De Bornier BM: Experimental studies on unconjugated bilirubin binding by human erythrocytes. Clin Chim Acta 186: 345–350, 1989
5. Nagaoka S, Cowger ML: Interaction of bilirubin with lipids studied by fluorescence quenching method. J Biol Chem 253: 2005–2011, 1978
6. Cestaro B, Cervato G, Ferrari S, Di Silvestro G, Monti D, Manitto P: Interaction of bilirubin with small unilamellar vesicles of dipalmitoylphosphatidylcholine. Ital J Biochem 32: 318–329, 1983
7. Vazquez J, Garcia-Calvo M, Valdivieso F, Mayor F, Mayor F Jr: Interaction of bilirubin with synaptosomal plasma membrane. J Biol Chem 263: 1255–1265, 1988
8. Leonard M, Noy N, Zakim D: The interaction of bilirubin with model and biological membranes. J Biol Chem 264: 5648–5652, 1989
9. Sheetz MP, Wang WP, Kreutzer DL: Polyphosphoinositides as regulators of membrane skeletal stability. Kroc Found Ser 16: 87–94, 1984
10. Haest CWM: Interaction between membrane skeleton proteins and the intrinsic domain of the erythrocyte membrane. Biochim Biophys Acta 694: 331–352, 1982
11. Rodgers W, Glaser M: Distribution of proteins and lipids in the erythrocyte membrane. Biochemistry 32: 12591–12598, 1993
12. Kumar A, Gupta CM: Red cell membrane abnormalities in chronic myeloid leukaemia. Nature 303: 632–633, 1983
13. Steck TL: The organization of proteins in the human red blood cell membrane. A review. J Cell Biol 62: 1–19, 1974
14. Palfrey HC, Waseem A: Protein kinase C in the human erythro-

cyte. Translocation to the plasma membrane proteins. J Biol Chem 260: 16021–16029, 1985
15. Rashid H, Owais M, Tayyab S: Bilirubin binding to normal and modified human erythrocyte membranes. Effect of phospholipases, neuraminidase, trypsin and $CaCl_2$ Mol Cell Biochem 228: 15–23, 2001
16. Kahlenberg A: Partial purification of a membrane protein from human erythrocytes involved in glucose transport. J Biol Chem 251: 1582–1590, 1972
17. Laemmli UK: Cleavage of structural proteins during the assembly of the head of bacteriophage T_4. Nature 227: 680–685, 1970
18. Lowry OH, Rosebrough NJ, Farr AJ, Randall RJ: Protein measurements with the Folin phenol reagent. J Biol Chem 193: 265–275, 1951
19. Fog J: Determination of bilirubin in serum as alkaline 'azobilirubin'. Scan J Clin Lab Invest 10: 241–245, 1958
20. Tayyab S, Ali MK: A comparative study on the extraction of membrane-bound bilirubin from erythrocyte membranes using various methods. J Biochem Biophys Meth 39: 39–45, 1999
21. Van Hoek AN, Wiener M, Bicknese S, Miercke L, Biwersi J, Verkman AS: Secondary structure analysis of purified functional CH1P28 water channels by CD and FTIR spectroscopy. Biochemistry 32: 11847–11856, 1993
22. Fairbanks G, Steck TL, Wallach DFH: Electrophoretic analysis of the major polypeptides of the human erythrocyte membrane. Biochemistry 10: 2606–2617, 1971
23. Steck TL, Kant JA: Preparation of impermeable ghosts and inside-out vesicles from human erythrocyte membranes. Meth Enzymol 31: 172–180, 1973
24. Jones MN, Nickson JK: Monosaccharide transport proteins of the human erythrocyte membrane. Biochim Biophys Acta 650: 1–20, 1981
25. Rashid H, Ali MK, Tayyab S: Differential accessibility of bilirubin to erythrocyte membrane vesicles bearing different structural features. Comp Biochem Physiol 127: 345–350, 2000
26. Marchesi VT, Andrews EP: Glycoproteins. Isolation from cell membranes with lithium diiodosalicylate. Science 174: 1247–1248, 1971
27. Makino S, Nakashima HJ: Behaviour of fragmented Band 3 from chymotrypsin-treated bovine erythrocyte membrane in non-ionic detergent solution. J Biochem 92: 1069–1077, 1982

Molecular and Cellular Biochemistry **246:** 179–185, 2003.

Arrest of cell cycle by Amida which is phosphorylated by Cdc2 kinase

Yehua Gan, Eiichi Taira, Yasuyuki Irie, Takahiro Fujimoto and Naomasa Miki

Department of Pharmacology, Osaka University Medical School, Yamada-oka, Suita, Osaka, Japan

Abstract

Amida was first isolated from a rat hippocampal cDNA library as an Arc-associated protein. Although previous studies have shown that Amida mRNA is predominantly expressed and developmentally regulated in rat testis and overexpression induces apoptosis, the function of Amida remains unclear. In this study, we found that overexpression of Amida inhibited cell growth. Flow cytometry analysis showed that Amida caused cell cycle inhibition in the S-phase and blocked cell cycle from entry into mitosis. Attempting to elucidate Amida effect on the cell cycle, we found that Amida was interacted with Cdc2 in mitosis and Amida's overexpression resulted in a decrease in Cdc2 kinase activity. In addition, Amida showed DNA-binding ability with DNA-affinity column chromatography. A region (aa, 76–189) between the two nuclear localization signals was found to be responsible for cell growth inhibition and DNA-binding activity, implying that DNA-binding activity may be necessary for Amida to repress cell cycle. Moreover, Amida was phosphorylated by Cdc2 kinase *in vitro* and Ser-180 of Amida was identified as the phosphorylation site. Furthermore, AmidaS180G (eliminate phosphorylation of Ser-180) showed stronger DNA-binding activity. Taken together, the data suggest that Amida may play an important role in cell cycle and may be partly regulated by Cdc2 kinase. (Mol Cell Biochem **246**: 179–185, 2003)

Key words: Amida, cell cycle, DNA-binding, Cdc2

Abbreviations: NLS – nuclear localization signal; PCR – polymerase chain reaction; GFP – green fluorescent protein; PAGE – polyacrylamide gel electrophoresis; PBS – phosphate-buffered saline; TBS – tris-buffered saline; PMSF – phenylmethylsulfonyl fluoride; DAPI – 4,6-diamidino-2-phenylindole

Introduction

Amida was first isolated from a rat hippocampal cDNA library by the yeast two-hybrid system as a novel protein associated with Arc [1]. Arc is an immediate early gene encoding an effector protein involved in activity-dependent plasticity of dendrites [2]. The human homologue of Amida has been recently identified as FB1 in childhood pre-B cell acute lymphoblastic leukemia [3]. Despite FB1 homology to Amida, neither protein has an established function. Previous studies have shown that Amida contains two nuclear localization signals (NLSs) which allows for transfection to the nucleus [2]. Amida mRNA is predominantly expressed and developmentally regulated in rat testis, suggesting that Amida may be involved in spermatogenesis [4]. FB1 localizes to chromosome 19q13.4. This chromosome region also contains a number of tumor suppressor genes [3]. Since the E2A/FB1 fusion appears to be a recurrent feature of pre-B acute lymphoblastic leukemia, it is suggested that it might have a role in the development and/or progression of leukemogenesis [3].

To verify this role, we examined whether or not Amida had any effect on mammalian cell cycle. We found that Amida inhibited S-phase progression and blocked cell cycle from entry into mitosis. To explain the possible mechanism of cell cycle arrest by Amida, we further found that Amida inhibited and interacted with Cdc2 kinase. In addition, we found

Address for offprints: N. Miki, Department of Pharmacology, Osaka University Medical School, Room A6, 2-2 Yamada-oka, Suita, Osaka 565-0871, Japan (E-mail: nmiki@pharma1.med.osaka-u.ac.jp)

that Amida had DNA-binding activity. Since Amida contains potential sites for Cdc2 kinase in the deduced amino acid sequence, we examined whether Amida is phosphorylated by Cdc2 kinase and the effect of the phosphorylation on DNA-binding activity of Amida.

Materials and methods

Cell culture and transfection

Cells were grown in DMEM (Sigma) with 10% fetal calf serum on 10-cm dishes in a humidified atmosphere containing 5% CO_2. Cells in 70% of confluence were transfected by 15 µl of lipofectamine reagent (Life Technologies, Inc.) with 3 µg of empty pEGFP-C1 mammalian expressing vector (containing green fluorescent protein, GFP) or the Amida insert or mutation of the Amida insert subcloned into the EGFP vector.

Colony-forming assay

To evaluate the effects of Amida on the cell cycle, COS-7 cells were transfected with GFP-Amida or GFP for 24 h and then harvested in single cell suspension by trypsin/EDTA. The cells were seeded into four 10-cm dishes at very low density and then cultured in DMEM containing 300 µg/ml G418 disulfate (nacalai tesque, Inc.) for 36, 72, 96 and 120 h. After washed off the cells undergoing apoptosis, the cells were then fixed with cold 70% ethanol and stained with DAPI (4, 6-diamidino-2-phenylindole). GFP-positive colonies were counted under a fluorescence microscope for cells transfected with GFP alone or GFP counterparts containing Amida or Amida mutants.

Flow cytometry

After transfection with GFP-Amida or GFP for 72 h, COS-7 cells were washed to remove cells undergoing apoptosis, harvested by trypsin/EDTA, and then fixed with cold 70% ethanol for 1 h on ice. The cells were stained with propidium iodide (10 µg/ml) after treated with 1 µg/ml RNase for 1 h. The cell suspensions were then analyzed with a fluorescence-activated cell sorter (FACS-Calibur, Becton Dickinson Co., Ltd.) [5].

H1 histone kinase assay in vitro

For preparations of Cdc2 kinase, COS-7 cells were transfected with GFP-Amida or GFP for 72 h and then lysed and sonicated in the lysis buffer [2]. The lysates were spun at 1,500 × g for 20 min at 4°C. Supernatants were precleared with 50 µl protein A-sepharose (Amersham Pharmacia Biotech) for 1 h at 4°C. Then, the supernatants were incubated with anti-Cdc2 antibody (Rochland, PA, USA) and 100 µl protein A-sepharose for 16 h at 4°C and subsequently centrifuged at 1,500 × g for 5 min. The immunoprecipitates were washed 3 times with the lysis buffer and twice with the kinase buffer (50 mM Tris-HCl, pH 7.5, 10 mM $MgCl_2$, 2 mM DTT, 1 mM EGTA, 0.01% Brij 35). Fifteen µl of the immunoprecipitates were incubated with 2 µg H1 histone (Sigma, type III-S) in 30 µl of kinase assay buffer containing 100 µM ATP and 10 µCi of [γ-^{32}P] ATP (6000 Ci/mmol, NEN Life Science Products) at 30°C for 30 min. The samples were then mixed with Laemmli sample buffer, boiled for 3 min, and then cooled on ice. The samples were then subjected to 10% SDS-PAGE and autoradiography.

Immunoprecipitation assay

Cells transfected with Amida could not be synchronized in M-phase because Amida arrested the cell cycle in S-phase and G2-phase. Therefore, supernatants from COS-7 cells transfected with GFP-Amida for 72 h were mixed with supernatants from M-phase HeLa cells that were collected by disassociating them from the dish through washing the dish repeatedly with the medium after treatment of 0.5 µg/ml nococdazole (Sigma) for 16 h and then centrifuged, or HeLa cells treated with 5 mM thymidine (Sigma) for 48 h, or asynchronous HeLa cells. The supernatants were precleared with 40 µl protein A-sepharose on ice for 1 h. Then the supernatants were immunoprecipitated with 1 µl anti-GFP antibody (Molecule Probe, Eugene, OR, USA) and 50 µl protein A-sepharose and then immunoblotted with anti-Cdc2 antibody, or immunoprecipitated with 2 µl anti-Cdc2 antibody and 50 µl protein A-sepharose and then immunoblotted with anti-GFP antibody, respectively.

Site-directed mutagenesis

Site-directed mutagenesis was initiated by annealing a mismatched oligonucleotide to the coding strand of the Amida cDNA. The 3′ primers for the point mutants of Amida were designed to start from four-base down-stream of the single ApaI site in the full-length Amida cDNA so that the single ApaI site was contained in these 3′ primers. 3′-primer for AmidaS167/180G is 5′-CACTGGGCCCCTCGCCAGGG-GCTGGCCGCTGGGGTCTAGGGTTGCAGTTGTCCTT-TGTGGTGGGCCAAGTCCCTC- 3′. The underlined codons (where the last C in the first underlined codon and the CC in the second underlined codon were the mismatch) replaced the original serine-180 and serine-167 by glycine respectively. 3′-primer for AmidaS180G is 5′-CACTGGGCCCCTCGC-CAGGGGCTGGGCCGCTGGGG-3′, where the last C in the underlined codon is mismatched. 3′-primer for AmidaS167G is 5′-CACTGGGCCCCTCGCCAGGGGCTGGGCTGCT-GGGGTCTAGGGTTGCAGTTGTCCTTGTGGTGGGCC-AAGTCCCTC-3′, where the CC in the underlined codon were mismatched. Those three kinds of 3′-primers were used with 5′-primer of Amida contained the first ATG down-stream of a single XhoI site, 5′-CCCCCTCGAGACATGGAG-TTGGAACAGAGAGAAG-3′, to produce PCR (polymerase chain reaction) fragments of the point mutants of Amida by Pyrobest Polymerase (TaKaRa, Japan). The full-length Amida cDNA (1–780 bp) has been inserted into pEGFP-C1 plasmid with XhoI and BamHI. Thus, the PCR fragments and the plasmid were incubated with XhoI and ApaI and purified. The purified PCR fragments (1–558 bp) and the plasmid containing only partial Amida cDNA (559–780 bp) were ligated to recover the full-length Amida cDNA with the point mutation. Each mutation was confirmed by sequencing.

Kinase assay in vitro

COS-7 cells transfected with GFP-Amida or GFP or the point mutants of GFP-Amida for 48 h were washed with PBS (–) (phosphate-buffered saline) twice and lysed by sonication in the lysis buffer [2]. The lysates were centrifuged at 15,000 × g for 20 min at 4°C. Supernatant was collected and treated with 50 µl Protein A-Sepharose and rotated at 4°C for 1 h and then centrifuged at 1,500 × g for 5 min at 4°C to eliminate non-specific binding materials. The supernatants were then added with 2 µl anti-GFP antibody and 50 µl Protein A-Sepharose for 16 h at 4°C and centrifuged. The precipitates were washed three times with the lysis buffer, twice with the kinase assay buffer. Then, they were incubated with 1.5 µl of p34 Cdc2 protein kinase composed of the catalytic subunit Cdc2 and its positive regulatory subunit cyclin B (New England BioLabs Inc.) in 30 µl of kinase assay buffer with 100 µM ATP and 10 µCi of [γ-^{32}P] ATP (6000 Ci/mmol; NEN Life Science Products) at 30°C for 30 min. The samples mixed with Laemmli sample buffer were boiled for 3 min and cooled on ice, then subjected to 10% SDS-PAGE and autoradiography.

DNA-binding assay

DNA-binding ability of Amida protein was examined by DNA-affinity column chromatography [4]. After transfection with GFP-Amida for 48 h, the COS-7 cells were sonicated in 0.5 TBS-T (12.5 mM, pH 7.5 Tris-HCl, 75 mM NaCl, 25 mM KCl, and 0.05% Tween 20) containing 1 mM PMSF (phenylmethylsulfonyl fluoride). The cell lysates were centrifuged at 15,000 × g for 20 min at 4°C. Then, the supernatants were applied to a column (10-ml bed volumes, Bio-Rad) filled with 1 ml of calf thymus native DNA-cellulose (Amersham Pharmacia Biotech) at 4°C. Washing fraction of 1 ml of 0.5 × TBS-T was collected after washing the column four times with 10 ml of 0.5 × TBS-T. The elution fractions were collected stepwisely with 1 ml of 0.1, 0.2, 0.6, and 1 M NaCl in 0.5 × TBS-T. The elution and washing fractions were immunoprecipitated and immunoblotted with anti-GFP antibody. The same protocol was used for Amida mutants, but the deletion mutants were eluted from the column with only 1 M NaCl.

Results

Inhibition of cell cycle by overexpression of Amida and identification of responsible region

We reported that overexpression of GFP-Amida induces apoptosis in 50% of transfectants [2]. However, we did not examine whether the remaining transfectants had any change of cell cycle. In our preliminary experiments, we transfected COS-7 cells with GFP for 24 h then harvested and seeded at low density and cultured for another 120 h in the presence of G418 (pEGFP-C1 vector contains a neomycin resistance gene). As expectedly, the transfected cells did not lose GFP during the time. In contrast, colonies consisted of GFP positive cells were observed under fluorescent microscopy. Therefore, we compared the effect of GFP-Amida fusion protein with GFP alone on cell cycle until 120 h after seeding the transfectants. The cell number of colonies consisted of GFP transfectants was approximately 20 at 120 h after seeding the transfected cells. However, the cell number of colonies consisted of GFP-Amida transfectants remained less than 3 at that time (Figs 1A and 1B). DAPI staining showed that after washing, the remaining GFP-Amida and its mutant positive cells had no signs of apoptosis. These data suggest that Amida has an ability to suppress cell growth. The inhibitory effect on cell growth by overexpression of Amida was reproduced in other cell lines such as C6, 293T, and HeLa cells (data not shown). Furthermore, a region (aa, 76–189) between the two NLSs of Amida was found to inhibit cell cycle, but C-terminus (aa, 190–259) and N-terminus (aa, 1–81) mutants of Amida did not suppress cell growth (Fig. 1B). The effect of Amida on cell cycle progression was examined by a flow cytometry in COS-7 cells transfected with GFP-Amida for 72 h (Fig. 1C). The COS-7 cells were subjected to GFP fluorescence (488 nm) and separated into fluorescence-positive group (TP) and fluorescence-negative group (TN) according to their intensities of GFP fluorescence. These two groups were used as the gates to analyze DNA contents. The fluorescence-negative cells (non-transfected cells) showed normal growth pattern (Fig. 1C, b). In contrast, the fluorescence-posi-

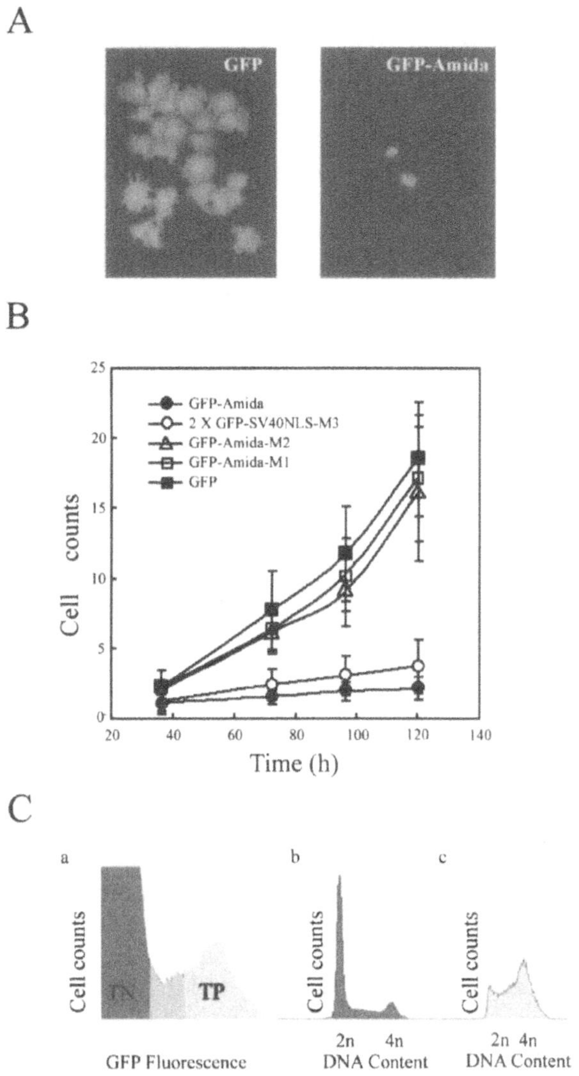

Fig. 1. Inhibition of cell growth by overexpression of Amida. (A) The representative colonies from COS-7 cells transfected with GFP or GFP-Amida. Pictures were photographed with a fluorescent microscope at 120 h after seeding the transfected cells. Scale bars, 10 µm. Note GFP diffused in the whole cell, but GFP-Amida only localized in nucleus. (B) Cell growth of transfected cells with Amida and the mutants. The COS-7 cells were transfected with GFP, GFP-Amida, GFP-Amida-M1 (aa, 1–81), GFP-Amida-M2 (aa, 190–259) or 2 × GFP- SV40NLS-Amida-M3 (aa, 76–189) (including double GFP and an SV40 nuclear localization signal), then cell numbers of 100 colonies were counted and averaged at the time point after seeding the transfected cells. The values are means of cell number and the bars represents S.D. (C) FACS analysis of COS-7 cells transfected with Amida. (a) The COS-7 cells were subjected to GFP fluorescence (488 nm) after transfection and separated into transfected group (TP) and non-transfected group (TN) according to their intensities of GFP fluorescence. These two groups were used as gates to analyze DNA contents. (b) The DNA content profile showed normal growth pattern from the gate of non-transfected cells (TN). (c) The DNA content profile of the gate of Amida transfected cells (TP) showed that cell cycle was inhibited continuously in the S-phase progression, accompanied with an inhibition of entry into mitosis.

tive cells (GFP-Amida transfected cells) showed that cell cycle was inhibited in S-phase progression and were accompanied by an inhibition of entry into mitosis (Fig. 1C, c). However, GFP alone showed no effect on cell cycle progression with flow cytometry (data not shown).

Inhibition of Cdc2 kinase activity in COS-7 cells transfected with Amida

Overexpression of p53 arrests cell cycle and affects Cdc2 kinase activity that was evaluated by H1 histone kinase assay *in vitro* [6]. To access the mechanism of cell cycle arrest by overexpression of Amida, we also measured the Cdc2 kinase activity in COS-7 cells transfected with GFP-Amida or GFP for 72 h. As shown in Fig. 2A, H1 histone phosphorylation by Cdc2 kinase from COS-7 cells transfected with GFP-Amida was less than cells transfected with GFP. This implies that Cdc2 kinase was inhibited by overexpression of Amida. Western blot showed no difference in protein level of Cdc2 between the cells transfected with GFP or GFP-Amida (Fig. 2B).

Association of Amida with endogenous Cdc2 at M-phase

As Amida arrested cell cycle and inhibited Cdc2 kinase activity, we examined whether Amida interacts with Cdc2. As

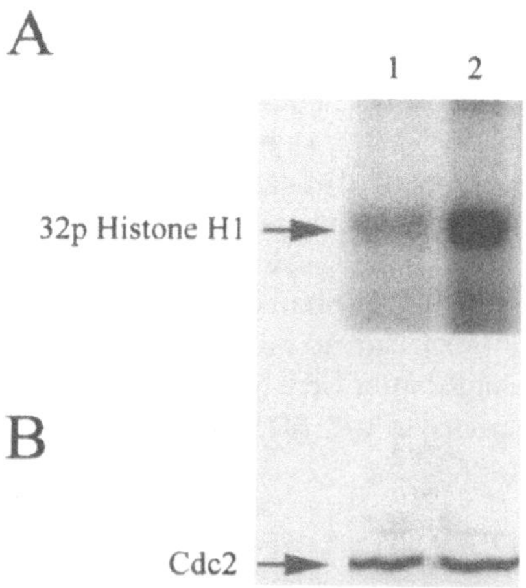

Fig. 2. Inhibition of Cdc2 kinase activity by overexpression of Amida. (A) Decrease of Cdc2 kinase activity in the COS-7 cells transfected with GFP-Amida. Two µg of H1 histone was phosphorylated in the presence [γ-^{32}p] ATP by Cdc2 which was immunoprecipitated from the lysates of COS-7 cells transfected with GFP-Amida (lane 1) or GFP (lane 2). (B) Western blot analysis of Cdc2 in COS-7 cells transfected with GFP-Amida (lane 1) or GFP (lane 2).

shown in Fig. 3, Amida interacted with endogenous Cdc2 from M-phase HeLa cells synchronized with nocodazole. GFP-Amida was detected in the pellets immunoprecipitated with anti-Cdc2 antibody and Cdc2 was detected in the pellets immunoprecipitated with anti-GFP antibody, reversibly. However, Amida interacted neither with Cdc2 from asynchronous HeLa cells nor with G1-phase HeLa cells synchronized with thymidine (data not shown).

Phosphorylation of Amida at serine-180 by Cdc2 kinase in vitro

Since Amida arrested cell cycle after transfection into COS-7 and contains five S/T-P-X-Z motifs in its deduced amino acid sequence that are potential phosphorylation sites for Cdc2 kinase [7], we examined whether Amida is phosphorylated by Cdc2 kinase. As shown in Fig. 4A, GFP-Amida, but not GFP, was phosphorylated by Cdc2 kinase. To identify the motif phosphorylated by Cdc2 kinase, three kinds of point mutants for two S-P-X-Z motifs in the cell growth inhibitory region of Amida were transfected and then immunoprecipitated with anti-GFP antibody. Only GFP-Amida (wild type) and GFP-AmidaS167G (Ser-167 changed into Gly) were phosphorylated after incubation with Cdc2/cyclin B complex, but neither GFP-Amida-S167/180G (Ser-167 and Ser-180 changed into Gly) nor GFP-AmidaS180G (Ser-180 changed into Gly) was phosphorylated (Fig. 4C). Therefore, Ser-180 in Amida was identified as the phosphorylation site of Cdc2 kinase. The expression of GFP-Amida and its mutants was shown by Western blot with anti-GFP antibody (Fig. 4C).

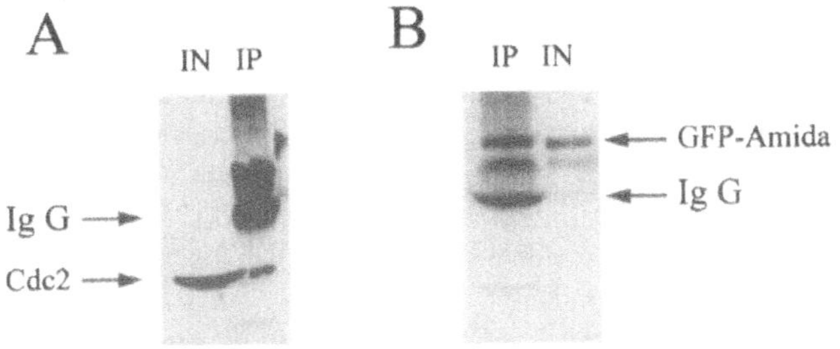

Fig. 3. Association between Amida and Cdc2 at M phase. The supernatant from COS-7 cells transfected with GFP-Amida for 48 h were mixed with that from M phase HeLa cells collected by disassociating them from the dish after treatment of nocodazole for 16 h and then immunoprecipitated with anti-GFP or anti-Cdc2 antibody, respectively. (A) Cdc2 was detected in the pellet (IP) immunoprecipitated by anti-GFP antibody. IN, input, 1/10 volume of the supernatant for immunoprecipitation. (B) GFP-Amida was detected in the pellet (IP) immunoprecipitated by anti-Cdc2 antibody. IN, input, 1/10 volume of the supernatant for immunoprecipitation.

DNA-binding activity of Amida and the identification of binding region

Since Amida is a nuclear protein and suppressed cell growth, DNA-binding ability of Amida was examined by DNA-affinity column chromatography. The lysate from COS-7 cells transfected with GFP-Amida was applied to the column and then washed and eluted. GFP-Amida was initially eluted with 0.1 M NaCl, greater concentrations of GFP-Amida were eluted with 0.2 and 0.6 M NaCl, but almost no GFP-Amida was eluted with 1 M NaCl (Fig. 5A). Next, the DNA-binding domain of Amida was determined by using deletion

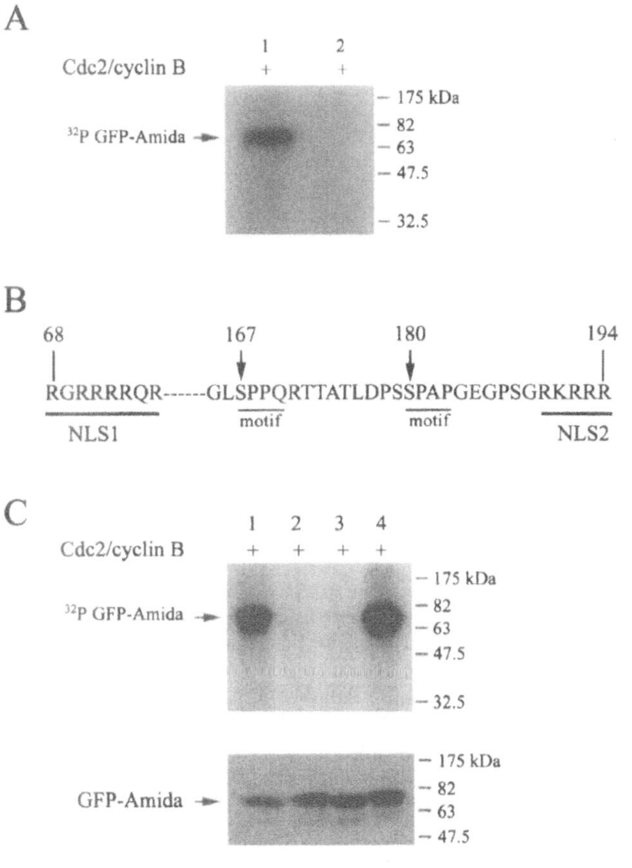

Fig. 4. In vitro phosphorylation of Amida at serine-180 by Cdc2 kinase. (A) Phosphorylation of Amida by Cdc2 kinase in vitro. The supernatants from COS-7 cells transfected with GFP-Amida (lane 1) or GFP (lane 2) were immunoprecipitated by anti-GFP antibody and then incubated with Cdc2/cyclin B complex in the presence of [γ-^{32}P] ATP. (B) Scheme for point mutation position (Ser-167 and Ser-180) in the motifs for Cdc2 kinase in the region between the nuclear localization signals (NLSs). (C) Phosphorylation of Amida at serine-180 by Cdc2 kinase. The supernatants from COS-7 cells transfected with GFP-Amida (lane 1), GFP-AmidaS167/180G (lane 2), GFP-AmidaS180G (lane 3) and GFP-AmidaS167G (lane 4) were immunoprecipitated by anti-GFP antibody and then incubated with Cdc2/cyclin B complex in the presence of [γ-^{32}P] ATP (upper panel). Western blot analysis of GFP-Amida and the mutants (lower panel).

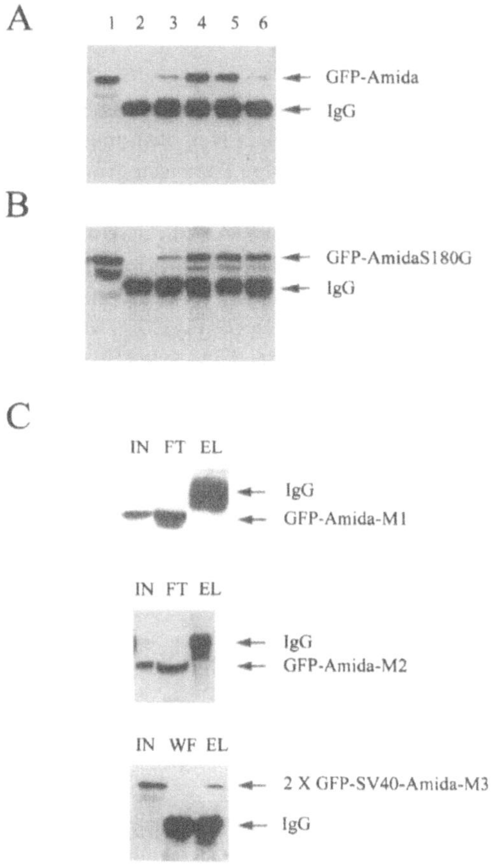

Fig. 5. DNA-binding activity of Amida and identification of binding domain. (A) DNA-binding activity of Amida. The supernatants of COS-7 cells transfected with GFP-Amida and its mutants were applied to the column filled with 1 ml calf thymus native DNA-cellulose and eluted with stepwise increasing concentrations of NaCl (lane 1, input; lane 2, washing fraction; lanes 3–6, elution fractions of 0.1, 0.2, 0.6 and 1.0 M NaCl). The elution and wash fractions of GFP-Amida were immunoprecipitated and immunoblotted with anti-GFP antibody. (B) AmidaS180G (eliminate phosphorylation of Ser-180) showed stronger DNA-binding activity than the wild type. GFP-AmidaS180G was still eluted with 1 M NaCl, as much as with 0.2 and 0.6 M NaCl (lane was indicated like panel A). (C) A region between the two NLSs of Amida (aa, 76–189) had DNA-binding ability. The column was eluted only with 1 M NaCl. The elution and flow through fractions of GFP-Amida-M1 (aa, 1–81) and GFP-Amida-M2 (aa, 190–259) or the elution and wash fractions of GFP-GFP-Amida-M3 (aa, 76–189) were immunoprecipitated and immunoblotted with anti-GFP antibody. IN – input; FT – flow through fraction; EL – elution fraction; WF – washing fraction.

mutants. The deletion mutant (M3, aa, 76–189) containing a region between the two NLSs of Amida was only detected in elution fraction, but not in washing fraction. However, the C-terminus mutant (M2, aa, 190–259) and N-terminus mutant (M1, aa, 1–81) of Amida were only detected in the fractions of flow through, but not in the elution fractions (Fig. 5C). The data indicate that the DNA-binding domain of Amida is located between the two NLSs (aa, 76–189), and may have an inhibitory effect on cell cycle.

Elimination of phosphorylation of Ser-180 affects Amida DNA-binding activity

Since Amida was phosphorylated at Ser-180 by Cdc2 kinase *in vitro* and exhibited DNA-binding activity, the effect of the phosphorylation of Ser-180 on DNA-binding activity of Amida was examined with GFP-AmidaS180G that was eliminated phosphorylation on Ser-180 *in vivo* after transfection. Like the wild type, GFP-AmidaS180G was initially eluted a little with 0.1 M NaCl, but unlike the wild type it was still eluted with 1 M NaCl, as much as with 0.2 and 0.6 M NaCl (Fig. 5B). The elution pattern of AmidaS180G is clearly different from that of the wild type. This implies that AmidaS180G has a stronger DNA-binding activity than the wild type.

Discussion

Previous studies have revealed that overexpression of GFP-Amida induces apoptosis in 50% of transfectants [2], but the fate of the remaining transfectants was unknown. In the present study, we have shown that overexpression of Amida was involved in the inhibition of cell cycle, mostly in S-phase progression accompanied by an inhibition of entry into mitosis (Fig. 1), if the transfectants did not undergo apoptosis. Therefore Amida may not only induce apoptosis but also arrest cell cycle. The domain responsible for Amida inhibition was mapped to a region between the two NLSs (Fig. 1B). The region was also revealed to be responsible for Amida DNA-binding activity (Fig. 5C). It is thus likely that the DNA-binding activity may be necessary for Amida to exert cell cycle inhibition. The results also imply that the region between the two NLSs is an important functional region of Amida.

The exact mechanism of Amida's inhibition of cell cycle is still unknown. However, according to the results of FACS, there exist several possible mechanisms for the involvement of Amida in cell cycle. One of the possible mechanisms is that Amida might affect Cdk2 or Cdk4, or both since overexpression of Amida appeared G1-phase peak lost and cell counts increased continuously in S-phase (Fig. 1C). We examined whether Amida also influenced Cdc2 activity as the results of FACS indicated an inhibition of entry into mitosis (Fig. 1C). We found that Cdc2 kinase activity was decreased in the cells transfected with Amida, but did not affect Cdc2 protein level (Fig. 2). Cdc2 kinase activity is critical for cell to progress from G2-phase into M phase [8]. Therefore the decrease of Cdc2 kinase is likely to be responsible for cell cycle blocked before the entry into mitosis by overexpression of Amida.

Amida was phosphorylated at Ser-180 by Cdc2 kinase *in vitro* (Fig. 4). Since overexpression of Amida arrested cell cycle before entry into mitosis and Amida contains five potential sites for Cdc2 kinase, we examined the phosphorylation of Amida by Cdc2 kinase and found that Amida was phosphorylated by Cdc2/cyclin B kinase in vitro (Fig. 4A). Furthermore, the phosphorylation site was identified as Ser-180 in the SPAP motif (Figs 4B and 4C), and this site could not be phosphorylated by Cdk2 (data not shown). The finding of association between Amida and endogenous Cdc2 at M phase (Fig. 3) further supported that this phosphorylation event may be highly possible *in vivo*.

Phosphorylation of Ser-180 by Cdc2 kinase might affect Amida DNA-binding activity. Protein phosphorylation and dephosphorylation can directly regulate many aspects of protein function, including cellular localization, protein stability, protein-protein interactions and DNA binding [9]. Since AmidaS180G (eliminated Ser-180 phosphorylation) showed stronger DNA-binding activity than the wild type (Figs 5A and 5B), it was suggested that the phosphorylation of Amida by Cdc2 kinase might decrease Amida DNA-binding activity. It has been speculated that one of the roles of Cdc2 phosphorylation of its substrates is to remove them from DNA and thus allow the chromosomes to become more highly condensed during mitosis [7]. Our suggestion is consistent with the previous observation that the DNA-binding activity will decrease when the SPXX motif (X, any amino acid) is phosphorylated [10]. Therefore, it is possible that the phosphorylation by Cdc2 kinase may release Amida from DNA to make DNA further condensation during M phase. Further study is needed to demonstrate that Ser-180 of Amida is phosphorylated by Cdc2 kinase *in vivo*, and to clarify mechanisms of Amida's inhibition of cell cycle.

References

1. Lyford G, Yamagata K, Kaufmann W, Barnes C, Sanders L, Copeland N, Gilbert D, Jenkins N, Lanahan A, Worley P: Arc, a growth factor and activity-regulated gene, encodes a novel cytoskeleton-associated protein that is enriched in neuronal dendrites. Neuron 14: 433–445, 1995
2. Irie Y, Yamagata K, Gan Y, Miyamoto K, Do E, Kuo C, Taira E, Miki N: Molecular cloning and characterization of Amida, a novel protein which interacts with a neuron-specific immediate early gene product Arc, contains novel nuclear localization signals, and causes cell death in cultured cells. J Biol Chem 275: 2647–2653, 2000
3. Brambillasca F, Mosna G, Colombo M, Rivolta A, Caslini C, Minuzzo M, Giudici G, Mizzi L, Biondi A, Privitera E: Identification of a novel molecular partner of the E2A gene in childhood leukemia. Leukemia 13: 369–375, 1999
4. Gan Y, Taira E, Irie Y, Tanaka H, Ichikawa H, Kumamaru E, Miki N: Amida predominantly expressed and developmentally regulated in rat testis. Biochem Biophys Res Commun 288: 407–412, 2002
5. Tanaka H, Yoshimura, Y, Nozaki M, Yomogida K, Tsuchida J, Tosaka Y, Habu T, Nakanishi T, Okada M, Nojima H, Nishimune Y: Identification and characterization of a haploid germ cell-specific nuclear protein kinase (Haspin) in spermatid nuclei and its effects on somatic cells. J Biol Chem 274: 17049–17057, 1999
6. Taylor W, DePrimo S, Agarwal A, Agarwal M, Schonthal A, Katula K, Stark G: Mechanisms of G2 arrest in response to overexpression of p53. Mol Biol Cell 10: 3607–3622, 1999
7. Moreno S, Nurse P: Substrates for p34cdc2: *In vivo* veritas? Cell 61: 549–551, 1990
8. Nurse P: Universal control mechanism regulating onset of M-phase. Nature 344: 503–508, 1990
9. Whitmarsh A, Davis R: Regulation of transcription factor function by phosphorylation. Cell Mol Life Sci 57: 1172–1183, 2000
10. Green G, Lee H, Poccia D: Phosphorylation weakens DNA binding by peptides containing multiple 'SPKK' sequences. J Biol Chem 268: 11247–1155, 1993

Molecular and Cellular Biochemistry **246:** 187–191, 2003.

Contribution of increased HMG-CoA reductase gene expression to hypercholesterolemia in experimental chronic renal failure

Michal Chmielewski,[1] Elzbieta Sucajtys,[1] Julian Swierczynski,[2] Boleslaw Rutkowski[1] and Wojciech Bogusławski[3]

Departments of [1] Nephrology, Transplantology and Internal Medicine; [2]Biochemistry; [3]Medical Chemistry, Medical University of Gdańsk, Gdańsk, Poland

Abstract

The aim of the present study was to examine hypothesis that the enhanced cholesterologenesis, found in rats with experimental chronic renal failure (CRF) resulted from the increased gene expression of 3-hydroxy-3-methylglutaryl coenzyme A (HMG-CoA) reductase – the rate limiting enzyme in the cholesterologenesis pathway, responsible for mevalonate synthesis. Wistar rats were used and experimental CRF was achieved by 5/6 nephrectomy model. We examined: (a) the changes in the rat liver microsomal HMG-CoA reductase activity, (b) the rat liver HMG-CoA reductase mRNA abundance in various times of day. Obtained data indicates that the increased activity of HMG-CoA reductase in the liver of rats with experimental CRF parallel enhanced mRNA level and suggests that enhanced cholesterol biosynthesis, observed in experimental CRF is at least in part due to the increased HMG-CoA reductase gene expression. The results also indicate that the physiological diurnal rhythm of HMG-CoA reductase activity is preserved in the course of experimental CRF. (Mol Cell Biochem **246**: 187–191, 2003)

Key words: cholesterol biosynthesis, HMG-CoA reductase, experimental chronic renal failure, hypercholesterolemia

Introduction

Hypercholesterolemia is a well known risk factor for cardiovascular complications. In the course of chronic renal failure (CRF) it contributes to the progression of the disease [1]. Therefore evaluation of mechanisms leading to hypercholesterolemia associated with CRF appears to be of importance. Recently we have shown increased rate of cholesterol biosynthesis both in livers and intestines of rats with experimental CRF. We have also documented that the physiological diurnal variability of cholesterologenesis is preserved in the course of experimental CRF [2].

Diurnal rhythm of cholesterol biosynthesis follows the changes in the gene expression of 3-hydroxy-3-methylglutaryl coenzyme A (HMG-CoA) reductase, the rate limiting enzyme in cholesterol biosynthesis [3]. Therefore in the case of enhanced cholesterologenesis one can suppose that the activity of HMG-CoA reductase paralleled by the level of its mRNA should be increased.

Indeed, Pandak *et al.* [4] showed higher hepatic HMG-CoA reductase activity in 75% nephrectomy rat CRF model, but the increase in activity was not associated with parallel increase in HMG-CoA reductase mRNA level and gene transcriptional activity. However diurnal rhythm of reductase activity has not been considered in these experiments.

Therefore the aim of the present study was to evaluate, if

Address for offprints: W. Bogusławski, Department of Medical Chemistry, Medical University of Gdańsk, 80-210 Gdańsk, Dębinki 1, Poland (E-mail: wbogus@amedec.amg.gda.pl)

the HMG-CoA reductase activity and its mRNA level are increased throughout the whole day, taking into consideration the diurnal rhythm of cholesterol biosynthesis.

Materials and methods

[3-^{14}C]HMG-CoA (specific activity 2216.3 MBq/mmol) and [^{3}H]mevalonate (sp.act. 9024 GBq/mmol) were obtained from the NEN Life Science Products Inc. (Boston, MA, USA). Unlabeled HMG-CoA, NADP^{+}, imidazole, Trizma-Base, glucose-6-phosphate, glucoso-6-phosphate dehydrogenase, DL-mevalonic acid lactone, dithiothreitol (DTT), were purchased from Sigma Chemical Co. (St. Louis, MO, USA). All other chemicals used were of analytical grade and Polish origin.

Induction of chronic renal failure

All experiments were performed under a protocol approved by the Ethical Committee of Medical University of Gdańsk (4/2000). Male Wistar rats (8–9 week old, weight 200–250 g at the beginning of the study) were used in all experiments. Animals were kept in individual cages with free access to water and food, and were maintained on a controlled lighting schedule, in which the room was illuminated from 6.00–18.00 daily. A standard diet for laboratory animals, 'Labofeed H'-Poland was used in all experiments. Experimental CRF was induced by our own modification of 5/6 nephrectomy model [5]. Animals were killed in the morning (between 7 and 9 a.m.) , in the evening (between 4 and 6 p.m.) and at midnight (between 11 p.m. and 1 a.m.), livers were removed and two 1 g pieces were taken for RNA determination, the remainder was used for preparation of microsomes.

Determination of urea, creatinine and cholesterol

Six weeks after the surgery rats were killed under a light anesthesia. Blood and livers were collected. Blood was centrifugated and serum used for the determination of urea, creatinine, and cholesterol levels. These parameters were determined using Hitachi 704 autoanalyser. Livers were used for microsomes isolation or immediately frozen in liquid nitrogen for total RNA isolation.

Evaluation of HMG-CoA reductase activity

Liver tissues were rinsed three times with 0.9% NaCl and three times in medium containing: 0.3 M sucrose, 10 mM mercaptoethanol and 15 mM Tris-HCl (pH 7.4) buffer. The tissues were drained on filter paper and homogenized manually with buffered solution in a glass Potter-Elvehjem homogeniser. The homogenate was centrifugated twice at 15,000 g for 10 min and the resulting supernatant was centrifugated at 105,000 g for 60 min. The microsomal pellet was resuspended in imidazole buffer containing: 1 mM EDTA, 250 mM NaCl, 5 mM DTT, and 50 mM imidazole-HCl (pH 7.4) buffer.

HMG-CoA reductase activity was assayed by the method of Shapiro *et al.* [6].

Isolation of total RNA and Northern blot analysis

The probe to detect HMG-CoA reductase mRNA was a 33-mer antisense oligonucleotide (5′-CTA TCC AGC GAC TGT GAG CGT GAA CAA GAA CCA-3′) (Gene Bank accession number D45862). The 31-mer antisense oligonucleotide (5′-CGC CTG CTG CCT TCC TTG GAT GTG GTA GCC-3′) was used as a probe for the 18S rRNA. The oligonucleotides were synthesized commercially (GENSET SA, Paris, France) with a single digoxigenin ligand at the 3′ end.

Total RNA was extracted from frozen tissue (approximately 1 g of tissue was used) by a guanidinium isothiocyanate-phenol/chloroform method [7] and finally dissolved in dimethyl pyrocarbonate-treated water. The RNA concentration was determined from the absorbance at 260 nm and all samples had an 260/280 nm absorbance ratio of about 2.0. RNA samples from control and nephrectomised rats were applied (10 µg/lane) to a 1% agarose gel containing 0.41 M formaldehyde and fractionated by horizontal gel electrophoresis. After electroforetic fractionation RNA was transferred overnight to a positively charged nylon membrane by capillary blotting and fixed with UV light. Prehybridization and hybridisation was performed as described recently [8]. Signals were scanned and quantified using the Sigma Scan software program (Jandel Scientific), the level of mRNA for HMG-CoA reductase was estimated. The values were normalised for the corresponding amount of 18S rRNA and the results were expressed in arbitrary units.

Radioactivity counting procedures

The scintillation fluid contained 200 mg of 1,4 di[2(5-phenyloxazolyl)] benzene (POPOP) and 4.0 g of 2,5-diphenyloxazole (PPO) and 60 g of naphtalene/liter of ethanol-ethylene-glycol-dioxane (10:2:100). All samples were analysed in Beckman liquid scintillation counter, and corrected for quenching by the external standard channels-ratio method.

Statistical analysis

Results are presented as means ± S.D. from the sample of all the rats studied. Statistical significance of differences between groups was assessed by one-way analysis of variance, followed by Student's t-test. Differences between groups were considered as significant when $p < 0.05$.

Results

Serum urea and creatinine were significantly higher in the CRF rats than in the sham operated animals (Table 1). Concentrations of serum cholesterol surpassed those found in the controls (Table 1). We haven't found any diurnal variation in the plasma cholesterol level in the control nor the CRF rats. Notably smaller average food intake and the mean body weight of the CRF rats as compared to the control group, were also determinants of a deep uremia (Table 2).

The HMG-CoA reductase mRNA levels are shown in Fig. 1 (part A). The values are presented with respect to 18S rRNA, and shown as a HMG-CoA reductase mRNA/18S rRNA ratio (part B). The results indicated on the fact that the diurnal rhythm of the changes in HMG-CoA reductase mRNA level was preserved in the course of experimental CRF. The values of the mRNA level were however higher in the CRF group.

The liver microsomal HMG-CoA reductase specific activity was increased in the CRF rats both during the day and in the night (Table 3). 3.8 and 3.0-fold increase in the specific activity of HMG-CoA reductase was observed in CRF rats in the evening and morning, respectively. However this difference didn't reach statistical significance in the midnight i.e. in the peak of the physiological diurnal rhythm of the enzyme's activity. In the control group the diurnal variability of HMG-CoA reductase activity was more apparent. It is important to note that differences in the specific activity of HMG-CoA reductase between CRF and control rats exceeded significantly the differences in the values of HMG-CoA reductase mRNA level.

Discussion

It is generally accepted that the HMG-CoA reductase is a rate limiting enzyme in the cholesterologenic metabolic pathway. Therefore increase in its activity would be expected to cause increase in the rate of cholesterologenesis. Indeed, our previously published results showed higher rate of cholesterol biosynthesis both in liver and intestine of CRF rats as compared to control animals [2].

The experiments of the present study were carried out in various times of day to compare the changes in HMG-CoA reductase activity in different periods of the physiological diurnal rhythm of cholesterologenesis. We observed higher activity of the enzyme both at light and at dark phases of the diurnal cycle in the CRF group. It is in accord with our earlier observation that diurnal rhythm of cholesterologenesis activity is preserved both in the liver and intestine of CRF rats [2].

It was suggested by Pandak *et al.* [4] that the rise in HMG-CoA reductase activity found in livers of rats with subtotal nephrectomy was not associated with parallel increase in HMG-CoA reductase mRNA level. The results of our experiments show that it is not the case.

Analysis of mRNA for HMG-CoA reductase shows that higher activity of this enzyme in CRF rats may result, at least in part, from the increased expression of HMG-CoA reductase gene. However it is important to note that in our 5/6 nephrectomy model animals were sacrificed 40 days after the second surgery in comparison to 21 days in the experiments of Pandak *et al.* That enabled the development of chronic renal failure with its metabolic consequences. However, it should be stressed that our results do not exclude suggestion of Pandak et al. that the observed increase in HMG-CoA reductase specific activity may be also due to post-transcriptional alteration in enzyme's catalytic activity.

Patients, as well as laboratory animals with CRF exhibit higher plasma mevalonate concentrations with a concomitant decrease in its urinary excretion [9, 10]. Higher plasma level of this direct product of HMG-CoA reductase activity paralels high concentrations of plasma cholesterol [11]. Increased expression of HMG-CoA gene in the course of CRF is therefore a surprising observation since in the presence of elevated mevalonate level HMG-CoA reductase activity and its mRNA level would be expected to be suppressed rather than stimulated [3]. Therefore it seems obvious that, as it was suggested by Pandak, normal feedback regulation of HMG-CoA reductase is altered in CRF rats. It is in accordance with the work of Kopito *et al.* [12] who found that even 4-fold rise of plasma mevalonate concentration, following bilateral nephrectomy did not suppress HMG-CoA reductase activity in rat's liver.

Table 1. Effect of experimental chronic renal failure (CRF) on serum creatinine, urea total and LDL cholesterol concentrations

Rats	Creatinine (mg/dl ± S.D.)	Urea (mg/dl ± S.D.)	Serum cholesterol (mg/dl ± S.D.)	LDL cholesterol (mg/dl ± S.D.)
Control (n = 14)	0.62 ± 0.02	54 ± 6.4	69 ± 10	28.8 ± 8.5
CRF rats (n = 14)	2.53 ± 1.23	270 ± 151.6	233 ± 61	97.3 ± 18.0
p	< 0.01	< 0.01	< 0.01	< 0.01

Table 2. Effect of experimental chronic renal failure (CRF) on body weight and average food intake

Rats	Body weight (g ± S.D.)	Food intake (g/day ± S.D.)
Control (n = 14)	430 ± 24	30.6 ± 1.9
CRF rats (n = 14)	351 ± 26	27.6 ± 4.5
p	< 0.01	< 0.02

Moreover, starvation is one of the physiological metabolic states which cause depression of liver cholesterologenesis. Despite that, in our experiments even significant decrease in food intake and body weight found in CRF rats was accompanied by higher activity of HMG-CoA reductase together with higher concentration of plasma and liver cholesterol as well as hepatic and intestinal cholesterologenesis [2].

At present it is difficult to answer which stages of the cellular cholesterol homeostasis are most affected in CRF rats.

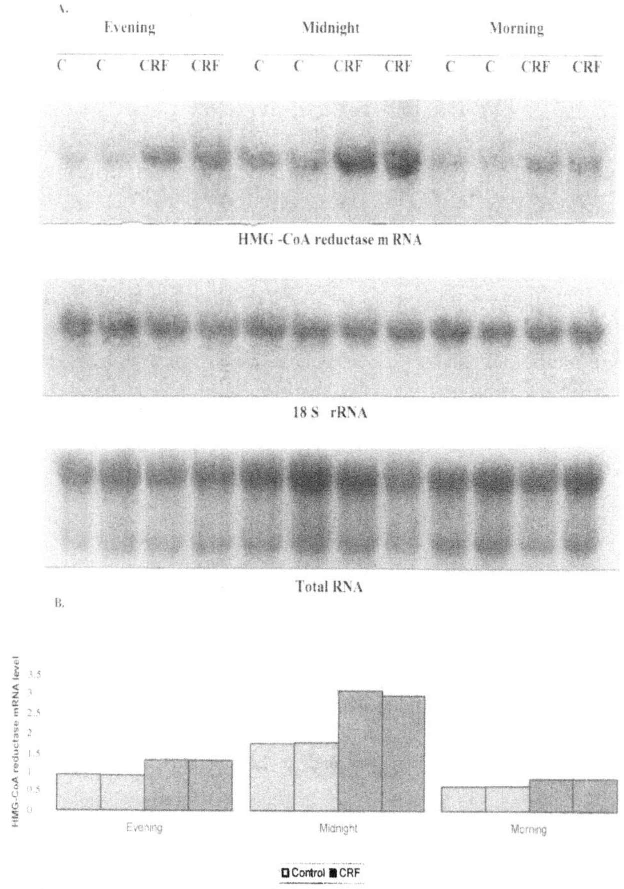

Fig. 1. (A) Northern blot analysis of liver HMG-CoA reductase mRNA (C – control rats; CRF – chronic renal failure rats). (B) Quantification of the HMG CoA reductase mRNA to 18S RNA ratio in control and CRF rats kept on the controlled dark/light schedule (values expressed in arbitrary units).

Table 3. Specific activity of rat liver microsomal HMG-CoA reductase in the sham operated (control) and CRF rats

Rats	HMG-CoA reductase specific activity (pmole/min × mg of microsomal protein ± S.D.)		
	Evening	Midnight	Morning
Sham operated (control) (n = 7)	180 ± 110	790 ± 160	30 ± 4
CRF (n = 7)	680 ± 190	1230 ± 450	90 ± 30
CRF/control	3.8	1.5	3
p	< 0.01	NS	< 0.01

Elevated activity of HMG-CoA reductase accompanying higher mRNA level, together with increased cellular and serum cholesterol concentrations (both LDL and HDL fractions increase about three fold as compared to controls) suggest that altered function of LDL receptor-mediated clearance of lipoprotein particles may be one of the steps affecting cellular cholesterol homeostasis in CRF rats.

It has been suggested that CRF generates an undefined stimulus for HMG-CoA reductase activity in the liver, which provides an additional enhancement of hepatic cholesterologenesis [13]. As was documented by Feingold and Grunefeld [14] intravenous administration of Tumor Necrosis Factor-alpha (TNF-α) stimulated hepatic sterol synthesis and increased HMG-CoA reductase activity in livers of rats. However, stimulation of sterologenesis by TNF-α was specifically localized to the liver, the administration of TNF-α did not stimulate cholesterol biosynthesis in the small intestine, adipose tissue, muscle nor skin. Moreover incorporation of mevalonate into cholesterol was not increased in the livers of TNF-treated animals. It is noteworthy that plasma concentrations of TNF-α as well as soluble TNF-α receptors have been found to be increased in chronically uremic patients [15].

On the other hand our previously published results [2] indicate that in CRF rats cholesterologenesis was stimulated both in the liver and intestinal tissues, and an over 2 fold stimulation of mevalonate incorporation into cholesterol has been found in the liver slices of CRF rats. Therefore it appears that TNF-α might be one of many factors responsible for the increase in HMG-CoA gene expression in the CRF rats.

In this paper we have provided evidence that increased mevalonate biosynthesis and in consequence cholesterologenesis in CRF rats may result at least in part from increased expression of HMG-CoA reductase on the gene level.

Acknowledgements

This work was supported by a grant from the Polish Committee for Scientific Research (KBN 4PO5A09919) and by

KBN grants via Medical University of Gdańsk (ST-4, ST-41, W 703).

References

1. Attman PO, Alaupovic P, Samuelsson O: Lipoprotein abnormalities as a risk factor for progressive nondiabetic renal disease. Kidney Int 71(suppl): S14–S17, 1999
2. Chmielewski M, Nieweglowski T, Swierczynski J, Rutkowski B, Boguslawski W: Diurnal rhythm of cholesterol biosynthesis in experimental chronic renal failure. Mol Cell Biochem 228: 33–37, 2001
3. Clarke CF, Fogelman AM, Edwards PA: Diurnal rhythm of rat liver mRNAs encoding 3-hydroxy-3-methylglutaryl coenzyme A reductase. Correlation of functional and total mRNA levels with enzyme activity and protein. J Biol Chem 259: 10439–10447, 1984
4. Pandak WM, Vlahcevic ZR, Heuman DM, Krieg RJ, Hanna JD, Chan JC: Post-transcriptional regulation of 3-hydroxy-3-methylglutaryl coenzyme A reductase and cholesterol 7 alpha-hydroxylase in rats with subtotal nephrectomy. Kidney Int 46: 358–364, 1994
5. Swierczynski J, Korczynska J, Szolkiewicz M, Karbowska J, Kochan Z, Nieweglowski T, Kusiak E, Rutkowski B: Low leptin mRNA level in adipose tissue and normoleptinemia in experimental chronic renal failure. Exp Nephrol 9: 54–59, 2001
6. Shapiro DJ, Nordstrom JL, Mitschelen JJ, Rodwell VW, Schimke RT: Micro assay for 3-hydroxy-3-methylglutaryl-CoA reductase in rat liver and in L-cell fibroblasts. Biochim Biophys Acta 370: 369–377, 1974
7. Chomczynski P, Sacchi N: Single-step method of RNA isolation by acid guanidinium thiocyanate-phenol-chloroform extraction. Anal Biochem 162: 156–159, 1987
8. Karbowska J, Kochan Z, Zelewski L, Swierczynski J: Tissue-specific effect of clofibrate on rat lipogenic enzyme gene expression. Eur J Pharmacol 370: 329–336, 1999
9. Scoppola A, De Paolis P, Menzinger G, Lala A, Di Giulio S: Plasma mevalonate concentrations in uremic patients. Kidney Int 51: 908–912, 1997
10. Scoppola A, Galiano P: The mevalonate pathway and renal diseases: experimental and clinical implications. Int J Artif Organs 21: 243–246, 1998
11. Parker TS, McNamara DJ, Brown CD, Kolb R, Ahrens EH Jr, Alberts AW, Tobert J, Chen J, De Schepper PJ: Plasma mevalonate as a measure of cholesterol synthesis in man. J Clin Invest 74: 795–804, 1984
12. Kopito RR, Weinstock SB, Freed LE, Murray DM, Brunengraber H: Metabolism of plasma mevalonate in rats and humans. J Lipid Res 23: 577–583, 1982
13. Feingold KR, Wiley MH, MacRae G, Kaysen G, Schoenfeld PY, Siperstein MD: The effect of uremia on circulating mevalonate metabolism in rats. Metabolism 32: 215–223, 1983
14. Feingold KR, Grunfeld C: Tumor necrosis factor-alpha stimulates hepatic lipogenesis in the rat *in vivo*. J Clin Invest 80: 184–190, 1987
15. Guarnieri G, Toigo G, Fiotti N, Ciocchi B, Situlin R, Giansante C, Vasile A, Carraro M, Faccini L, Biolo G: Mechanisms of malnutrition in uremia. Kidney Int 62(suppl): S41–S44, 1997

Molecular and Cellular Biochemistry **246:** 193–196, 2003.

Anti-oxidant effect of flavonoids on the susceptibility of LDL oxidation

Gholam Ali Naderi,[1] Seddigheh Asgary,[2] Nizal Sarraf-Zadegan[3] and Hamid Shirvany[1]

Departments of [1]Biochemistry; [2]Pharmacognosy; [3]Cardiology, Isfahan Cardiovascular Research Center, Isfahan University of Medical Sciences, Isfahan, Iran

Abstract

In vitro studies have demonstrated increased atherogenicity of oxidized low-density lipoprotein (ox-LDL) compared to native LDL. Oxidative modification of LDL alters its structure allowing LDL to be taken up by scavenger receptors on macrophage, endothelial, and smooth muscle cells, leading to the formation of lipid-laden foam cells, the hallmark of early atherosclerotic lesions. The susceptibility of LDL to *in vitro* oxidation was assessed essentially by the technique described by Esterbauer *et al.* LDL oxidation were monitored by change in 234-absorbance in the presence and absence of pure flavonoids. Morin, genistein, apigenin and biochanin A, naringin and quercetin were used at different concentration. These flavonoids significantly inhibit *in vitro* LDL oxidation, genistein, morin and naringin have stronger inhibitory activity against LDL oxidation than biochanin A or apigenin. This study show that flavonoids prevent *in vitro* LDL oxidation and probably would be important to prevent atherosclerosis. (Mol Cell Biochem **246**: 193–196, 2003)

Key words: antioxidant, LDL oxidation, flavonoid, natural product, atherosclerosis

Introduction

There is a large body of evidence that oxidative processes may be involved in the pathogenesis of atherosclerosis [1]. Oxidative modification of low-density lipoprotein enhances uptake by macrophages and cellular accumulation of cholesterol (foam cells) [1, 2]. Many other mechanisms may explain the increased atherogenicity of oxidized LDL cholesterol [2].

On the other hand, epidemiological data regarding the preventive role of dietary or serum anti-oxidants such as selenium, vitamin E, and β carotene or carotenoids are much more extensive [3–6].

Flavonoids are major components of flowering plants. They have received considerable attention because of their possible beneficial effects as dietary components in the prevention of human diseases [7–10]. Notably, a recent ACS publication devoted to food phytochemicals focused on the anti-oxidant and antitumor properties of flavonoids [11]. The possible beneficial effect of red wine flavonoids inhibiting low density lipid oxidation in the blood is also under active investigation. In this study the *in vitro* effects of some pure flavonoids, on the susceptibility of LDL to oxidation was examined.

Material and methods

Chemical

Morin, querctein, genistein, naringin were purchased from Merck chemical company and apigenin and biochanin A from Aldrich. 1.5 mg of each flavonoid were dissolved separately in 1 ml of DMSO.

In vitro *oxidation of LDL*

LDL was prepared by preparative ultracentrifugation in the presence of EDTA [12]. LDL was desalted by dialysis against

Address for offprints: G. Ali Naderi, Isfahan Cardiovascular Research Center, Amin Hospital, Ibn-sina Street, Isfahan, Iran (E-mail: isfcarvasrc@hotmail.com)

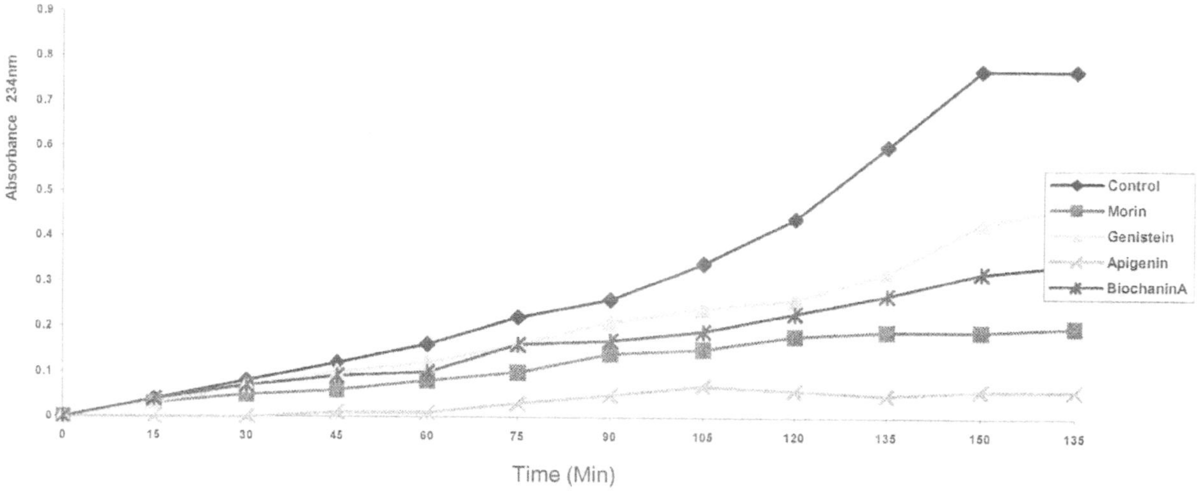

Fig. 1. Effects of *in vitro* addition of genistein, biochanin A, morin, apigenin (0, 50 μg/ml) to LDL (15 μg/ml) on the susceptibility of low density lipoprotein to 5 μM CuSO4-induced oxidation. Oxidation was determined by monitoring the change at 234 nm due to increase in conjugated diene formation for 135 min.

phosphate buffered saline under nitrogen and then adjusted to a protein concentration of 200 mg/L. They were supplemented with CuSO4 at a final concentration of 15 μg/ml.

The kinetic of LDL oxidation was followed by determining the absorption of conjugated dienes at 234 nm as a function of time [13].

Results

Figure 1 shows the effect of genistein, biochanin A, morin and apigenin on the susceptibility of LDL to copper-induced oxidation compared with oxidation of LDL as a control.

Morin and genistein showed a significant inhibitory effect on LDL oxidation over the time of incubation, but biochanin A and apigenin were less.

Naringin also showed a significantly inhibitory effect on LDL oxidation compared to control (Fig. 2). Quercetin at different concentration affected the susceptibility of LDL to oxidation. At high concentrations (250 and 500 ng/ml), there was a prooxidant effect and in contrast at lower concentrations (50, 100 ng/ml) quercetin produced an antioxidant effect (Fig. 3).

Discussion

In this study we showed copper-catalyzed oxidation of LDL preincubated with flavonoids was significantly inhibited as assessed by monitoring over 135 min the changes in absorption at 234 nm caused by a decrease in conjugated diene formation.

Other studies reported similar effects of flavonoids on LDL oxidation. For example, Mangiapane *et al.* [14] reported that catechin completely inhibited the copper catalyzed oxidation of LDL at 20 mg/l. Miura *et al.* [15] reported that the lag time before the onset of conjugated diene formation in porcine LDL was prolonged > 100 min in the presence of 0.5 μmol/l epigallocatechin gallate or theaflavin compared with only 21 min in the absence of flavonoids. In this study the antioxidative activity of several flavonoids was compared. The ability to prolong lag time before copper induced oxidation was observed occurred the following order: sesaminol > quercetin > epigallocatechin gallate > theaflavin > myricetin > butylated hydroxytoluene > α-tocopherol [16]. Our study showed that genistein, morin and naringin have stronger inhibitory activity against LDL oxidation than biochanin A or apigenin. Morin has more hydroxyl groups, which are considered to be necessary for radical scavenging activity, than apigenin or biochanin A. This may be one reason for the more marked effects of morin.

Quercetin has binary effect on LDL oxidation. At high concentrations (250, 500 ng/ml) the flavonoids acted as a prooxidant, and at low concentrations (50, 100 ng/ml) it was an antioxidant this biphasic effect may be due to its structural properties or an ability to generate free radicals at high concentration. Flavonoids have been shown to react with superoxide anion [17], hydroxyl radicals and lipid peroxy radicals [18].

Flavonoids have been reported to chelate iron [17] and copper [19] and this may partly explain their antioxidant effects. There is growing evidence to suggest that oxidized LDL

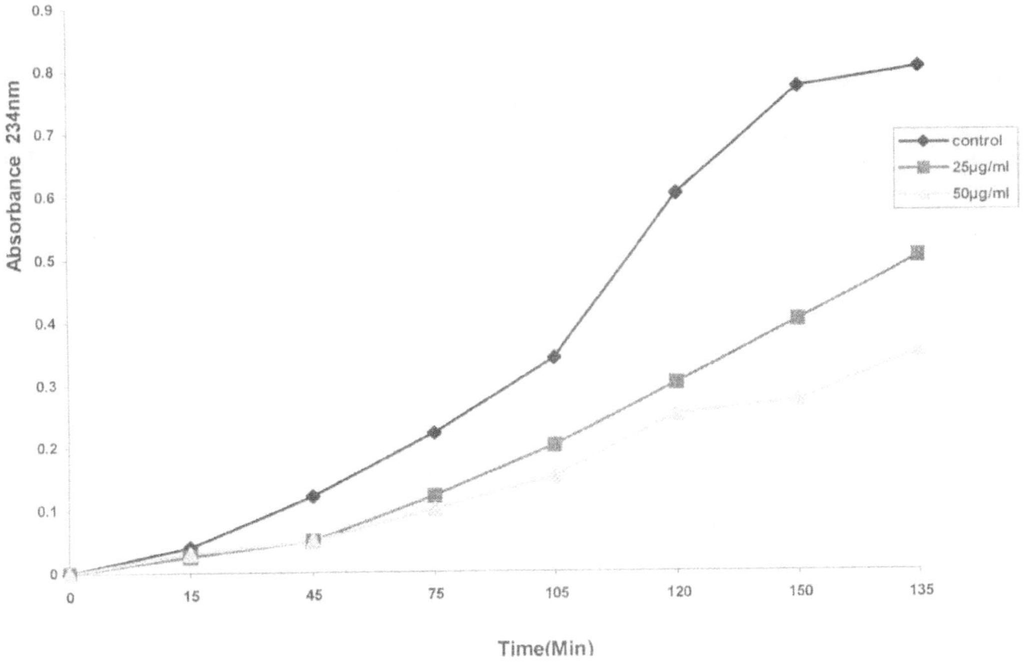

Fig. 2. Effects of *in vitro* addition of naringin (0, 25, 50 μg/ml) to LDL (15 μg/ml) on the susceptibility of low density lipoprotein to 5 μM CuSO4-induced oxidation. Oxidation was determined by monitoring the change at 234 nm due to increase in conjugated diene formation for 135 min.

might be one of the modified LDL species responsible for cholesterol loading of the macrophage [2]. We showed that certain flavonoids have marked inhibitory effects on LDL oxidation. Dietary flavonoid intake has been reported to be inversely associated with the incidence of coronary artery disease. Thus, it is possible that flavonoids reduce the rate of formation of oxidized LDL and inhibit the growth of atherogenic plaques.

In conclusion, we clearly showed that flavonoids especially morin and genistein added to LDL *in vitro* protected LDL from oxidation. These flavonoids may have favorable effects in atherosclerosis.

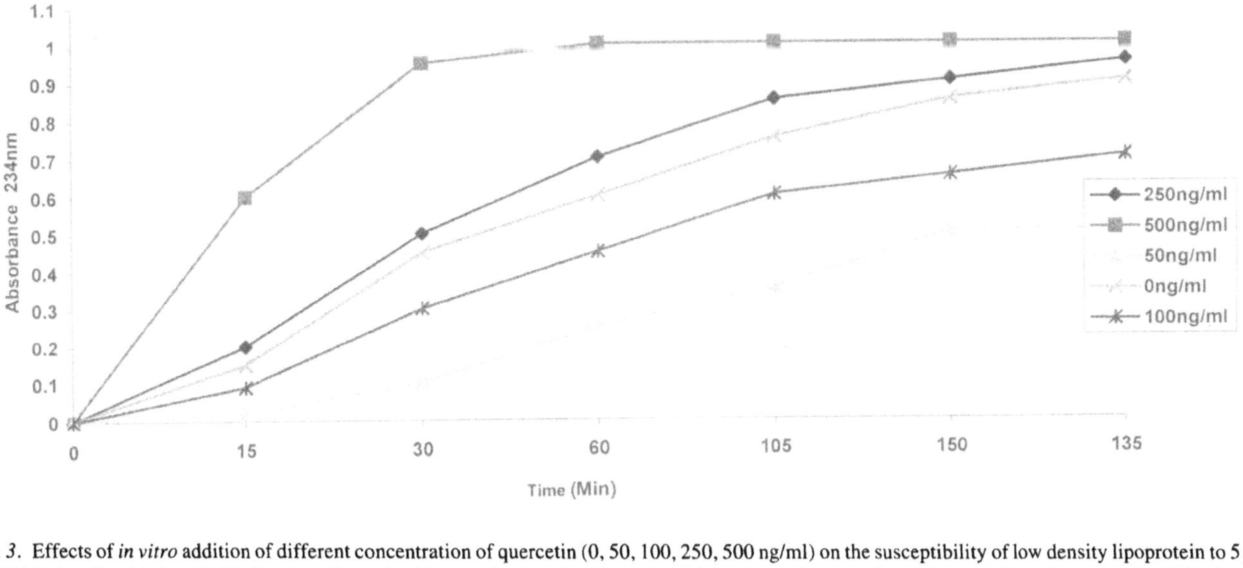

Fig. 3. Effects of *in vitro* addition of different concentration of quercetin (0, 50, 100, 250, 500 ng/ml) on the susceptibility of low density lipoprotein to 5 μM CuSO4-induced oxidation. Oxidation was determined by monitoring the change at 234 nm due to increase in conjugated diene formation for 135 min.

References

1. Witzum JL: The oxidation hypothesis of atherosclerosis. Lancet 344: 793–798, 1994
2. Witzum JL, Steinberg D: Role of oxidized low density lipoprotein in atherogeneis. J Clin Invest 88: 1785–1792, 1991
3. Geg KF, Brubacher GB, Stahelin HB: Plasma levels of anti-oxidant vitamins in relation to ischemic heart disease. A J Clin Nutr 45: 1368–1377, 1987
4. Gaziano JM, Manson JE, Buring JE, Hennekens CH: Dietary anti-oxidants and cardiovascular disease. Ann NY Acad Sci 669: 249–259, 1992
5. Steinbrg D: Workshop participants. Anti-oxidants in the prevention of human atherosclerosis. Circulation 85: 2338–2344, 1992
6. Susukawa M, Ishikawa T, Yoshuda H, Nakamura H: Effect of *in vivo* supplemention with low dose vitamin E on susceptibility of low density lipoprotein and high density lipoprotein to oxidative modification. J Am Coll Nutr 14: 16–52, 1995
7. Hertog MGL, Feskens EJM, Hollman PCH, Katan MB, Komnow D: Dietary antioxidant flavonoids and risk of coronary heart disease: The Zutphen Elderly study. Lancet 345: 1007–1011, 1993
8. Hertog MGL, Kromhout D, Aravanis C *et al.*: Flavonoids intake and long-term risk of coronary heart disease and cancer in the seven countries study. Arch Intern Med 155: 381–386,1995
9. Renaul S, Lorgeril M: Wine, alcohol, platelets and the French paradox for coronary heart disease. Lancet 339: 1523–1526 1992
10. Wagner H: In: J.L. Beal, E. Reinhard (eds). Natural Products as Medicinal Agents. Hippokrates-Verlag, Stuttgart, 1981, p 217
11. Kuana MT, Sawo JO, Ho CT, Rosen CT (eds).: Food Phytochemicals for Cancer Prevention. Acs Symposium Series, vols. 546/547. Acs, Washington DC, 1994
12. Lindgren FT, Jensen LC, Wills RD, Freeman NK: Flotation rates, molecular weights, and hydrated densities of the low density lipoproteins. Lipids 4: 337–344, 1969
13. Esterbauer H, Striegl G, Puhl H, Waeg G, Striegl G: Continuous monitoring of *in vitro* oxidation of human low density lipoprotein. Free Radic Res Commun 6: 67–75, 1989
14. Mangiapane H, Thomson J, Salter A, Brown S, Bell GD, White DA: The inhibition of the oxidation of low-density lipoprotein by (+)catechin. A naturally occurring flavonoid. Biochem pharmacol 43: 445–450, 1992
15. Miura S, Watanable J, Sano M: Effects of various natural anti-oxidants on the Cu^{++}-mediated oxidative modification of low density lipoprotein. Biol Pharm Bull 18: 1–4, 1995
16. De whalley CV, Rankin SM, Hovlt JRS, Jessup W: Flavonoids inhibit the oxidative modification of low-density lipoproteins by macrophages. Biochem pharmacol 39: 1743–1750, 1990
17. Afanasev IB, Dorozhko AL, Brodoski AV, Kostyuk VA, Potapovitch AL: Chelating and free radical scavenging mechanisms of inhibitory action of rutin and quercetin in lipid peroxidation. Biochem Pharmacol 38:1763–1769, 1989
18. Husain SR, Cillard J, Cillard P: Hydroxyl radical scavenging activities of flavonoids. Phytochemistry 26: 2489–2491, 1987
19. Thompson M, Williams CR, Elliol CEP: Stability of flavonoid of copper (11) and flavonoid anti-oxidant activity. Anal Chim Acta 85: 375–381, 1976

Molecular and Cellular Biochemistry **246:** 197–201, 2003.

Index to Volume 246

Developments in Molecular and Cellular Biochemistry

Series Editor: Naranjan S. Dhalla, Ph.D., M.D. (Hon.), FACC

1. V.A. Najjar (ed.): *Biological Effects of Glutamic Acid and Its Derivatives.* 1981 ISBN 90-6193-841-4
2. V.A. Najjar (ed.): *Immunologically Active Peptides.* 1981 ISBN 90-6193-842-2
3. V.A. Najjar (ed.): *Enzyme Induction and Modulation.* 1983 ISBN 0-89838-583-0
4. V.A. Najjar and L. Lorand (eds.): *Transglutaminase.* 1984 ISBN 0-89838-593-8
5. G.J. van der Vusse (ed.): *Lipid Metabolism in Normoxic and Ischemic Heart.* 1989 ISBN 0-7923-0479-9
6. J.F.C. Glatz and G.J. van der Vusse (eds.): *Cellular Fatty Acid-Binding Proteins.* 1990 ISBN 0-7923-0896-4
7. H.E. Morgan (ed.): *Molecular Mechanisms of Cellular Growth.* 1991 ISBN 0-7923-1183-3
8. G.J. van der Vusse and H. Stam (eds.): *Lipid Metabolism in the Health and Diseased Heart.* 1992 ISBN 0-7923-1850-1
9. Y. Yazaki and S. Mochizuki (eds.): *Cellular Function and Metabolism.* 1993 ISBN 0-7923-2158-8
10. J.F.C. Glatz and G.J. van der Vusse (eds.): *Cellular Fatty-Acid-Binding Proteins, II.* 1993 ISBN 0-7923-2395-5
11. R.L. Khandelwal and J.H. Wang (eds.): *Reversible Protein Phosphorylation in Cell Regulation.* 1993 ISBN 0-7923-2637-7
12. J. Moss and P. Zahradka (eds.): *ADP-Ribosylation: Metabolic Effects and Regulatory Functions.* 1994 ISBN 0-7923-2951-1
13. V.A. Saks and R. Ventura-Clapier (eds.): *Cellular Bioenergetics: Role of Coupled Creatine Kinases.* 1994 ISBN 0-7923-2952-X
14. J. Slezák and A. Ziegelhöffer (eds.): *Cellular Interactions in Cardiac Pathophysiology.* 1995 ISBN 0-7923-3573-2
15. J.A. Barnes, H.G. Coore, A.H. Mohammed and R.K. Sharma (eds.): *Signal Transduction Mechanisms.* 1995 ISBN 0-7923-3663-1
16. A.K. Srivastava and J.-L. Chiasson (eds.): *Vanadium Compounds: Biochemical and Therapeutic Applications.* 1995 ISBN 0-7923-3763-8
17. J.M.J. Lamers and P.D. Verdouw (eds.): *Biochemistry of Signal Transduction in Myocardium.* 1996 ISBN 0-7923-4067-1
18. E.-G. Krause and R. Vetter (eds.): *Biochemical Mechanisms in Heart Function.* 1996 ISBN 0-7923-4118-X
19. R. Vetter and E.-G. Krause (eds.): *Biochemical Regulation of Myocardium.* 1996 ISBN 0-7923-4259-3
20. G.N. Pierce and W.C. Claycomb (eds.): *Novel Methods in Molecular and Cellular Biochemistry of Muscle.* 1997 ISBN 0-7923-4387-5
21. F.N. Gellerich and S. Zierz (eds.): *Detection of Mitochondrial Diseases.* 1997 ISBN 0-7923-9925-0
22. P.K. Singal, V. Panagia and G.N. Pierce (eds.): *The Cellular Basis of Cardiovascular Function in Health and Disease.* 1997 ISBN 0-7923-9974-9
23. S. Abdel-aleem and J.E. Lowe (eds.): *Cardiac Metabolism in Health and Disease.* 1998 ISBN 0-7923-8104-1
24. A.K. Srivastava and B. Posner (eds.): *Insulin Action.* 1998 ISBN 0-7923-8113-0
25. V.A. Saks, R. Ventura-Clapier, X. Leverve, A. Rossi and M. Rigoulet (eds.): *Bioenergetics of the Cell: Quantitative Aspects.* 1998 ISBN 0-7923-8118-1
26. G.N. Pierce, H. Rupp, T. Izumi and A. Grynberg (eds.): *Molecular and Cellular Effects of Nutrition on Disease Processes.* 1998 ISBN 0-7923-8171-8
27. K. Ahmed, E. Chambaz and O.G. Issinger (eds.): *Molecular and Cellular View of Protein Kinase CK2.* 1998 ISBN 0-7923-8208-0

Developments in Molecular and Cellular Biochemistry

28. M.V. Cohen, J.M. Downey, R.J. Gelpi and J. Slezák (eds.): *Myocardial Ischemia and Reperfusion*. 1998 ISBN 0-7923-8173-4
29. D.A. Bernlohr and L. Banaszak (eds.): *Lipid Binding Proteins within Molecular and Cellular Biochemistry*. 1998 ISBN 0-7923-8223-4
30. R. Albarez-Gonzalez (ed.): *ADP-Ribosylation Reactions: From Bacterial Pathogenesis to Cancer*. 1998 ISBN 0-7923-8235-8
31. S. Imai and M. Endo (eds.): *Muscle Physiology and Biochemistry*. 1998 ISBN 0-7923-8264-1
32. D.K. Das (ed.): *Stress Adaptation, Prophylaxis and Treatment*. 1999 ISBN 0-7923-8406-7
33. H. Rupp and B. Maisch (eds.): *Control of Gene Expression by Catecholamines and the Renin-Angiotensin System* ISBN 0-7923-7981-0
34. X. Shi, V. Castranova, V. Vallyathan and W.G. Perry (eds.): *Molecular Mechanisms of Metal Toxicity and Carcinogenesis*. 2001 ISBN 0-7923-7498-3
35. K. Ahmed, J.E. Allende and O.-G. Issinger (eds.): *Protein Kinase CK2 – From Structure to Regulation*. 2001 ISBN 0-7923-7666-8
36. R.K. Sharma (ed.): *Guanylate Cyclase*. 2002 ISBN 0-7923-7682-8
37. V. Vallyathan, V. Castranova and S. Shi (eds.): *Oxygen/Nitrogen Radicals: Cell Injury and Disease*. 2002 ISBN 1-4020-7085-3
38. J.F.C. Glatz (ed.): *Cellular Lipid Binding Proteins*. 2002 ISBN 1-4020-7220-1
39. E. Kardami, L. Hryshko and N. Mesaeli (eds.): *Cardiac Cell Biology*. 2003 ISBN 1-4020-7296-1
40. J.F. Clark (ed.): *Guanidino Compounds in Biology and Medicine*. 2003 ISBN 1-4020-7297-X
41. P. Zahradka, J. Wigle and G.N. Pierce (eds.): *Vascular Biochemistry*. 2003 ISBN 1-4020-7398-4

KLUWER ACADEMIC PUBLISHERS – DORDRECHT / BOSTON / LONDON

GPSR Compliance

The European Union's (EU) General Product Safety Regulation (GPSR) is a set of rules that requires consumer products to be safe and our obligations to ensure this.

If you have any concerns about our products, you can contact us on ProductSafety@springernature.com

In case Publisher is established outside the EU, the EU authorized representative is:

Springer Nature Customer Service Center GmbH
Europaplatz 3
69115 Heidelberg, Germany

Batch number: 10755943

Printed by Printforce, the Netherlands